12th International Conference on Numerical Modeling of Space Plasma Flows (ASTRONUM 2017)

Journal of Physics: Conference Series Volume 1031

Saint-Malo, France
26 - 30 June 2017

Editors:

Nikolai V. Pogorelov
Edouard Audit
Gary P. Zank

ISBN: 978-1-5108-6406-1
ISSN: 1742-6588

Printed from e-media with permission by:

Curran Associates, Inc.
57 Morehouse Lane
Red Hook, NY 12571

Some format issues inherent in the e-media version may also appear in this print version.

Printed by Curran Associates, Inc. (2018)

For permission requests, please contact the Institute of Physics at the address below.

Institute of Physics
Dirac House, Temple Back
Bristol BS1 6BE UK

Phone: 44 1 17 929 7481
Fax: 44 1 17 920 0979

techtracking@iop.org

Additional copies of this publication are available from:

Curran Associates, Inc.
57 Morehouse Lane
Red Hook, NY 12571 USA
Phone: 845-758-0400
Fax: 845-758-2633
Email: curran@proceedings.com
Web: www.proceedings.com

Table of contents

Volume 1031

12th International Conference on Numerical Modeling of Space Plasma Flows: ASTRONUM-2017

26–30 June 2017, Saint-Malo, France

Accepted papers received: 10 May 2018
Published online: 1 June 2018

Preface

Papers

Turbulence and Cosmic Ray Transport

Astrophysical Flows

Space Plasma Flows

Numerical Methods

IOP Conf. Series: Journal of Physics: Conf. Series **1031** (2018) 011001 doi:10.1088/1742-6596/1031/1/011001

PREFACE

The Center for Space Plasma and Aeronomic Research (CSPAR) at the University of Alabama in Huntsville (UAH) and Maison de la Simulation, a joint laboratory at the French Alternative Energies and Atomic Energy Commission (CEA), National Center for Scientific Research (CNRS), University of Paris-Sud, and University of Versailles, organized the 12th International Conference on Numerical Modeling of Space Plasma Flows (ASTRONUM-2017) on June 26—30, 2017 in Saint-Malo, France.

The Program Committee consisted of Tahar Amari (CNRS Ecole Polytechnique, France), Edouard Audit (CEA/CNRS Maison de la Simulation, Gif-sur-Yvette, France, co-chair), Amitava Bhattacharjee (Princeton University, USA), Phillip Colella (Lawrence Berkeley National Laboratory, USA), Anthony Mezzacappa (University of Tennessee, Knoxville, USA), Ewald Müller (Max-Planck-Institute for Astrophysics, Garching, Germany), Nikolai Pogorelov (University of Alabama in Huntsville/CSPAR, USA, chair), Kazunari Shibata (Kyoto University, Japan), James Stone (Princeton University, USA), Jon Linker (Predictive Science, Inc., USA), and Gary P. Zank (University of Alabama in Huntsville, USA).

The conference attracted 82 scientists from 21 countries representing different branches of the plasma simulation community. The distinctive feature of this conference is a combination of diverse research topics, all of which are essential for performing high-resolution, continuum mechanics and particle, simulations of physical phenomena in space physics and astrophysics. Among such topics were software packages for modeling and analyzing plasma flows; advanced numerical methods for space and astrophysical flows; large-scale fluid-based, kinetic, and hybrid simulations; turbulence and cosmic ray transport; and magnetohydrodynamics. The applications discussed included cosmology and galaxy formation, supernova explosions, physics of the Sun-heliosphere-magnetosphere environments, the interstellar medium and star formation, stellar physics, experimental plasma physics, astrophysical accretion, numerical methods for ideal and non-ideal, relativistic and nonrelativistic MHD, etc. The proceedings volume is structured so that it covers all of these topics.

The structure of the ASTRONUM conference series is based on the idea that modelers working in seemingly distant fields should have an opportunity to share their scientific achievements with the broad community of computational scientists performing numerical experiments. As in previous ASTRONUM meetings, we were interested in physical systems that are coupled across a multiplicity of spatial and temporal that incorporate diverse physical processes.

The contributors to this volume are both young researchers and renowned experts in space physics and astrophysics, applied mathematics, and computational physics. This book describes the application of numerical methods and the algorithms themselves, allowing us to discuss the challenges that theory imposes on numerical schemes for solving partial differential equations describing collisional and collisionless processes in space and astrophysical plasmas.

We would like to thank the participants who submitted their papers to Proceedings of ASTRONUM-2017 and especially to those who reviewed manuscripts thus ensuring the high quality of this publication. We also are grateful to Valérie Belle, Axelle Nouhaud, and Bruno Thooris for the excellent management of the conference.

The book will be useful to graduate and postgraduate students majoring in space physics, astrophysics, numerical, engineering, and applied mathematics. It is also aimed at specialists in applied mathematics, and various fields of physics that involve flows of partially ionized plasmas at both the collisional and collisionless levels.

Nikolai V. Pogorelov, Edouard Audit, and Gary P. Zank

April 27, 2018

Peer review statement

All papers published in this volume of *Journal of Physics: Conference Series* have been peer reviewed through processes administered by the proceedings Editors. Reviews were conducted by expert referees to the professional and scientific standards expected of a proceedings journal published by IOP Publishing.

The Rate of Three-dimensional Hall-MHD Reconnection

Andrey Beresnyak
Research Support Instruments, Latham, MD 20706

Abstract. There is a discrepancy between reconnection rates observed in a) various plasma environments, b) plasma PIC simulations and c) resistive fluid simulations. Careful observations of solar flares show reconnection rates between $0.001 - 0.1v_A$ (see, e.g., [1]), plasma PIC and Hall-MHD simulations tend to show faster reconnection rate of around $0.1v_A$ and pure MHD resistive simulations tend to converge to $0.015v_A$. The common explanation of the difference between b and c [2] is that plasma reconnection is inherently different from resistive MHD reconnection due to the different mechanisms of how individual field lines break and reconnect. This would imply a universal rate of $0.1v_A$ in electron-proton plasma. In this paper, I report three-dimensional Hall-MHD simulations with resolution up to 2304×4608^2. Reconnection is mediated by self-excited turbulence in the current layer and is indeed faster than MHD reconnection. However, the reconnection rate goes down as the layer width increases. For the average layer width between two and ten ion skin depths (d_i), we recover "classic" value of $0.1v_A$, however, the trend of the rate to decrease may bring it to the fluid value for widths of around $170d_i$, although reaching this width is not possible with current numerical resources. In any case, based on these simulations, now we are certain that the $0.1v_A$ value has been favored before due to the limited range of scales available in numerical simulations.

1. Introduction

Magnetic reconnection is a topological rearrangement of the magnetic field lines, leading to the release of magnetic energy, which is associated with solar X-ray flares [3, 4]. Magnetic field lines are supposed to be frozen into the well-conducting plasma, so quiet laminar reconnection [3, 5] is extremely slow and cannot explain observed phenomena. This prompted research into collisionless reconnection [6, 7, 8]. The stochasticity of magnetic field lines due to ambient turbulence [9, 10, 11] leads to fast reconnection, also the tearing instability of the thin current sheet [12, 13] was proposed as a driver of resistivity-independent reconnection, which was shown to be consistent with two-dimensional simulations [14, 15].

Recent three-dimensional MHD simulations [16] further investigated the development of instability of the current layer and found that unlike 2D case the layer is not dominated by plasmoids but has fully developed turbulence with motion on all scales between the layer width and the dissipation scale. It was confirmed that reconnection is fast (independent on resistivity) and [16] interpreted this as a consequence of turbulence locality. In this picture, the rate was determined not by the critical Lundquist number, but rather by inherent properties of turbulence, such as the ratio of magnetic to kinetic energies and the Kolmogorov constant.

Reconnection rate is an important parameter to understand many space and solar physics phenomena. For example, reconnection releases up to 6×10^{33} erg in gigantic X-ray flares on the Sun. The dayside reconnection in front of the Earth magnetosphere allows fast particles from

Run	$n_x \cdot n_y \cdot n_z$	d_i	dt	L/d_i	d_i/d	ν_4
R0	384×768^2	0.32	2.25e-5	20	39	9.00e-08
R1	768×1536^2	0.16	1.06e-5	39	39	9.33e-09
R2	1536×3072^2	0.08	5.30e-6	79	39	9.26e-10
Q0	576×1152^2	0.052	5.00e-5	121	9.5	9.50e-09
Q1	1152×2304^2	0.026	2.75e-5	242	9.5	9.43e-10
Q2	2304×4608^2	0.013	1.25e-5	483	9.5	9.35e-11

Table 1. Parameters in our Hall-MHD simulations, where grid size d, ion skin depth d_i, box size L, time step dt and hyperdiffusivity ν_4 are in code units, such that $L = 2\pi$, and Alfvén crossing time of L is also 2π. Finally, the code has $\rho = 1$ and the code unit of mass is absent.

the solar wind to penetrate within closed field lines and often affect satellites. The nightside reconnection results in a contraction of the magnetotail, causing auroral substorms.

The understanding of magnetic reconnection is important not only in the Sun and heliosphere but also in a wide range of astrophysical objects. The push toward larger resolutions and larger numbers of particles in PIC and classical hybrid simulations have resulted in an impressive progress in understanding plasma reconnection on scales up to several hundred ion skin depths [7, 17]. At the same time, MHD fluid simulations also became commonplace [18, 19, 20], moving from 2D to 3D setups [10, 16, 21] and breaking through the critical Lundquist number required for spontaneous reconnection (10^4).

Many astrophysical objects, such as the interstellar medium in our Galaxy, feature relatively high level of ambient turbulence and the reconnection is supposed to be fast [9] due to the existing magnetic field stochasticity. In highly magnetized environments, e.g., the solar surface or the pulsar wind nebulae, the level of ambient turbulence is much lower, and the turbulence, spontaneously generated by the current sheet and fueled by the reconnecting field itself is of great interest. Magnetospheric observations which show an enhanced level of turbulence inside current sheets [22, 23], are supportive of this spontaneous reconnection picture.

Despite the success of MHD reconnection, it has been argued that the plasma reconnection rate could be higher and the value around $0.1v_A$ obtained by most authors in PIC and Hall-MHD simulations has been brought as an argument that plasma reconnection is qualitatively different due to the different mechanisms of how individual field lines break and reconnect. In this case, studying MHD limit in reconnection would be of little practical use, because all relevant MHD fluids in space physics and astrophysics are actually plasmas. This will also bring into question each MHD study having unresolved current layers, ostensibly due to an improperly resolved inflow condition at the discontinuity.

The question remain, however, why the observed solar reconnection rates are often much lower than $0.1v_A$? May plasma simulations not have yet achieved sufficient scale separation between fluid scales and the ion skin depth? This separation in nature is quite large, a typical length of a large-scale reconnection sheet in the solar corona is around $L = 10$ Mm. Thus, the width of the outflow in the reconnection region can be estimated as $d = rL \sim 0.4$ Mm, where $r \approx 0.04$ is the reconnection rate. This corresponds to 3×10^4 ion skin depths. It is also important to note that observations indicate that coronal reconnection is of the Y-point type, not a single X-point type. This means that important physics happens along the full 10 Mm length of the current layer and across the full width of 10^4 ion skin depths. Such vast dynamical range of scales is out of reach of the modern PIC or hybrid simulations.

The spontaneously evolving turbulent current layer in MHD develops turbulence that pretty much resembles generic MHD turbulence [16]. In this case, we expect turbulence to be local and the large-scale effects to be disconnected from the small-scale plasma effects and the MHD rate

Figure 1. Left: The all-periodic box reconnection setup with two current layers between the regions of reconnecting field $\pm B_{y0}$. Upper right: the surface of mixing of the electron fluid (the boundary between mixed and unmixed fluid), which is the surface where reconnection occurs, the color of the surface indicates the intensity of mixing. Yellow tubes are the magnetic field lines. Bottom right, we zoom into the surface of the electron fluid mixing and see complex and multi-scale nature of reconnection. The large-scale ridges visible on the surface on the left are characteristic angular outflows of Hall reconnection, while small scales shows a lot of structure pertaining to fast electron MHD turbulence.

of ~ 0.015 of Alfvén speed v_A should be universal.

The study of reconnection has split over various pathways, often creating confusion regarding the approximation being made. In the big picture of things, we subdivide reconnection into small-scale or plasma-scale and large-scale cases. (see, e.g., [24] for an overview of various reconnection regimes).

In the first case, the length of the reconnection layer is at most a few tens of ion skin depths and the layer thickness is of order or smaller than ion skin depth. This is the case which is commonly realized in plasma experiments. It can also be applied, with some caution, to the reconnection in the Earth magnetosphere which determines its connection to the solar wind. We believe that this regime has already been extensively studied with PIC codes, see, e.g. [7, 17].

We would like to concentrate on the second case, large-scale reconnection, where the length of the current layer is at least a hundred ion skin depths. Such cases include astrophysical reconnection, where it is important for understanding star formation, flares from pulsars and acceleration in high energy sources, such as blazars or GRBs. The other example is the reconnection in the solar corona, where the typical length of a large reconnection layer is 10

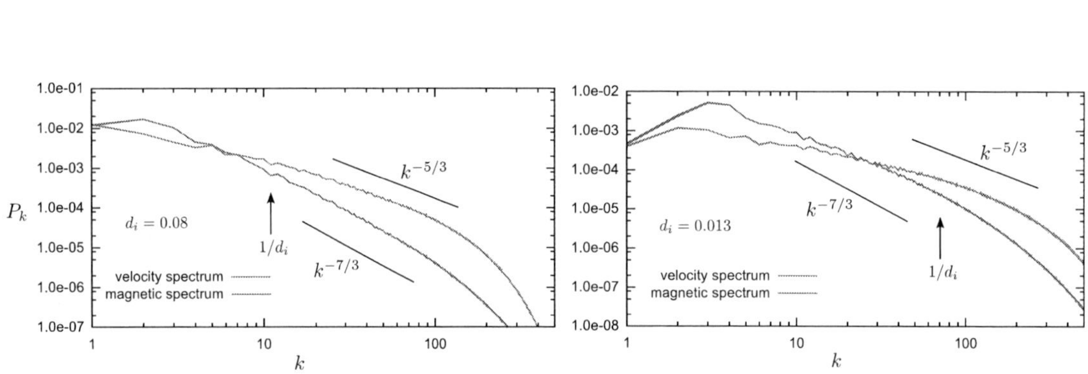

Figure 2. Kinetic and magnetic spectra in simulation R2 and Q2 (see Table 1). On small scales magnetic perturbations are primarily right-circularly-polarized whistlers, which have a steep spectrum of $k^{-7/3}$. While slower velocity perturbations are ion cyclotron waves, the nature of their cascade is still debated.

Mm. In this large-scale case, it has become clear recently that 2D physics is not sufficient to explain the physics of fast reconnection correctly. For example, while 3D results usually show fast reconnection, in the 2D case the three papers [25], [26], [27] all show different results: the first two reported slow reconnection with scalings $(d_i/L)^{3/2}$ and $(d_i/L)^{1/2}$ correspondingly, while the last – fast reconnection with $0.1v_A$. While 2D reconnection show plasmoids, the 3D case often have turbulent flux ropes.

The key role of turbulence in 3D reconnection still has to be studied in great detail. In this project, we decided to concentrate on the spontaneous reconnection that does not have any imposed turbulence in it. This may correspond to solar events, which look very quiet before erupting.

2. Previous MHD findings

Earlier, I simulated spontaneous reconnection in resistive MHD with the main results presented in [16]. The power spectra of turbulent perturbations resembled decaying magnetic turbulence with magnetic contribution dominating over kinetic on large scales and approximate equipartition on small scales. The total energy spectral slope was between -1.5 and -1.7, roughly consistent with Goldreich-Sridhar [28] scaling. Such spectral slopes are indicative of a turbulent cascade, alternatively called local-in-scale turbulence. The concept of such turbulence was introduced by Kolmogorov [29] who showed that the energy transfer between large and small scales is dominated by so-called local interactions, i.e. nonlinear interations between two scales separated by a factor of two or so. Scale-locality is a key ingredient in theories of turbulent reconnection. Indeed full scale-locality will imply that large-scale quantities, such as reconnection rate and dissipation rate per unit area should be independent on microphysics. Unlike the 2D case where large plasmoids couple large and small scales directly, the 3D turbulence we observe appears to be local-in-scale.

The dynamics of large scales, which determines such properties as the bulk reconnection and dissipation rates, will be disconnected from the dynamics on small scales, plasma parameters, and microscopic dissipation. We were arguing in [16] that this would imply that any large-scale spontaneous reconnection case will produce the same rate, i.e., $0.015v_A$.

3. Hall-MHD Equations

Hall-MHD is a simple framework for electron-proton plasmas. We use one-fluid incopressible Hall-MHD that nevertheless has a defining feature of dispersive plasmas below the ion skip depth d_i – the highly dispersive whistler waves that was argued to result in the higher reconnection

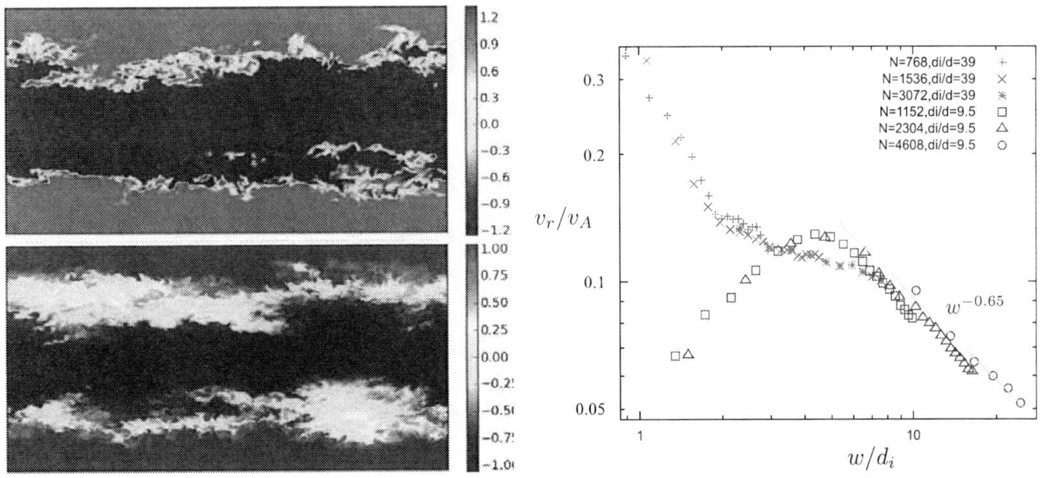

Figure 3. Left: we evolve passive scalars associated with ions (top) and electrons (bottom). Here we show xz slices of the data cubes, i.e. the slices parallel to the mean field B_x. Right: Dimensionless reconnection rate as a function of the mixing layer's width for all simulations in Table 1. The rate shows "classic" value of 0.1 for moderate width and then started to decline. We conjecture that for $w \gg d_i$ the rate will converge to the fluid value of 0.015. Note that the rate is largely insensitive to the size of the datacube. It does depend of the ratio of d_i/d (red vs blue), but only for early times when $w < 3..6d_i$.

rate because they open an X-point to a constant angle [30, 31].

Although the Hall-MHD is not capturing all elements of plasma dynamics, it had been verified to accelerate reconnection [2]. We hope, therefore, that Hall-MHD will allow us to uncover the puzzle of the discrepancy between MHD rates and plasma rates.

The incompressible Hall-MHD equations with explicit dissipation are

$$\begin{aligned}
\partial_t \mathbf{v} &= \hat{S}(-\omega \times \mathbf{v} + \mathbf{j} \times \mathbf{b}) - \nu_n \nabla^4 \mathbf{v}, && (1)\\
\partial_t \mathbf{b} &= \nabla \times ((\mathbf{v} - d_i \mathbf{j}) \times b) - \nu_n \nabla^4 \mathbf{b}, && (2)\\
\partial_t \phi_i &= \mathbf{v} \cdot \nabla \phi_i - \nu_n \nabla^4 \phi_i, && (3)\\
\partial_t \phi_e &= (\mathbf{v} - d_i \mathbf{j}) \cdot \nabla \phi_e - \nu_n \nabla^4 \phi_e. && (4)
\end{aligned}$$

We also show evolution equations for passive scalars ϕ_i, ϕ_e for ion and electron fluids respectively. Here we introduced the current density $\mathbf{j} = \nabla \times \mathbf{B}$ and vorticity $\omega = \nabla \times \mathbf{v}$. We renormalized magnetic field to velocity units $\mathbf{b} = \mathbf{B}/\rho^{1/2}$ (the absence of 4π is due to Heaviside units), and used the solenoidal projection operator $\hat{S} = (1 - \nabla\Delta^{-1}\nabla)$ to get rid of pressure.

The forth-order diffusivity is essential in the Hall-MHD case to terminate the whistler cascade. While natural forth-order diffusive terms exist in plasmas, such as an extra 4th order magnetic diffusivity due to the scalar electron viscosity, we treat this diffusivity as just a tool to terminate the cascade.

4. Simulations and results

We are following our earlier paper on spontaneous reconnection in resistive MHD [16] regarding initial setup, which is the thin planar current layer and tiny initial perturbations (see Fig. 1). Thus we will be able to compare results between MHD and Hall-MHD case directly.

We ran two series of simulations with different ratio of the ion skin depth d_i to the grid size d. See Table 1. Cases R0-R2, have "physically motivated" d_i, i.e., we assumed that the whistler cascade terminates on electron scales $\sim d_e = d_i\sqrt{m_e/m_i} \approx d_i/42$. This is actually the observed range of the steeper part of the spectrum of the dispersive cascade in the solar wind. This condition is approximately satisfied with $d_i N = 40(2\pi)$

Cases Q0-Q2 have d_e - d_i range of scales diminished by a factor of 4, e.g. using 4 times smaller d_i. Here we are motivated to go to the largest possible box sizes, in units of d_i. This also broadly corresponds to PIC simulations with the reduced m_i/m_e ratio.

All simulations were evolved to around $t = 4$ in code units at which point reconnection may be influenced by the finite box size (see Fig. 3).

We introduce w, the average width of the mixed scalar, i.e. the volume of the mixed scalar in one layer, divided by the XY cross-section of the datacube, $4\pi^2$ in our case. We determined the reconnection rate by the rate of mixing of the electron passive scalar as

$$v_r = dw/dt. \tag{5}$$

On Figures 1-3 we show that current layer indeed become fully turbulent. Figure 2 shows energy spectra of the velocity and magnetic perturbation which are broadly consistent with MHD turbulence on large scales and whistler turbulence for B on small scales and ion-cyclotron turbulence for v on small scales. Figure 1, right panel, show mixing surface of the passive scalar (a threshold value of passive scalar of 0.95), which looks like multi scale mixing surface characteristic for turbulence. Figure 3, left panel, indicates that mixing of electron fluid is more intense than of the ion fluid, which is characteristic for plasma reconnection. We found, however, that on large timescales the mixing rates of ion and electron fluids tend to converge, which is indicative of MHD regime. Figure 3, right panel, shows the reconnection rate as a function of average layer width. It is amazing that despite the difference in simulation setups the values for the rate largely converge for $w > 3d_i$. We also confirm that the finite box size is not a problem for $t = 4$, as the simulation in smaller box, say Q1 is reproduced by the simulation in larger box Q2.

The overall conclusion, perhaps greatly surprising that $0.1v_A$ is not a universal number, but rather a constant trend over ranges of scales currently explored by numerics. Starting with $w > 7d_i$ the rate is decreasing. We speculate that given the observed trend, the rate may approach MHD fluid rate at $w \approx 170d_i$, which would require box sizes five times larger than we used in this study, i.e., around 20000^3, which is currently not possible.

5. Acknowledgements
An award of computer time was provided by the Innovative and Novel Computational Impact on Theory and Experiment (INCITE) program. This research used resources of the Argonne Leadership Computing Facility, which is a DOE Office of Science User Facility supported under Contract DE-AC02-06CH11357. I'm grateful to Dmitry Pekurovsky who constantly improved performance of his P3DFFT library, which allowed me to perform larger simulations.

6. Appendix: Code detail.
1CARU5 is a pseudospectral C++ code which solves hydrodynamic, MHD, Reduced MHD, Hall MHD and Electron MHD equations. Some details may be found in [32, 33]. The pseudospectral code solves above equations as the ordinary differential equation in time for each spacial Fourier harmonic, the "pseudo" coming from the fact that nonlinear term is calculated in real space, and then converted back to Fourier space. Unlike finite difference schemes pseudospectral scheme does not suffer from either dispersion of dissipation errors. The dissipation is explicit and divergence-free condition for velocity and magnetic field are done with simple algebraic

operations in Fourier space. For time integration we use leapfrog scheme which is time-reversible and numerical dissipation is absent, because nonlinear term, calculated in this manner, preserve both energy and cross-helicity.

By testing the code, we verified that the following standard prescriptions for the dissipation and time step are sufficient for the simulation to stay accurate and stable. The forth-order diffusivity coefficient ν_4 is determined by

$$\left(\frac{\nu_4^3}{\epsilon d_i^2}\right)^{1/8} = \frac{3.3}{N}, \tag{6}$$

where ϵ is the dissipation rate, which in our case was taken 0.08 in code units and N refers to the number of grid points in the larger dimension (which is assumed to have the size 2π in code units). This condition terminates whistler cascade on grid scale, where 3.3 was found empirically for sufficient numerical accuracy. The time step dt was chosen by CFL number of 0.1 for the whistler mode:

$$dt = 0.1(2\pi/N)^2/v_A d_i. \tag{7}$$

Hall-MHD code with two passive scalars does 18 direct, and 8 inverse Fourier transforms per time step, 26 in total. Since in Hall-MHD velocity evolves on the longer timescale than the magnetic field, it is possible to save some computing time by sub-stepping (see, e.g., [34]). The small-scale evolution of b is faster than v by a factor of around $d_iN/(2\pi)$. We implemented substepping routine when certain v evolution steps are skipped, e.g., in cases Q0-Q3 we will be doing 3 magnetic steps per 1 kinetic, which gives a substepping reduction of SR=4/6=0.67 of the original number of transforms and in the cases R0-R2 we will have 6 magnetic steps per 1 kinetic, SR=7/12=0.58. With substepping improvement, Hall-MHD code is about as fast as hydro code, per time step. Despite massive amounts of FFT transform, each requiring all-to-all communication, 1CARU5 scaled well up to 256k cores on IBM BG/Q platform and reached the peak of 0.3 Pflops on ALCF Mira, mostly because of the highly optimized parallel FFT library P3DFFT [35].

References

[1] Dere K P 1996 *The Astrophysical Journal* **472** 864 URL `http://stacks.iop.org/0004-637X/472/i=2/a=864`
[2] Birn J, Drake J F, Shay M A, Rogers B N, Denton R E, Hesse M, Kuznetsova M, Ma Z W, Bhattacharjee A, Otto A and Pritchett P L 2001 JGR **106** 3715–3720
[3] Sweet P A 1969 Ann. Rep. A&A **7** 149
[4] Dungey J W 1961 *Physical Review Letters* **6** 47–48
[5] Parker E N 1957 JGR **62** 509–520
[6] Drake J F, Swisdak M, Che H and Shay M A 2006 Nature **443** 553–556
[7] Daughton W, Roytershteyn V, Karimabadi H, Yin L, Albright B J, Bergen B and Bowers K J 2011 *Nature Physics* **7** 539–542
[8] Che H, Drake J F and Swisdak M 2011 Nature **474** 184–187
[9] Lazarian A and Vishniac E T 1999 Astrophys. J. **517** 700–718 (*Preprint* `arXiv:astro-ph/9811037`)
[10] Kowal G, Lazarian A, Vishniac E T and Otmianowska-Mazur K 2009 Astrophys. J. **700** 63–85
[11] Eyink G L, Lazarian A and Vishniac E T 2011 Astrophys. J. **743** 51 (*Preprint* `1103.1882`)
[12] Biskamp D 1986 *Physics of Fluids* **29** 1520–1531
[13] Loureiro N F, Schekochihin A A and Cowley S C 2007 *Physics of Plasmas* **14** 100703 (pages 4)
[14] Huang Y M and Bhattacharjee A 2010 *Physics of Plasmas* **17** 062104 (*Preprint* `1003.5951`)
[15] Loureiro N F, Samtaney R, Schekochihin A A and Uzdensky D A 2012 *Physics of Plasmas* **19** 042303 (*Preprint* `1108.4040`)
[16] Beresnyak A 2017 Astrophys. J. **834** 47 (*Preprint* `1301.7424`)
[17] Karimabadi H, Dorelli J, Roytershteyn V, Daughton W and Chacón L 2011 *Physical review letters* **107** 025002
[18] Birn J, Artemyev A, Baker D, Echim M, Hoshino M and Zelenyi L 2012 *Space science reviews* **173** 49–102
[19] Birn J, Runov A and Hesse M 2014 *Journal of Geophysical Research: Space Physics* **119** 3604–3616

[20] Samtaney R, Loureiro N, Uzdensky D, Schekochihin A and Cowley S 2009 *Physical review letters* **103** 105004
[21] Oishi J S, Mac Low M M, Collins D C and Tamura M 2015 ApJ **806** L12 (*Preprint* 1505.04653)
[22] Matsumoto H, Deng X H, Kojima H and Anderson R R 2003 GRL **30** 1326
[23] Cattell C, Dombeck J, Wygant J, Drake J F, Swisdak M, Goldstein M L, Keith W, Fazakerley A, André M, Lucek E and Balogh A 2005 *Journal of Geophysical Research (Space Physics)* **110** A01211
[24] Ji H and Daughton W 2011 *Physics of Plasmas* **18** 111207
[25] Wang X, Bhattacharjee A and Ma Z 2001 *Physical review letters* **87** 265003
[26] Fitzpatrick R 2004 *Physics of Plasmas* **11** 937–946
[27] Huba J and Rudakov L 2004 *Physical review letters* **93** 175003
[28] Goldreich P and Sridhar S 1995 Astrophys. J. **438** 763–775
[29] Kolmogorov A 1941 *Akademiia Nauk SSSR Doklady* **30** 301–305
[30] Sonnerup B U Ö 1979 *Space Plasma Physics: The Study of Solar-System Plasmas. Volume 2* p 879
[31] Shay M A, Drake J F and Swisdak M 2007 *Physical Review Letters* **99** 155002 (*Preprint* 0704.0818)
[32] Beresnyak A and Lazarian A 2009 Astrophys. J. **702** 460–471 (*Preprint* 0904.2574)
[33] Beresnyak A and Lazarian A 2009 Astrophys. J. **702** 1190–1198 (*Preprint* 0812.0812)
[34] Huba J 2005 *Proceedings of ISSS* vol 7 pp 26–31
[35] Pekurovsky D 2012 *SIAM Journal on Scientific Computing* **34** C192–C209

IOP Conf. Series: Journal of Physics: Conf. Series **1031** (2018) 012002 doi:10.1088/1742-6596/1031/1/012002

Three-dimensional simulations of solar wind turbulence with the hybrid code CAMELIA

L. Franci[1,2], P. Hellinger[3], M. Guarrasi[4], C. H. K. Chen[2], E. Papini[1], A. Verdini[1], L. Matteini[5], and S. Landi[1,6]

[1]Dipartimento di Fisica e Astronomia, Università degli Studi di Firenze, Firenze, Italy
[2]School of Physics and Astronomy, Queen Mary University of London, London, UK
[3]Astronomical Institute, CAS, Prague, Czech Republic
[4]SuperComputing Applications and Innovation Department, CINECA, Bologna, Italy
[5]LESIA-Observatoire de Paris, Meudon, France
[6]INAF - Osservatorio Astrofisico di Arcetri, Firenze, Italy

E-mail: `franci@arcetri.astro.it`

Abstract. We investigate the spectral properties of plasma turbulence from fluid to sub-ion scales by means of high-resolution three-dimensional (3D) numerical simulations performed with the hybrid particle-in-cell (HPIC) code CAMELIA. We produce extended turbulent spectra with well-defined power laws for the magnetic, ion bulk velocity, density, and electric fluctuations. The present results are in good agreement with previous two-dimensional (2D) HPIC simulations, especially in the kinetic range of scales, and reproduce several features observed in solar wind spectra. By providing scaling tests on many different architectures and convergence studies, we prove CAMELIA to represent a very efficient, accurate and reliable tool for investigating the development of the turbulent cascade in the solar wind, being able to cover simultaneously several decades in wavenumber, also in 3D.

1. Introduction

Turbulence in magnetized collisionless plasmas, such as the solar wind, is one of the major challenges of space physics and astrophysics. Both the anisotropic flow of energy toward smaller scales (cascade) and the damping of energy at dissipative scales are poorly understood. Solar wind turbulent fluctuations are generated at large magneto-hydrodynamic (MHD) scales and dissipated at scales where particle kinetics dominates. Indeed, in situ measurements show spectra of the plasma and electromagnetic fields with a power-law scaling spanning several decades in frequency, with a spectral break in the magnetic and density spectrum at proton scales, separating the MHD inertial range cascade from a second power-law interval at kinetic scales (see [1] for a recent summary). A further change in the spectral properties occurs at the electron scales, although distinguishing between an exponential cut-off [2] or a power law [3] is not straightforward and thus no universal behavior is observed.

Very large numerical resources are required to investigate the whole turbulent cascade, since at least two full decades in wavevectors across the transition needs to be covered simultaneously. Full Particle-In-Cell (PIC) simulations represent the most comprehensive numerical tool for simulating the plasma dynamics up to electron spatial and temporal scales, e.g., [4, 5, 6, 7, 8, 9, 10, 11, 12, 13]. However, due to computational limitations, they typically employ a limited accuracy (e.g., small resolution, small number of particles, small ion-to-electron mass ratio).

Run	Grid	Δx	$L\,[d_i]$	B^{rms}/B_0	$k_0 d_i$	$k_{inj} d_i$	ppc
A	512^3	0.25	128	0.40	0.05	0.25	2048
B	256^3	0.25	64	0.38	0.10	0.30	2048
C	256^3	0.25	64	0.38	0.10	0.30	1024
D	256^3	0.25	64	0.38	0.10	0.30	512
E	2048^2	0.125	256	0.24	0.025	0.28	8000

Table 1. List of simulations and their main different parameters

Alternatively, reduced models have been largely used in the last decade to simulate plasma turbulence at kinetic scales, both in 2D and 3D, e.g., Hall-MHD [14, 15, 16, 17, 18], Electron-MHD [19, 20, 21], Gyrokinetic [22, 23, 24, 25], finite Larmor radius-Landau fluid [26, 27, 28], hybrid Vlasov-Maxwell [29, 30, 31, 32, 33, 34, 35, 36], and Hybrid Particle-In-Cell (HPIC) [37, 38, 39, 40, 41, 42, 43, 44, 45].

In particular, very high-resolution 2D simulations, performed with the HPIC code CAMELIA, recently fully covered the transition between fluid and kinetic scales [46, 47, 48]. Moreover, they produced extended turbulent spectra with well-defined power laws for the plasma and electromagnetic fields, in agreement with solar wind observations. Such results have been recently extended to 3D [49].

2. Numerical setup

The simulations presented here have been performed using the 3D HPIC code CAMELIA (see Sec. 5 for details). The characteristic spatial unit is the ion (proton) inertial length, $d_i = v_A/\Omega_i$, v_A being the Alfvén speed, while the temporal unit is the inverse ion gyrofrequency, $1/\Omega_i$. We compare four simulations, all employing a periodic cubic grid with spatial resolution $\Delta x = 0.25\,d_i$, with two different numbers of grid points (512^3 and 256^3) and, consequently, two different box sizes ($L_{box} = 128\,d_i$ and $64 d_i$), and with three different numbers of (macro)particle-per-cell, ppc, representing protons (512, 1024, and 2048). In all four cases, the equilibrium system is a homogeneous plasma, with uniform density and temperature, embedded in a mean magnetic field, $\boldsymbol{B}_0$. The plasma beta, i.e., the ratio of the plasma pressure to the magnetic pressure, is $\beta = 0.5$ for both the protons and the electrons. This system is perturbed with magnetic and ion bulk velocity fluctuations, purely perpendicular to $\boldsymbol{B}_0$, consisting of a superposition of Fourier modes of equal amplitude and random phases in the range $k_0 < k < k_{inj}$ ($k = \sqrt{k_x^2 + k_y^2 + k_z^2}$). The minimum wavenumber, $k_0 = 2\pi/L_{box}$, is $0.05\,d_i^{-1}$ for the large box and $0.10\,d_i^{-1}$ for the small one, while the maximum injection scale is $k_{inj} = 0.25\,d_i^{-1}$ and $0.30\,d_i^{-1}$, respectively. A more detailed description of the initialization and of the physical and numerical parameters can be found in [46, 47]. The main differences between the four 3D simulations are summarized in Tab. 1 (runs A-D), where we also recall the parameters of a 2D HPIC simulation with the same β (run E), presented in [46, 47].

3. Results

We start with the results of run A, already presented in [49], focusing on the spectral properties of th electromagnetic and plasma fluctuations at the time when the turbulent cascade has fully developed. For the definitions of the 1D omnidirectional spectra, P_{1D}, the 1D reduced perpendicular and parallel spectra with respect to the global mean field $\boldsymbol{B}_0$, $P_{1D,\perp}$ and $P_{1D,\|}$, and for the description of the filtering procedure used to obtain them from the 3D power spectra, please refer to [49].

In the top panels of Fig. 1, we show a comprehensive overview of the 1D omnidirectional power spectra of all fields for the 3D run A (left), to be compared with the results of the 2D run E [46, 47] (right). Additionally, characteristic power laws are drawn with dashed black lines as reference. The main

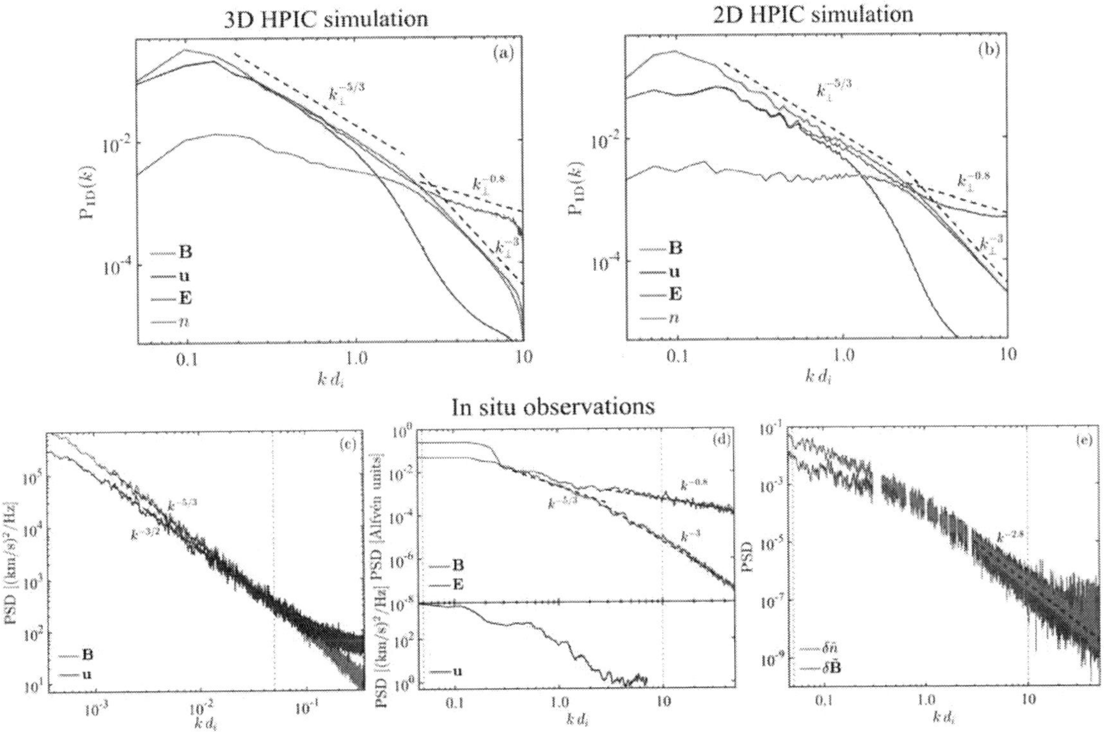

Figure 1. Qualitative comparison of the 1D omnidirectional spectra of the magnetic field (red), ion bulk velocity (blue), electric field (green), and density (purple) between the 3D run A and the 2D run E (panels (a) and (b), respectively). Additionally, the bottom panels show results of solar wind and magnetosheath observations from [50] (c), [51] (d), and [52] (e). A light blue and an orange dashed lines mark the smaller and larger wavenumbers covered by the 3D HPIC simulation, respectively.

similarities between the 3D and the 2D cases are: (i) the double power-law behavior of the magnetic fluctuations, with a spectral index close to $-5/3$ at MHD scales and a steepening (~ -3) at sub-ion scales, (ii) the strong coupling between magnetic and density fluctuations at sub-ion scales, with the same amplitude and similar slopes, (iii) the flattening of the electric field spectrum in the kinetic range, with a spectral index close to -0.8, (iv) the sharp decline of the ion bulk velocity spectrum, which quickly reaches the noise level. All these features are in broad agreement with solar wind observations [50, 51, 52], as shown in the bottom panels of Fig. 1, where a light blue and a orange dashed lines mark the smaller and larger wavenumbers covered by the 3D HPIC simulation, respectively. A particular attention needs to be paid to the ion bulk velocity spectrum, which is the most difficult to measure, in both simulations and data, due to noise (numerical and instrumental, respectively). Indeed, the flattening toward small scales observed in Fig. 1c is only due to noise, reached already at $kd_i \sim 0.1$ (see the discussion in [50]). With better measurement in the kinetic range, a steeper slope can be estimated from Fig. 1d (~ -3.5, see [51]). Such slope is much steeper than the one in the MHD range and steeper than the slopes of the spectra of all other fields at sub-ion scales. Although the range of scales here is not large enough to compare in details with simulations, there is actually reasonably good agreement, at the level we can measure. In particular, the ion bulk velocity spectrum drops around $kd_i \sim 1$ and exhibits the steepest slope. As far as the MHD scales are concerned, differences arise for the velocity, electric field, and density spectra. In particular, in the 3D case the magnetic and velocity fluctuations are more strongly coupled, since they are almost equal amplitude, in Alfvén units, up to $kd_\mathrm{i} \lesssim 1$. A higher level of density

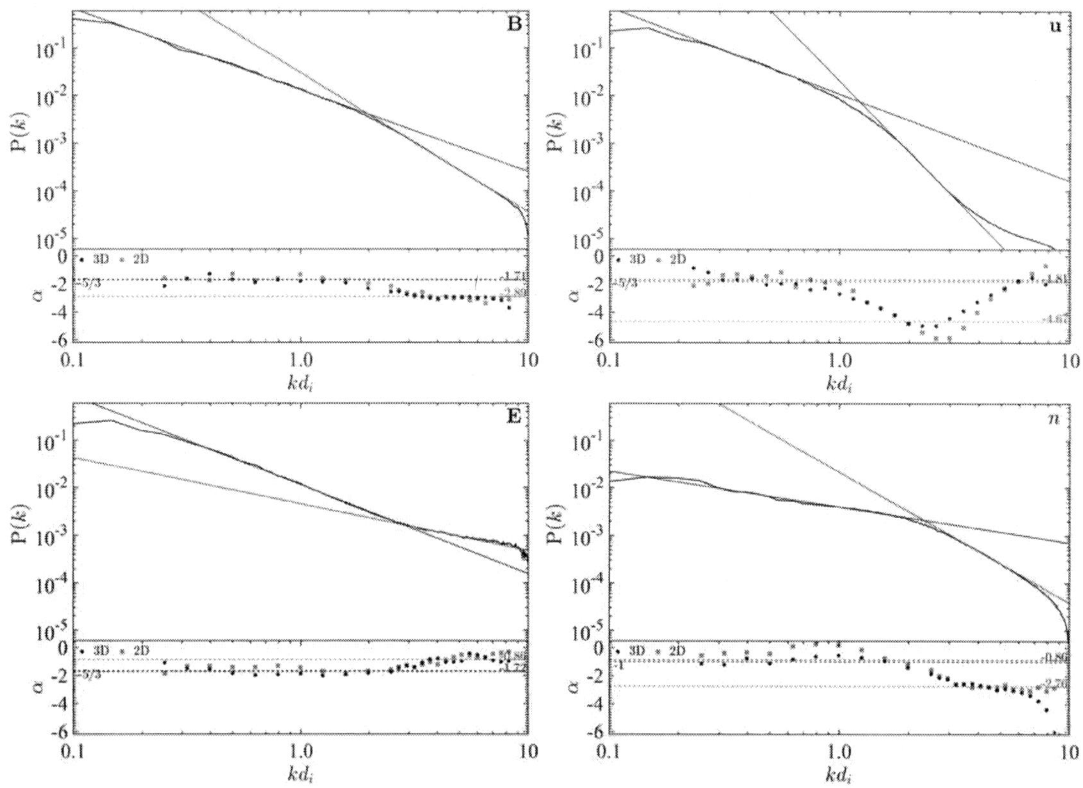

Figure 2. 1D omnidirectional power spectra of the magnetic field (top left), ion bulk velocity (top right), electric field (bottom left), and density (bottom right) for run A. Power-law global fits are reported for the MHD range (blue) and the kinetic range (red). In the bottom region of each panel, the value of the local spectral index, α (black), is compared with the results of the two global fits in corresponding colors and with the local spectral index of the 2D run E (green).

fluctuations is also observed. Unexpectedly, a better agreement is recovered between observations and the 2D simulation for what concerns the ion bulk velocity spectrum. The latter exhibits a lower level of fluctuations than the 3D case, and therefore a sizeable residual energy (difference between magnetic and kinetic energy). This discrepancy between 2D and 3D might be due to the different geometry or to the different setting, e.g., to the initial level of fluctuations or resolution (see Tab. 1). Indeed, the 3D physical domain is a factor of 2 smaller than the 2D one and this, and this might constrain the dynamics at the largest scales.

In Fig. 2, we show in separate panels the 1D omnidirectional spectra of the magnetic field, $\boldsymbol{B}$ (top left), ion bulk velocity, $\boldsymbol{u}$ (top right), electric field, $\boldsymbol{E}$ (bottom left), and density, n (bottom right) for run A. We fit each spectrum with two power laws, one the MHD range (blu straight line) and one in the kinetic range (red), as already done for 2D HPIC simulations [48]. Additionally, we perform local power-law fits over many small intervals in the range $k_\perp d_i \in [0.25, 10]$. The values of the local spectral index, α, are shown with black dots in the bottom part of each panel. Two horizontal dashed line represent the slopes of the two global fits, with the respective colors. We also directly compare α for the 3D run A (black) and the 2D run E (green).

The magnetic field spectrum exhibits two clear power-law intervals spanning almost two decades in wavevector, with a spectral index of $\sim -5/3$ and ~ -2.9 at MHD and sub-ion scales, respectively, and a

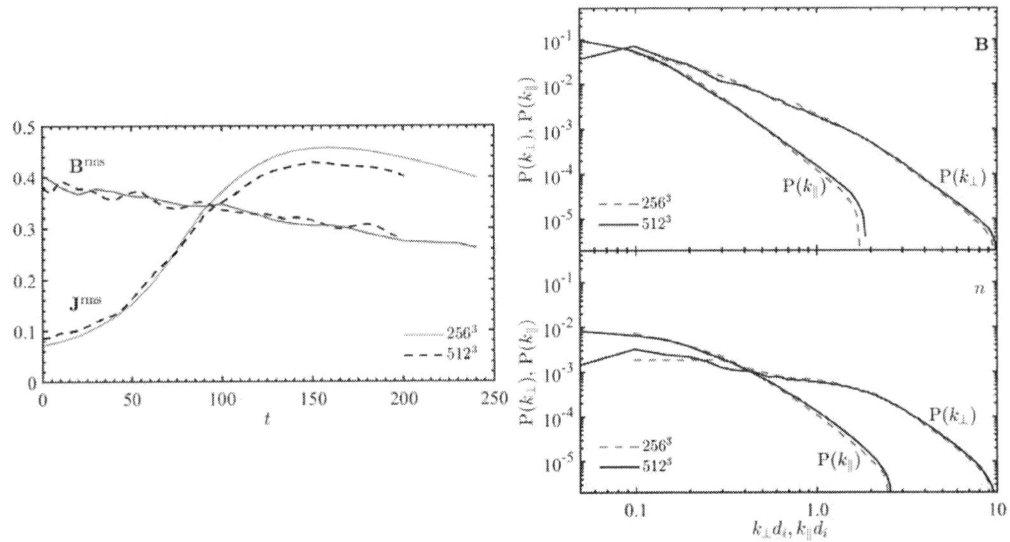

Figure 3. Comparison between run A and B, employing different numbers of grid points. Left panel: time evolution of rms of the magnetic field, *B*rms and of current density, $J^{\rm rms}$. Right panels: 1D reduced perpendicular and parallel spectra of the magnetic (red, top) and density (purple, bottom) fluctuations.

transition at $kd_{\rm i} \gtrsim 2$. The velocity spectrum shows a power-law-like behaviour at large scales, although less extended than the magnetic field's, with a spectral index close to $-5/3$ in 3D and slightly closer to $-3/2$ in 2D. For $kd_{\rm i} \gtrsim 1$, the spectrum drops very rapidly, reaching the ppc noise level at slightly smaller scales. The hint of a power-law shape can be inferred, significantly steeper than the magnetic field's. However, the small extent in k prevents us from providing either a clear evidence of a power law with respect to an exponential drop, nor a common value of the slope between the 3D and the 2D runs. The electric field spectrum shows an extended Kolmogorov-like power law at MHD scales and flattens toward a spectral index of ~ -0.8 around ion scales, consistently with the generalized Ohm's law [47]. Finally, the density spectrum is almost flat at intermediate scales, with a slope ~ -0.9, although it seems to be slightly steeper at the largest scales. A transition is clearly observed around ion scales, followed by a power law with a spectral index of ~ -2.8.

4. Role of numerical parameters

A very high accuracy is required to investigate the turbulent cascade from large MHD scales to small kinetic ones. In particular, in order to quantitatively compare numerical results with solar wind observations (e.g., the spectral indices of electromagnetic and plasma fluctuations and the scale of the MHD-kinetic spectral break) one needs to cover simultaneously at least two full decades in wavevectors across the transition. Moreover, artificial effects due to the finite spatial resolution, the finite box size and the finite number of particles have to be carefully checked, so that the numerical results can be considered robust and reliable. While all this has been feasible in 2D in recent years [47], it is not trivial in 3D. Here we provide such an analysis, in support of the results recently presented in [49].

4.1. Box size

We investigate the effects of the box size by running two simulations with the same setting and numerical parameters, except for different grid sizes, 512^3 (run A) and 256^3 (run B). Consequently, the box size changes from $128\,d_{\rm i}$ to $64\,d_{\rm i}$. The left panel of Fig. 3 shows that the use of a smaller box does not affect considerably the time evolution of the system. The rms value of the magnetic fluctuations, $B^{\rm rms}$ (red

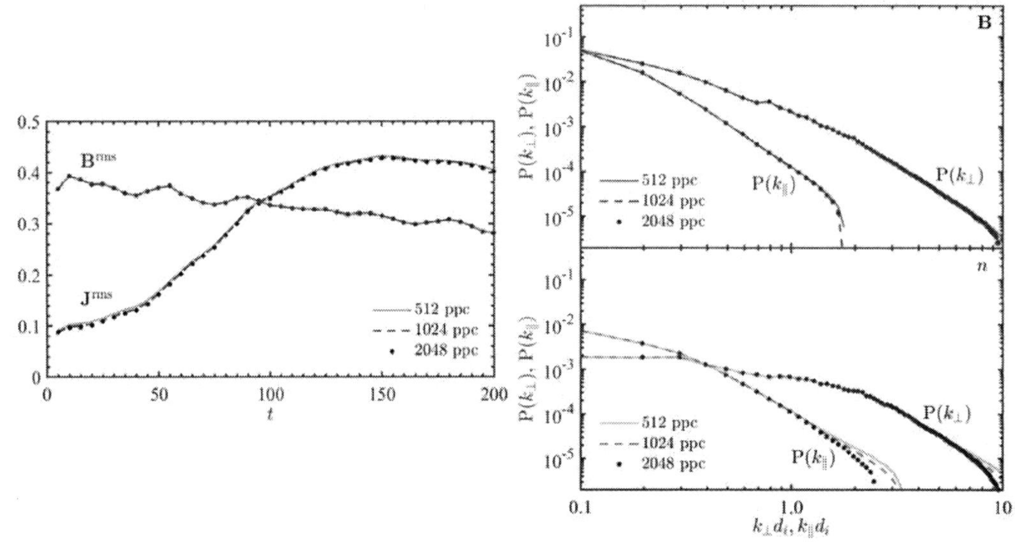

Figure 4. The same as in Fig. 3, but for the comparison between run B, C, and D, employing different numbers of ppc.

line), exhibits very small variations around the same average value, with the same slow decrease, while the rms of the current density, J^{rms} (black), shows the very same behavior up to $t \sim 100$ but then it reaches a slightly smaller peak value. This difference is mainly due to the fact that for run B the same resistivity as for run A was set, while it should be slightly decreased in order to take into account the small increase in the minimum wavenumber and in the injection scale. In the right panel of Fig. 3, we also compare the reduced perpendicular and parallel spectra of the total magnetic fluctuations (top) and the density fluctuations (bottom) between runs A and B. The differences in $P^{\boldsymbol{B}}_{1D,\perp}$ and $P^{n}_{1D,\perp}$ are indeed negligible. Even $P^{\boldsymbol{B}}_{1D,\|}$ and $P^{n}_{1D,\|}$ show no significant differences, except for a very small deviation at the smallest parallel scales, that are not suppressed by the filtering procedure, i.e., for $k_\| \, d_i \gtrsim 1$.

4.2. *Number of particles (ppc)*

We investigate the effects of the number of particles, in order to determine the optimal number which allows for obtaining reliable results at sub-ion scales. We use two simulations employing the small box and the same parameters as in run B, except for different numbers of ppc, i.e., 1024 (run C) and 512 (run D) instead of 2048. The left panel of Fig. 4 shows that the use of fewer particles, at least a factor of 4, has no effect on the evolution of global quantities: both B^{rms} and J^{rms} exhibit the same behavior over time. The same consideration holds for the spectral properties of the magnetic fluctuations (top right panel of Fig. 4). The only, almost negligible, differences arise in $P^{n}_{1D,\perp}$ (bottom right) at the scales corresponding to the grid spacing and, correspondingly, in $P^{n}_{1D,\|}$ at scales $k_\| \, d_i \gtrsim 1$. Note that the filtering procedure already removes most of the artificial features due to the noise, making the use of a very large number of ppc unnecessary. As a consequence, for an intermediate-beta case such as the one presented here, 512 ppc could be already enough in order to investigate the development of the kinetic cascade and the spectral properties. However, if one intends to investigate the evolution of the proton temperature (not discussed here), the use of a larger number of ppc would likely be advisable, as previously shown in 2D [47].

5. The hybrid particle-in-cell (HPIC) code CAMELIA

5.1. Overview and applications

CAMELIA[1] (Current Advance Method Et cycLIc leApfrog) is a HPIC code, where the electrons are considered as a massless, charge neutralizing fluid, whereas the ions are described as macroparticles, i.e., statistically-representative portions of the distribution function in the phase space. Up to 2016, the code was only parallelized through the Message Passing Interface (MPI) Library using spatial and/or particle decompositions, with the former being available in all spatial directions. In early 2017, a hybrid parallelization using both MPI and the Open MultiProcessing (OpenMP) Interface has been implemented, which fully exploits the latest architectures for High Performance Computing, including many-integrated cores Intel Knights Landing processors. Different formats can be chosen for the output files, i.e., ASCII, binary, or HDF5, including parallel binary (MPI-IO) and parallel HDF5. A checkpoint/restart procedure is implemented, with a HDF5 file written by each MPI process, assuring a wide portability across different platforms. The boundary conditions are periodic in all directions. Alternatively, in 2D, one can choose reflecting boundary conditions on one side and open boundary conditions, with a continuous injection of particles, on the other side (e.g., allowing for the generation of shock waves).

CAMELIA has two major extensions, being able to model (i) the effects of a slow expansion, using the Hybrid Expanding-Box (HEB) model (Hellinger et al., 2003; Hellinger and Travnicek, 2005), and (ii) the effects of Coulomb collisions, using the Langevin representation (Hellinger and Travnicek, 2010, 2015). In the last decade, it has been extensively used to investigate the spectral and heating properties of solar wind turbulence from MHD to sub-ion scales in 2D [46, 47], recovering a good agreement with solar wind observations. Such findings have also been validated by comparing with the results obtained by numerical simulations performed with the Eulerian hybrid Vlasov-Maxwell code HVM [53]. A particular focus has been devoted to the correlation between vorticity and proton temperature [54], to the dependence of the ion-scale spectral break on the plasma beta [48], and to the role of magnetic reconnection as a trigger for the sub-ion scale cascade [55]. The study of the spectral behavior of electromagnetic and plasma fluctuations has recently been extended to 3D [49]. Furthermore, the coexistence of kinetic instabilities with strong plasma turbulence has been investigated by means of 2D HEB simulations including the effects of the solar wind expansion [42, 56].

5.2. Numerical scheme

CAMELIA is based on the Current Advance Method and Cyclic Leapfrog (CAM-CL) code of Matthews [57]. The system is governed by the Vlasov-fluid equations, comprising the equations of motions for individual ions, and the electron fluid equations. The ions are described by a PIC model, a technique used to solve a certain class of partial differential equations where particles (or fluid elements) in a Lagrangian frame are tracked in continuous phase space, whereas moments of the distribution such as densities and currents are computed simultaneously on Eulerian (stationary) mesh points. The central engine of the code is a CAM-CL algorithm, which integrates the differential equations in a manner that is explicit in time and spatially local. The plasma has two time-independent components: ion macro-particles with position and velocities and the magnetic field specified at the nodes of a regular computing grid. The displacement current is neglected in Maxwell's equations, so there is no equation for the time-evolution of the electric field, which is just a function of the ion moments (interpolated at grid points from particle data), the magnetic field, and the electron temperature. (Bi, tri)linear interpolation is used for moment collection and evaluation of the Lorentz force at particle positions. An explicit form of the time-dependent differential equation is used. A CAM method is used to advance the ion current density. The original leapfrog particle advance has been replaced by the more precise Boris' algorithm [58], which has been proved to have an excellent long-term accuracy [59]. This requires the fields to be known at half time step ahead of the particle velocities, which is achieved by advancing the current density to this time

[1] Official CAMELIA website: http://terezka.asu.cas.cz/helinger/camelia.html

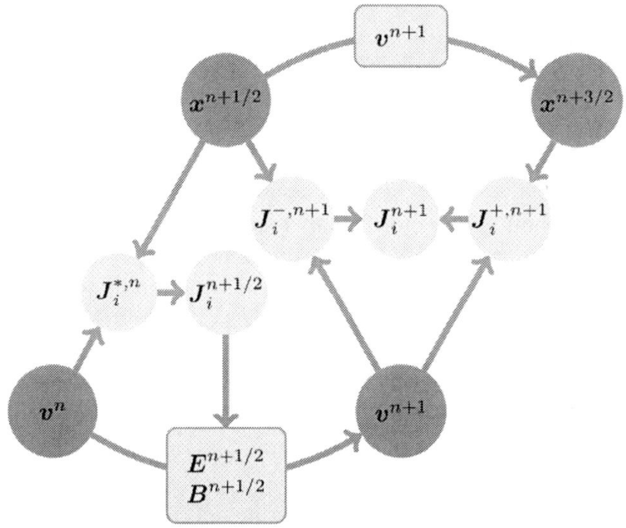

Figure 5. Schematic diagram of the code CAMELIA. The particle positions and velocities are taken at the different times, separated by half the (particle) time step $\Delta t/2$, v^n and $x^{n+1/2}$, where the subscript n refers to time t and $n+1/2$ to time $t+\Delta t/2$. At the beginning, there are the current density, $J_i^{n+1/2} = J(x^{n+1/2})$, and the "free-streaming" ionic current, $J_i^{*,n} = J_i(x^{n+1/2}, v^n)$, as well as $J^n = J_n(x_n)$ and $J_i^n = J_i(x_n, v_n)$. The magnetic field is advanced (using two copies via cyclic leapfrog) from B_n to $B^{n+1/2}$ with $E(J^n, J_i^n, B)$. The Current Advance Method advances $J_i^{*,n}$ to $J_i^{n+1/2}$. Using $B^{n+1/2}$ and $E^{n+1/2} = E(J^{n+1/2}, J_i^{n+1/2}, B^{n+1/2})$, the particles are advenced using the Boris' scheme to v^{n+1} and $x^{n+3/2}$ and moments are collected, $J^{n+3/2} = J(x^{n+3/2})$; J^{n+1} is obtained as an average of $J^{n+1/2}$ and $J^{n+3/2}$. Forward and backward "free-streaming" ionic currents, $J_i^{-,n+1} = J_i(x^{n+1/2}, v^{n+1})$ and $J_i^{+,n+1} = J_i(x^{n+3/2}, v^{n+1})$, are collected and their average give J_i^{n+1}. Finally, $B^{n+1/2}$ is advanced to B^{n+1}.

	Peak Performance	Clock	Cores/node	RAM/node	Network
SuperMike-II	146 TFlops	2.6 GHz	16	32 GB	40 Gbit/s Infiniband
Fermi	2 PFlops	1.6 GHz	16	16 GB	5D Torus
Cartesius	1.8 PFlops	2.6 GHz	24	64 GB	100 Gbit/s Infiniband
Marconi-A1	2 PFlops	2.3 GHz	36	128 GB	Intel OmniPath
Marconi-A2	11 PFlops	1.4 GHz	68	16+96 GB	Intel OmniPath

Table 2. Main parameters of the different HPC systems used for scaling tests

step, with only one computational pass through the particle data at each time step (see Fig. 5). Cyclic leapfrog is used to advance the magnetic field with a sub-stepping using two copies of the magnetic field.

5.3. Performance

CAMELIA has been optimized and run on many thousands of cores on High Performance Computing (HPC) systems with different architectures. Here we provide scaling tests performed on:

- SuperMike-II (Intel Sandy Bridge) @ Louisiana State University, Baton Rouge, U.S. (2015)

- Fermi (IBM Blue Gene/Q) @ CINECA, Bologna, Italy (2016)
- Cartesius (Intel Haswell) @ SURFsara, Amsterdam, The Netherlands (2017)
- Marconi-A1 (Intel Broadwell) @ CINECA, Bologna, Italy (2017)
- Marconi-A2 (Intel Knights Landing) @ CINECA, Bologna, Italy (2017)

The main features of these HPC systems are summarized and compared in Tab. 2.

For all systems, we performed test simulations with seven different problem sizes (three for each machine) for measuring the strong-scaling performances and four different sizes (one or more for each machine) for the weak scaling (see Fig. 6- 10). In CAMELIA, we evolve more than 30 different field components (taking into account both physical variables and additional temporary auxiliary grid variables), plus the 3 position components and the 3 velocity components of each particle. Since we always put thousands of particles in each grid cell, the particles positions and velocities represent the dominant contribution to the memory requirement, which is of the order of 4 Bytes (single precision) $\times$ 6 variables $\times$ number of ppc $\times$ number of cells. In Tab. 3, we list all the different problem sizes, indicating the total number of particles in the whole simulation domain and the corresponding total memory requirement. A color has been associated to each problem size, so that a qualitative comparison of code performances for a fixed sized between different machines can be easily done by eye. Note that the same global size can result from different configurations, i.e., number of grid points $\times$ number of ppc. These are explicitly indicated in the legend of each plot.

In Fig. 6-10, we report all the scaling tests that we performed on the above mentioned systems, both for strong scaling (left panels) and weak scaling (right panels), versus the number of cores. Until 2016, only a pure MPI parallelization was implemented, so the number of cores for SuperMike-II, Fermi, Cartesius and Marconi-A1 also corresponds to the number of MPI processes. Since early 2017, a hybrid MPI+OpenMP version is also available, allowing us to fully exploit the Intel Many Integrated Core Architecture. Test simulations on Marconi-A2 were run with 32 MPI processes per node, i.e., half the number of cores per node (since 4 over 68 were left available for the operating system and I/O operations) and 8 OpenMP threads per MPI task. Therefore, the code scalability has been measured up to 16384 MPI tasks $\times$ 8 OpenMP threads/MPI task = 131072 total threads.

Fig. 6-10 show that both the strong and the weak scalability are quite good in all the analyzed architectures. In particular, the weak scalability is very promising. The parallel efficiency on standard Intel x86 architectures is near to 1 at least up to 2048 cores. Both in the Haswell and Broadwell cases (Cartesius and Marconi-A1) this efficiency is also slightly greater than 1, and this is probably due to a more optimal use of the memory with an high number of cores. Furthermore, the weak scalability on Blue Gene/Q architectures is also more promising. Indeed, in almost all the analyzed cases (with the exception of the smallest one) the parallel efficiency in near to 1 at least up to 8192 cores. This increase of parallel performance is likely due to a combination of factors, the most important ones being the high speed Torus 5D Blue Gene/Q network and the ratio between the network (and memory, of course) bandwidth and the CPU clock. The weak scaling tests on F also clearly show that, for the same problem size, CAMELIA is more efficient in running 3D simulations than 2D, due to the more favorable ratio between computations and communications when parallelizing the computational domain into cubes instead of rectangles. Finally for what concerns the parallel performances on KNL machines, the weak parallel efficiency is close to 1, as expected, up to 4096 cores (i.e. the same number of nodes as in the Broadwell case), which correspond to 16384 total OpenMP threads, while it is lower for larger number of cores, mainly due to the fact that a smaller workload per core has been employed.

Size	Particles	RAM	Color
XS	$\sim 2.1 \times 10^9$	48 GB	violet
S	$\sim 8.6 \times 10^9$	192 GB	indigo
M	$\sim 1.7 \times 10^{10}$	384 GB	blue
L	$\sim 3.4 \times 10^{10}$	768 GB	green
XL	$\sim 6.7 \times 10^{10}$	1.5 TB	yellow
XXL	$\sim 1.4 \times 10^{11}$	3.0 TB	orange
XXXL	$\sim 2.7 \times 10^{11}$	6.0 TB	red

Table 3. Problem sizes for strong scaling tests

Size	Particles/MPI	RAM/MPI	Color
S	$\sim 4.2 \times 10^7$	96 MB	indigo
M	$\sim 1.7 \times 10^7$	384 MB	blue
L	$\sim 3.4 \times 10^7$	768 MB	green
XL	$\sim 6.7 \times 10^7$	1.5 GB	yellow

Table 4. Problem sizes for weak scaling tests

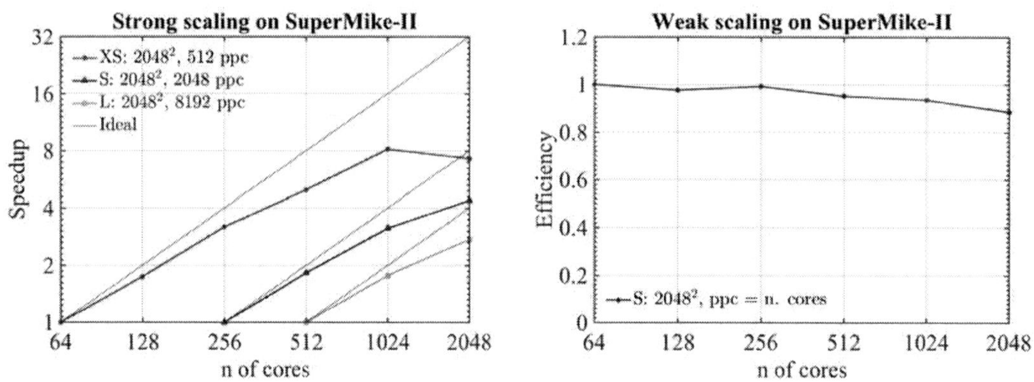

Figure 6. Code performances of CAMELIA's pure MPI version on the Sandy Bridge architecture, i.e., "SuperMike-II": strong scaling (left panel) and weak scaling tests (right panel).

6. Conclusions

CAMELIA is a 3D hybrid kinetic numerical code, suitable for investigating the development of plasma turbulence and its interplay with kinetic instabilities over a wide range of scales. Indeed, it allows for the simultaneous modelling of the MHD and the sub-ion-scale dynamics, fully capturing the transition between the two regimes. Being highly performing on many different HPC systems, CAMELIA allows us to perform state-of-the-art 2D HPIC simulations in a walltime of a few hours and 3D HPIC simulations within a day. In the last few years, the results of the HPIC simulations performed with the code CAMELIA showed a remarkable agreement with solar wind observations, both in 2D and 3D.

As far as 3D simulations are concerned, we showed that a 256^3 computational domain with a spatial resolution of $d_i/4$ and 512 particle-per-cell may be accurate enough for investigating the development of the turbulent cascade simultaneously from MHD to sub-ion scales in an intermediate- or small-beta

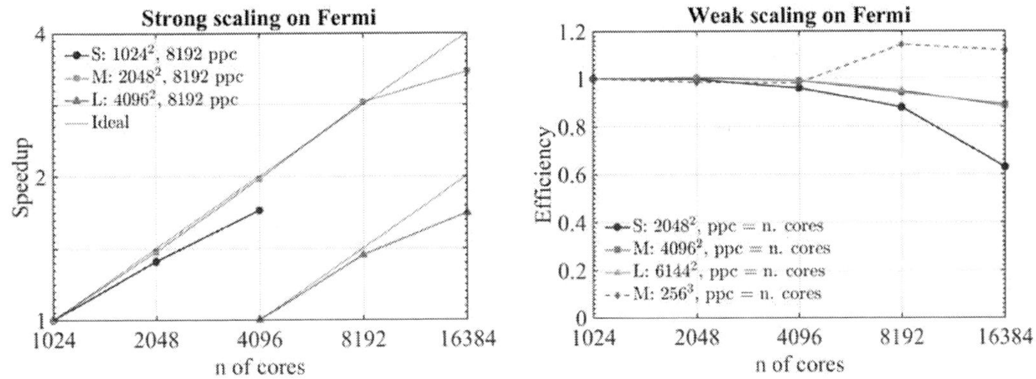

Figure 7. The same as in Fig. 6, but for an IBM Blue Gene/Q architecture, i.e., “Fermi”.

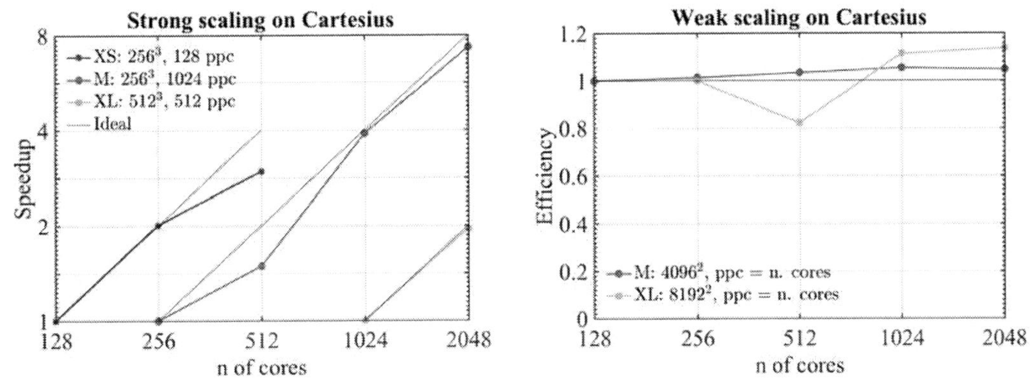

Figure 8. The same as in Fig. 6, but for an Intel Haswell architecture, i.e., “Cartesius”.

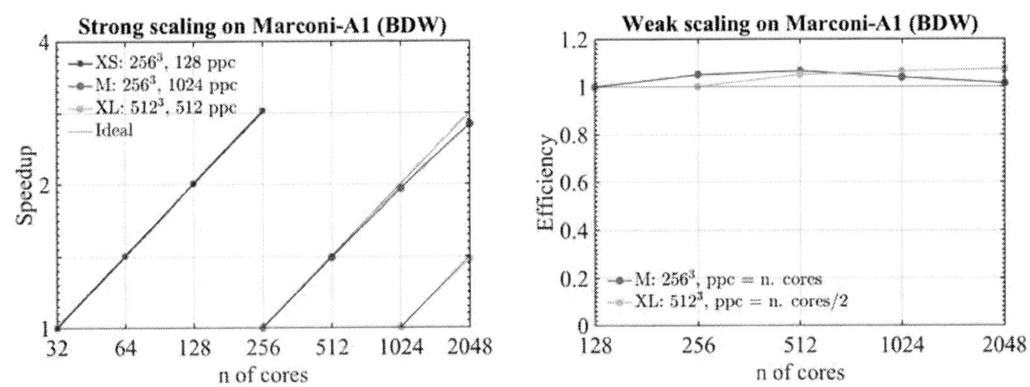

Figure 9. The same as in Fig. 6, but for an Intel Broadwell architecture, i.e., “Marconi-A1 (BDW)”.

plasma. This “reduced” setting can therefore be safely employed for future parameter studies and convergence studies in three dimensions. However, based on [47, 48], we can reasonably expect that a larger number of particles would be mandatory for properly investigating the evolution of the proton

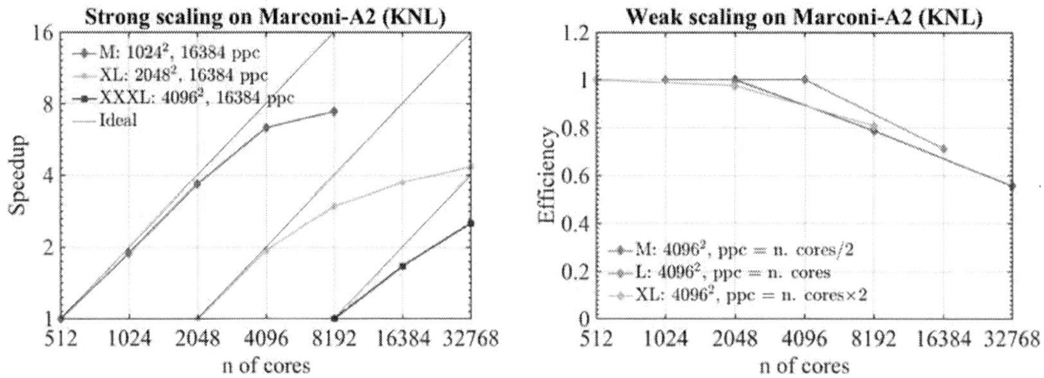

Figure 10. Code performances of CAMELIA's hybrid MPI+OpenMP version on an Intel Knights Landing architecture, i.e., "Marconi-A2.

temperature, or the spectral properties in systems with $\beta \gtrsim 1$.

7. Acknowledgments
The authors thank T. Tullio for useful comments and suggestions. LF was funded by Fondazione Cassa di Risparmio di Firenze through the project "Giovani Ricercatori Protagonisti". PH acknowledges GACR grant 18-08861S. CHKC is supported by an STFC Ernest Rutherford Fellowship. We acknowledge PRACE for awarding us access to resource Cartesius based in the Netherlands at SURFsara through the DECI-13 (Distributed European Computing Initiative) call (project HybTurb3D), and the CINECA award under the ISCRA initiative, for the availability of high performance computing resources and support (grants HP10C877C4, HP10CVCUF1, HP10B2DRR4, HP10BUUOJM). We warmly thank Frank Löffler for providing HPC resources through the Louisiana State University (allocation hpc hyrel14). Data deposit was provided by the EUDAT infrastructure (https://www.eudat.eu/).

References
[1] Chen C H K 2016 *Journal of Plasma Physics* **82** 535820602
[2] Alexandrova O, Lacombe C, Mangeney A, Grappin R and Maksimovic M 2012 *Astrophys. J.* **760** 121
[3] Sahraoui F, Huang S Y, Belmont G, Goldstein M L, Rétino A, Robert P and De Patoul J 2013 *Astrophys. J.* **777** 15
[4] Svidzinski V A, Li H, Rose H A, Albright B J and Bowers K J 2009 *Phys. Plasmas* **16** 122310
[5] Saito S, Gary S P and Narita Y 2010 *Phys. Plasmas* **17** 122316
[6] Chang O, Gary P S and Wang J 2011 *Geophys. Res. Lett.* **38** L22102
[7] Camporeale E and Burgess D 2011 *Astrophys. J.* **730** 114
[8] Gary S P, Chang O and Wang J 2012 *Astrophys. J.* **755** 142
[9] Wan M, Matthaeus W H, Karimabadi H, Roytershteyn V, Shay M, Wu P, Daughton W, Loring B and Chapman S C 2012 *Phys. Rev. Lett.* **109** 195001
[10] Karimabadi H, Roytershteyn V, Wan M, Matthaeus W H, Daughton W, Wu P, Shay M, Loring B, Borovsky J, Leon2ardis E, Chapman S C and Nakamura T K M 2013 *Phys. Plasmas* **20** 012303
[11] Wan M, Matthaeus W H, Roytershteyn V, Karimabadi H, Parashar T, Wu P and Shay M 2015 *Phys. Rev. Lett.* **114** 175002
[12] Gary S P, Hughes R S and Wang J 2016 *Astrophys. J.* **816** 102
[13] Wan M, Matthaeus W H, Roytershteyn V, Parashar T N, Wu P and Karimabadi H 2016 *Phys. Plasmas* **23** 042307
[14] Shaikh D and Shukla P K 2009 *Phys. Rev. Lett.* **102** 045004
[15] Shaikh D and Zank G P 2009 *MNRAS* **400** 1881–1891
[16] Martin L N, De Vita G, Sorriso-Valvo L, Dmitruk P, Nigro G, Primavera L and Carbone V 2013 *Phys. Rev. E* **88** 063107
[17] Rodriguez Imazio P, Martin L N, Dmitruk P and Mininni P D 2013 *Phys. Plasmas* **20** 052506
[18] Gómez D, Martín L N and Dmitruk P 2013 *Advances in Space Research* **51** 1916–1923
[19] Cho J and Lazarian A 2009 *Astrophys. J.* **701** 236–252
[20] Shaikh D 2009 *MNRAS* **395** 2292–2298

[21] Meyrand R and Galtier S 2013 *Phys. Rev. Lett.* **111** 264501
[22] Howes G G, Tenbarge J M, Dorland W, Quataert E, Schekochihin A A, Numata R and Tatsuno T 2011 *Phys. Rev. Lett.* **107** 035004
[23] TenBarge J M and Howes G G 2013 *Astrophys. J. Lett.* **771** L27
[24] Hatch D R, Jenko F, Bratanov V and Navarro A B 2014 *Journal of Plasma Physics* **80** 531âĂŞ551
[25] Li T C, Howes G G, Klein K G and TenBarge J M 2016 *Astrophys. J. Lett.* **832** L24
[26] Passot T, Henri P, Laveder D and Sulem P L 2014 *Eur. Phys. J. D* **68** 207
[27] Sulem P L and Passot T 2015 *Journal of Plasma Physics* **81** 325810103
[28] Kobayashi S, Sahraoui F, Passot T, Laveder D, Sulem P L, Huang S Y, Henri P and Smets R 2017 *Astrophys. J.* **839** 122
[29] Servidio S, Valentini F, Califano F and Veltri P 2012 *Phys. Rev. Lett.* **108** 045001
[30] Greco A, Valentini F, Servidio S and Matthaeus W H 2012 *Phys. Rev. E* **86** 066405
[31] Perrone D, Valentini F, Servidio S, Dalena S and Veltri P 2014 *Eur. Phys. J. D* **68** 209
[32] Valentini F, Servidio S, Perrone D, Califano F, Matthaeus W H and Veltri P 2014 *Phys. Plasmas* **21** 082307
[33] Servidio S, Valentini F, Perrone D, Greco A, Califano F, Matthaeus W H and Veltri P 2015 *Journal of Plasma Physics* **81** 325810107
[34] Valentini, F, Vásconez, C L, Pezzi, O, Servidio, S, Malara, F and Pucci, F 2017 *Astron. Astrophys.* **599** A8 URL `https://doi.org/10.1051/0004-6361/201629240`
[35] Cerri S S and Califano F 2017 *New Journal of Physics* **19** 025007
[36] Cerri S S, Servidio S and Califano F 2017 *Astrophys. J. Lett.* **846** L18
[37] Verscharen D, Marsch E, Motschmann U and Müller J 2012 *Phys. Plasmas* **19** 022305
[38] Vasquez B J and Markovskii S A 2012 *Astrophys. J.* **747** 19
[39] Comişel H, Verscharen D, Narita Y and Motschmann U 2013 *Phys. Plasmas* **20** 090701
[40] Parashar T N, Vasquez B J and Markovskii S A 2014 *Phys. Plasmas* **21** 022301
[41] Vasquez B J, Markovskii S A and Chandran B D G 2014 *Astrophys. J.* **788** 178
[42] Hellinger P, Matteini L, Landi S, Verdini A, Franci L and Trávníček P M 2015 *Astrophys. J. Lett.* **811** L32
[43] Ozak N, Ofman L and Viñas A F 2015 *Astrophys. J.* **799** 77
[44] Parashar T N, Matthaeus W H, Shay M A and Wan M 2015 *Astrophys. J.* **811** 112
[45] Parashar T N and Matthaeus W H 2016 *Astrophys. J.* **832** 57
[46] Franci L, Verdini A, Matteini L, Landi S and Hellinger P 2015 *Astrophys. J. Lett.* **804** L39
[47] Franci L, Landi S, Matteini L, Verdini A and Hellinger P 2015 *Astrophys. J.* **812** 21
[48] Franci L, Landi S, Matteini L, Verdini A and Hellinger P 2016 *Astrophys. J.* **833** 91
[49] Franci L, Landi S, Verdini A, Matteini L and Hellinger P 2018 *Astrophys. J.* **853** 26
[50] Chen C H K, Bale S D, Salem C S and Maruca B A 2013 *Astrophys. J.* **770** 125
[51] Chen C H K and Boldyrev S 2017 *Astrophys. J.* **842** 122
[52] Chen C H K, Boldyrev S, Xia Q and Perez J C 2013 *Phys. Rev. Lett.* **110** 225002
[53] Cerri S S, Franci L, Califano F, Landi S and Hellinger P 2017 *Journal of Plasma Physics* **83** 705830202
[54] Franci L, Hellinger P, Matteini L, Verdini A and Landi S 2016 *American Institute of Physics Conference Series* (*American Institute of Physics Conference Series* vol 1720) p 040003
[55] Franci L, Cerri S S, Califano F, Landi S, Papini E, Verdini A, Matteini L, Jenko F and Hellinger P 2017 *Astrophys. J. Lett.* **850** L16
[56] Hellinger P, Landi S, Matteini L, Verdini A and Franci L 2017 *Astrophys. J.* **838** 158
[57] Matthews A P 1994 *J Comput Phys* **112** 102 – 116 ISSN 0021-9991 URL `http://www.sciencedirect.com/science/article/pii/S0021999184710849`
[58] Boris J P 1970 *Proceeding of Fourth Conference on Numerical Simulations of Plasmas*
[59] Qin H, Zhang S, Xiao J, Liu J, Sun Y and Tang W M 2013 *Phys. Plasmas* **20** 084503

IOP Conf. Series: Journal of Physics: Conf. Series **1031** (2018) 012003 doi:10.1088/1742-6596/1031/1/012003

X-Ray Emitting Gas in Circumstellar Nebulae

Jane Arthur

Instituto de Radioastronomía y Astrofísica, UNAM Campus Morelia, Apartado Postal 3-72, 58090 Morelia, Michoacán, México

E-mail: j.arthur@irya.unam.mx

Abstract. Recent high-resolution 2D radiation-hydrodynamic numerical simulations of the formation and evolution of hot bubbles around evolved stars are described. The simulations take into account the evolution of the stellar parameters such as ionizing photon rate, wind velocity and mass-loss rate for a range of initial stellar masses. For low-mass stars, a planetary nebula with a lifetime of a few thousand years forms around the central hot star, while for massive stars the result is a Wolf-Rayet nebula, which has a lifetime of tens of thousands of years. In both cases, instabilities in the fast wind-slow wind interaction zone produce clumps and filaments in the swept-up shell of nebular material. Turbulent mixing and thermal conduction at the corrugated interface can produce quantities of intermediate temperature and density gas between the hot, shocked wind bubble, and the swept-up photoionized nebular material, which can emit in soft, diffuse X-rays. Sampling of the resultant theoretical spectra helps to make meaningful comparisons with recent observations of planetary nebulae.

1. Introduction

Circumstellar nebulae are formed when a fast stellar wind sweeps up gas that was expelled during a previous stage of the star's evolution and the dense, compressed shell is then photoionized by energetic photons from the hot central object. Although optical nebulae characterize this interaction, interior to the dense, photoionized shell is a bubble of hot, shocked fast wind plasma, which, in principle, should be observable in X-rays.

In general terms, after core hydrogen burning is finished, a star leaves the main sequence and its subsequent evolution proceeds rapidly. The star moves to the red on Hertsprung-Russell diagram and expels its hydrogen-rich envelope before heading to the blue as the hot helium-rich core becomes exposed. This process occurs for both low-mass ($M_{\rm init} < 8M_\odot$) and very high-mass ($M_{\rm init} > 30M_\odot$) stars. For low-mass stars, the end result is a planetary nebula (PN), while the very high-mass equivalent is a Wolf-Rayet nebula (WRN).

In both low-mass and very high-mass scenarios, the details of the mass loss immediately prior to the fast wind stage are crucial for the formation and appearance of the nebula. For low-mass stars, the evolution from the asymptotic giant branch to the white dwarf cooling track is reasonably well understood. Whatever the initial mass of these stars ($< 8M_\odot$), the final mass must be less than a Chandrasekhar limiting mass ($\sim 1.4M_\odot$) and so the progenitor can lose more than half its mass through thermal pulses and superwind in the final stages. In fact, only stars with initial mass less than about $3.5M_\odot$ form observable PN; higher progenitor masses lead to objects with optically thick envelopes.

The late stages of very high-mass stellar evolution are less well understood. Stellar winds are important throughout the lifetime of these stars, even on the main sequence. Stellar rotation

IOP Conf. Series: Journal of Physics: Conf. Series **1031** (2018) 012003 doi:10.1088/1742-6596/1031/1/012003

promotes mixing in the stellar interior, which enhances mass loss at the surface and thus extends the main sequence lifetime of these objects. A massive star will lose half its mass in dense, slow winds during a red supergiant (RSG) or luminous blue variable (LBV) stage.

The timescales for PN before the fast wind switches off and the star settles onto the white dwarf cooling track are very short, typically less than 10,000 yrs and the size scales are small < 0.5 pc. However, PN are quite numerous and reasonably bright objects. WRN are much larger > 3 pc and because there is still ongoing fusion in the hot central object, the lifetimes are much longer. However, observable WR wind-blown nebulae are few in number and they are lower luminosity than PN because, being such extended objects, the densities are lower.

Hot gas is expected to fill both PN and WRN but X-ray observations present a puzzling picture. The CHANPLANS program is a statistically significant study of X-ray emission from PNe within 1.5 kpc of the Sun, which to date has detected 30% of target PNe in diffuse X-rays [1, 2]. On the other hand, only 4 WRN have been detected in X-rays [3, 4, 5]. Curiously, the derived X-ray temperature of the emitting gas for all these objects is in the narrow range 1–3 $\times 10^6$ K, even though the stellar wind velocities would suggest that the hot, shocked gas should have temperatures in excess of 10^7 K.

In this paper, we discuss numerical models of PN and WRN and the formation and evolution of hot bubbles in these objects. We examine the X-ray properties of the hot gas and possible reasons for the apparently low temperatures.

2. Formation of Circumstellar Nebulae

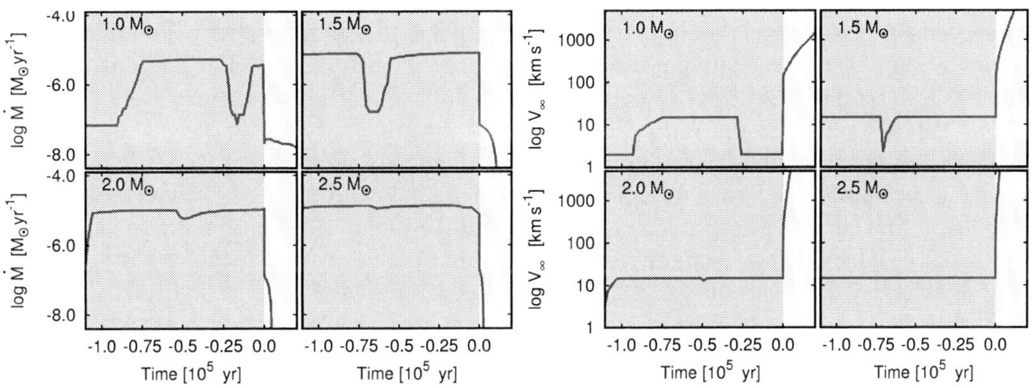

Figure 1. Evolution of fast stellar wind parameters for evolved, low-mass stars of 4 different progenitor masses. Left panel: mass-loss rate. Right panel: wind velocity. The red shaded area represents the AGB stage, while the blue shaded area starts once $T_{\rm eff} > 10^4$ K.

Circumstellar nebulae are shaped both by the details of the mass-loss prior to the onset of the fast wind, as well as by the acceleration timescale of the fast wind.

For low-mass progenitors near the top of the asymptotic giant branch (AGB), the superwind is characterized by slow, dense ($v_\infty < 30$ km s^{-1},$\dot{M} \sim 10^{-4} M_\odot$ yr^{-1}) episodic mass loss due to radial pulsations in the expanded stellar envelope. The ejected material forms multiple shells around the star. The transition to central star of a PN is accompanied by a large reduction in the mass-loss rate ($\dot{M} \sim 10^{-8} M_\odot$ yr^{-1}) and corresponding increase in the stellar wind velocity, which can reach > 1000 km s^{-1} on a timescale of a thousand years (see Figure 1). The stellar ionizing photon rate follows a similar trend to the mass-loss rate and drops off steeply. PN are only detectable for a few thousand years because the expansion means the gas becomes very diffuse and also the photons illuminating the nebula drop off. The effective temperature of the star is a good observational proxy for time during this stage.

The evolution of very massive stars depends very much on the ingredients of the stellar evolution code and different groups of researchers have produced very different results. For example, Figure 2 shows results for post-main-sequence mass loss for a progenitor $40M_{\odot}$ star obtained by the Geneva group, where the difference between the two models is that one assumes an initial rotation velocity of 300 km s^{-1} (at the equator) for the star, while the other model has no rotation [6]. The mass-loss rates and timescales are widely different between the models. The intensity and duration of the slow wind phase determines how far away from the star and how dense the shells of material swept up by the fast wind can be.

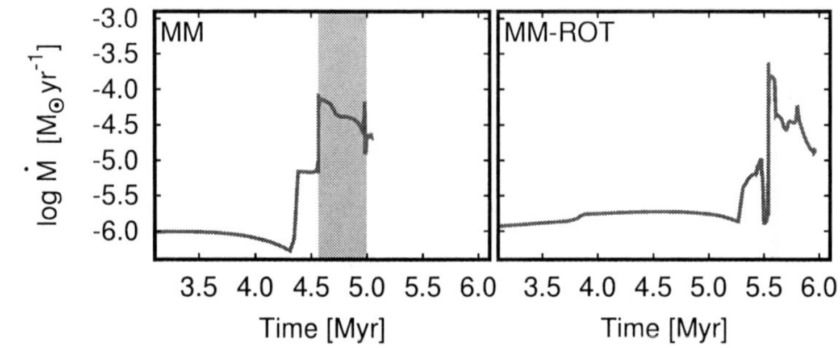

Figure 2. Mass-loss rate for post-main-sequence evolution of a $40M_{\odot}$ star obtained from Geneva Group stellar evolution models with (MM-ROT) and without (MM) rotation [6]. Shaded areas indicate slow, dense wind phase.

3. Numerical Simulations

We have performed numerical simulations in 1D spherical symmetry and 2D axisymmetry of the interaction of the time-evolving fast stellar wind with the circumstellar medium sculpted by previous slow, intense mass-loss episodes for both high-mass and low-mass stars [7, 8, 9, 10]. These calculations include the radiative transfer of the time-varying ionizing photons from the central star. Radiative cooling with appropriate chemical abundances is taken into account. The calculations are performed both with and without thermal conduction, which affects the density and temperature distributions at the edge of the hot bubble. Conduction has been postulated as a means of reducing the temperature in the hot bubble in order to achieve the temperatures indicated by X-ray observations [11, 12]. Our conduction models and pure hydrodynamic models are taken to represent the two extremes of behaviour. Magnetic fields in PN are extremely weak but conduction by thermal electrons will be inhibited in the direction perpendicular to the field lines; the real behaviour could be expected to fall somewhere in between our 2 extremes.

The initial conditions for these calculations are set by the evolution in the slow, dense wind stage. For full details of the code, see [7, 8, 10].

3.1. Planetary Nebulae

We start with PN, in particular, the results for a $1.5M_{\odot}$ initial mass star. Figure 3 shows the number density of the ionized gas and temperature distribution after 7400 yrs of post-AGB evolution during the 2D axisymmetric simulation. Results are shown for simulations with and without thermal conduction. Instabilities develop early in the interaction between the fast wind and the earlier slow, dense wind material, leading to the formation of clumps and filaments in the swept-up shell. Photoevaporated flows from the heads of the inward-pointing filaments carve out cavities in the interior of the bubble, and interacting non-radial flows generate turbulent mixing regions. Heat diffusion by electrons from the hot bubble into the surrounding swept-up shell evaporates cold, dense material into the bubble interior. Both processes lead to mixing at the interface, which has a large surface area due to the corrugations.

Figure 4 shows the density and profiles of 1D spherically symmetric simulations of the same PN, together with angle-averaged profiles obtained from the 2D axisymmetric simulations where

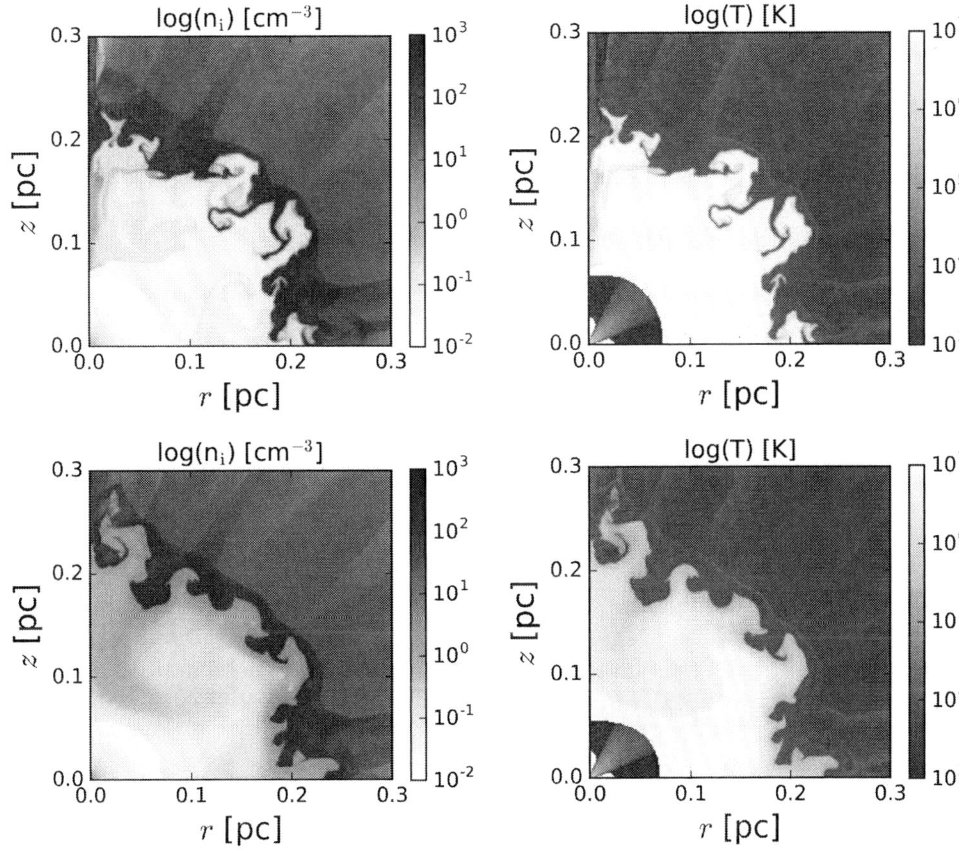

Figure 3. Number density in the ionized gas and temperature distribution after 7400 yrs of post-AGB evolution. Top panels: simulation without conduction. Bottom panels: simulation with thermal conduction.

all simulations have the same numerical resolution and input physics. It can be seen that thermal conduction reduces the temperature throughout the hot bubble and evaporation raises the density close to the contact discontinuity. On the other hand, instabilities do not occur and there are no turbulent, hydrodynamic mixing regions. The density of the swept-up shell of the 1D simulations is higher, and the shell is narrower. In this $1.5 M_\odot$ case, however, the shell remains optically thin. In the 2D simulations, opacity differences in the clumpy unstable shell lead to ray patterns in the photoionized envelope gas beyond due to shadowing effects.

The simulations proceed until the stellar wind mechanical luminosity falls off and back-filling occurs in the nebula as the photoionized material expands inwards.

3.2. Wolf-Rayet Nebulae

The essence of a WRN is the interaction of a fast wind with a dense circumstellar medium produced during the red supergiant (RSG) or luminous blue variable (LBV) immediately preceding the transition to Wolf-Rayet star. Although the exact details vary, the scenario is the same regardless of the stellar evolution model adopted. The main difference is the distance the slow wind material has managed to travel from the star before the fast wind switches on. The interaction consists of an initial stage, in which the fast wind drives a shock into the dense circumstellar medium, which sweeps up a dense shell of material that quickly becomes unstable

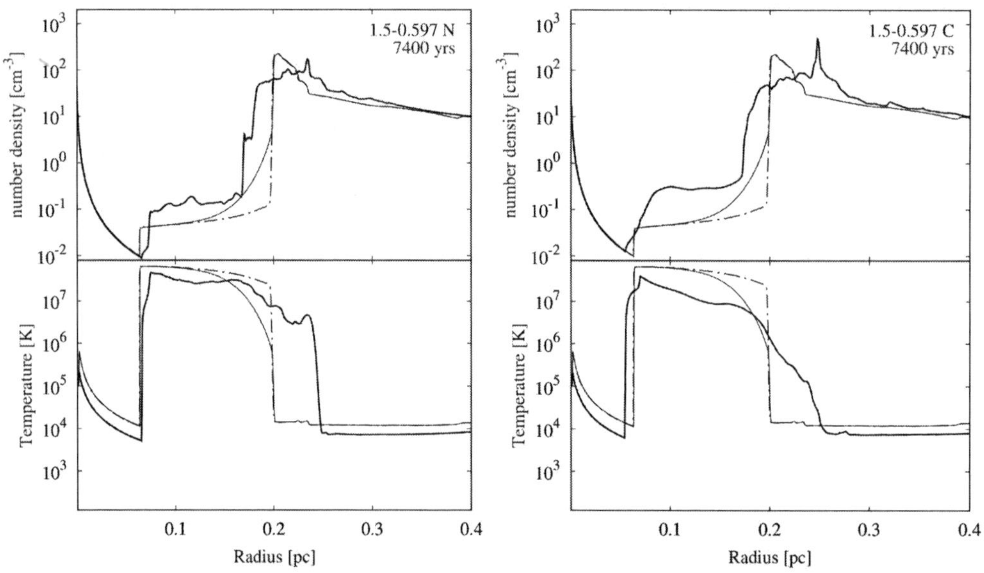

Figure 4. Density and temperature profiles of 1D spherically symmetric simulations (thin lines — with conduction; dashed lines — without conduction) together with angle-averaged profiles from 2D axisymmetric simulations (thick lines) of planetary nebula evolution for a $1.5M_{\odot}$ initial mass star. Left panel: 2D simulation without conduction. Right panel: 2D simulation with thermal conduction. All simulations shown after 7400 years of post-AGB evolution.

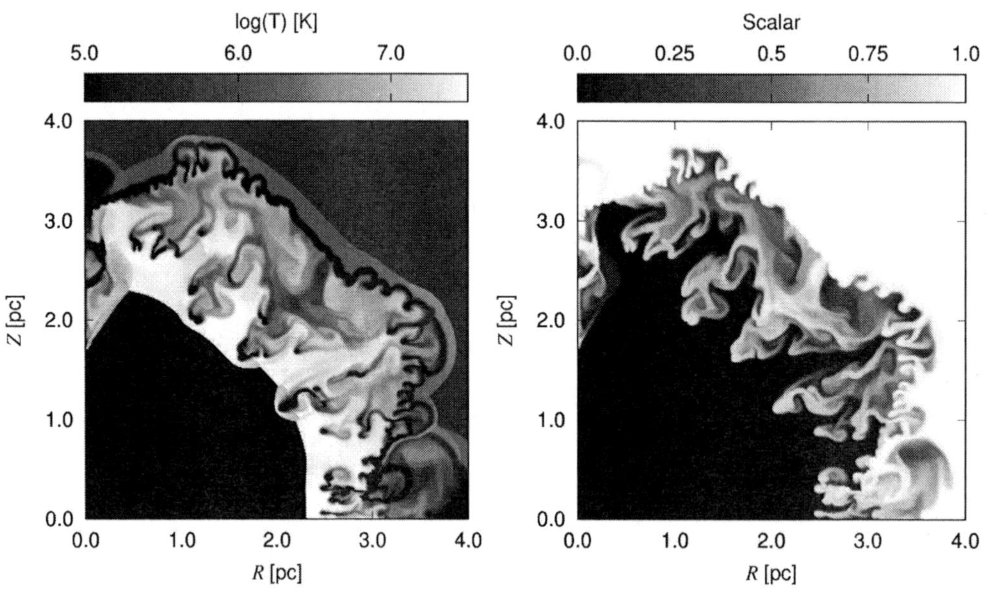

Figure 5. Left panel: logarithm of temperature for a typical Wolf-Rayet nebula simulation (without thermal conduction) after 20,000 yrs of evolution. Right panel: value of scalar tracing fast-wind material (scalar value 0) and ambient material (scalar value 1).

(initially due to the thin shell instability and then the shadowing instability) and breaks up in to clumps and filaments. Later, the ragged swept-up shell interacts with the edge of the circumstellar material and breaks out into the low-density interstellar medium beyond. The size scales and time scales of WRN are both about an order of magnitude greater than those of PN. The mass-loss rate of Wolf-Rayet stars is at least 3 orders of magnitude greater than that of the central stars of planetary nebulae, and so the fast wind material in WRN will have a relatively more important contribution to the X-ray emission of the diffuse gas.

Figure 5 shows the temperature distribution in a typical simulation, together with the distribution of a scalar, which has value 0 in the fast wind and value 1 in the ambient material. The initial condition was a spherically symmetric density distribution generated by a uniform, slow, dense wind from the RSG stage that ejected a total of $15 M_\odot$ of material over 200,000 yrs. The wind speed of 15 km s^{-1} puts the rim of the dense circumstellar material at a distance 2.3 pc from the star. The fast wind has a constant velocity of 1600 km s^{-1}, a mass-loss rate of $7 \times 10^{-5} M_\odot$ yr^{-1}, and the ionizing photon rate is 2.6×10^{49} s^{-1}. The simulation is shown 20,000 yrs after the onset of the fast wind. The clumps and filaments are the result of instabilities in the swept-up shell, which have then collided with the dense rim of the circumstellar material and broken out into the diffuse, main-sequence wind bubble beyond.

4. X-ray Emission

The X-ray emission corresponding to the hot bubbles produced in the simulations can be calculated easily using the idea of Differential Emission Measure (DEM) if we assume that the planetary (or Wolf-Rayet) nebula is optically thin to X-ray photons.

4.1. Differential Emission Measure

In this work, we define the Differential Emission Measure, $DEM(T_\mathrm{b})$, of gas in temperature bin T_b of width ΔT_b as follows:

$$\mathrm{DEM}(T_\mathrm{b}) = \sum_{k, T_k \in T_\mathrm{b}} n_\mathrm{H} n_\mathrm{e} \Delta V_k \,, \qquad (1)$$

where the summation is over all computational cells, k, of volume ΔV_k. The hydrogen number density and electron density are n_H and n_e, respectively. We consider 100 temperature bins equally spaced in $\log T$ between $\log T = 5.0$ and $\log T = 9.0$ with bin width 0.04 dex.

Figure 6 shows the DEM histograms at four different times for the simulations described in Section 3. In each profile, the maximum temperature corresponds to the immediate postshock temperature at the inner wind shock ($T_\mathrm{s} \simeq 3 m_\mathrm{H} v_\infty^2/(32k)$ in fully ionized gas). The wind velocity increases with time (see Figure 1), and so the maximum temperature increases with time.

The 1D DEM profiles without conduction are very different to the other results. In this case, the DEM profile consists of discrete values at lower temperatures and a continuous block at higher temperatures. The continuous block corresponds to the postshock temperatures in the accelerating wind. The low temperature discrete values correspond to a numerical effect at the contact discontinuity and will change with numerical resolution. The 1D results with thermal conduction show how heat diffusion spreads out the temperature range of the DEM profile. The 2D results show how mixing due to non-radial flows around the clumps and filaments formed by instabilities fully populates the temperature range. The plateau at $\log T < 5.7$ in the 2D case with conduction corresponds to the conduction layer close to the interface.

Figure 7 shows the DEM distribution of the Wolf-Rayet simulation described earlier. There is a clear separation between the contributions of the fast wind, which dominates the high-temperature range $\log T > 6.5$, and the ambient material, which dominates the range $\log T < 6.5$. Even though the constant fast-wind velocity would suggest a uniform post-shock temperature, the flow of the fast wind around the clumps and filaments leads to a spread of temperatures. By the same token, interacting shock waves in the ablated material around the clumps and filaments raise the temperature in the nebular gas.

IOP Conf. Series: Journal of Physics: Conf. Series **1031** (2018) 012003 doi:10.1088/1742-6596/1031/1/012003

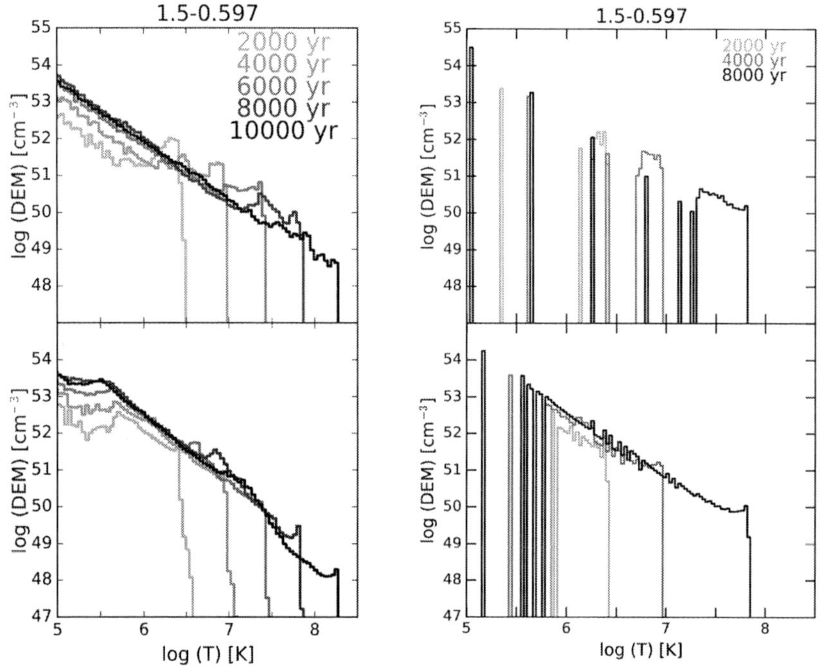

Figure 6. Differential emission measure histograms for the planetary nebula $1.5M_{\odot}$ initial mass numerical simulations. Left: 2D axisymmetric simulation. Right: 1D spherically symmetric simulation. In each case the upper panel shows the results without conduction and the lower panel shows the results for the simulation with thermal conduction.

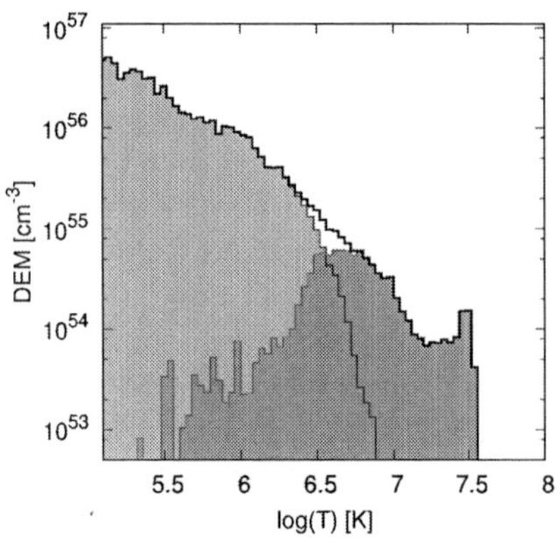

Figure 7. DEM distribution of the Wolf-Rayet nebula simulation shown in Figure 5. The contribution of the fast wind is shown in red and that of the ambient material is shown in blue.

4.2. Emission Coefficient

The emission coefficient is the combined result of all the atomic processes contributing to the X-ray emission in a desired energy band for gas at a given temperature. The emission coefficient as a function of temperature is defined by

$$\epsilon(T) = \int_{E_1}^{E_2} \epsilon(T, E) dE \tag{2}$$

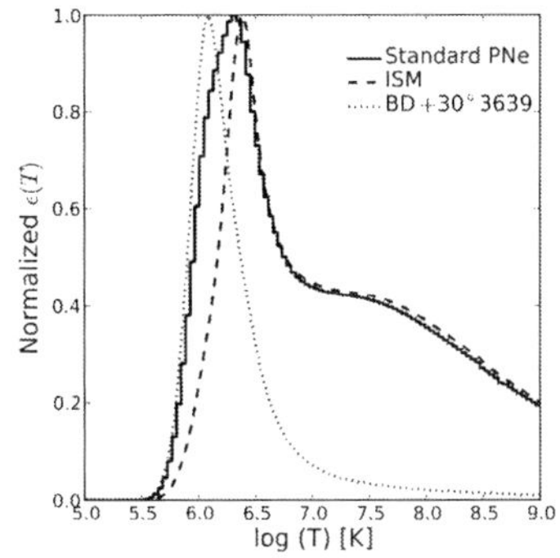

Figure 8. Emission coefficient for ISM, generic galactic planetary nebula and the hydrogen-poor planetary nebula BD+30°3639.

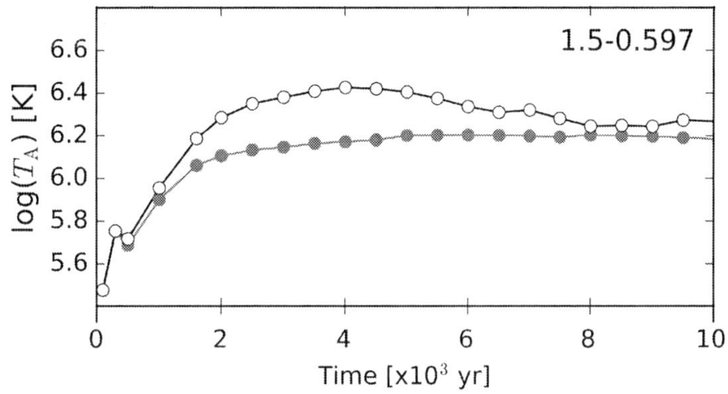

Figure 9. Average temperature of hot gas as a function of post-AGB time for the $1.5M_\odot$ axisymmetric simulations.

with $E_1 = 0.3$ keV and $E_2 = 2.0$ keV corresponding to the soft energy band of the *XMM-Newton* and *Chandra* telescopes. We used the CHIANTI code [13, 14, 15] to calculate the emission coefficient for chemical abundances appropriate to PN and other astrophysical objects, shown in Figure 8. It is noteworthy that the emission coefficient has a sharp peak at about $\log T = 6.3$ ($T = 2.0 \times 10^6$ K).

The emission-weighted average temperature is defined by

$$T_{\rm A} = \frac{\int T\epsilon(T)\mathrm{DEM}(T)dT}{\int \epsilon(T)\mathrm{DEM}(T)dT} \tag{3}$$

and gives a reasonable estimate to the characteristic temperature derived from X-ray observations. Even though the DEM profiles cover a wide range of temperatures, the sharp peak of the emission coefficient filters out values that are far from the peak value. Figure 9 shows how the average temperature of our simulations changes with time. The simulations with thermal conduction achieve a constant mean temperature, $\log T_{\rm A} \simeq 6.2$, after only 2000 years. Those without thermal conduction have a slightly higher mean temperature with more variation. The variation can be understood in terms of the DEM profile, which shows a bump close to the maximum temperature at early times that is not seen in the DEM profiles with conduction.

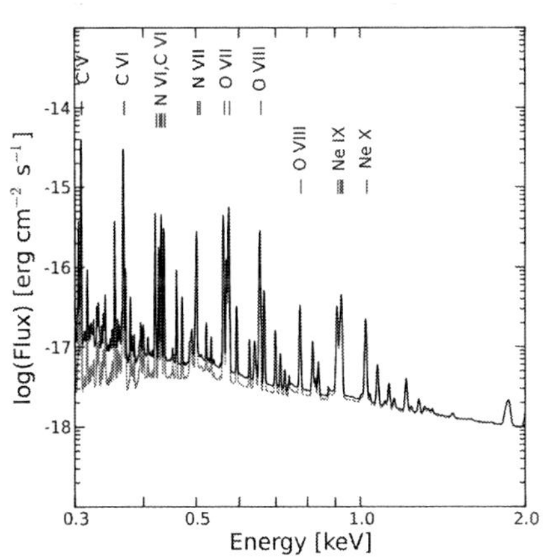

Figure 10. Synthetic spectrum (black line) from the planetary nebula $1.5M_\odot$ initial mass simulation after 8,000 yrs with principal emission lines indicated. The red line shows the effect of an absorbing neutral column density $N_{\rm H} = 8 \times 10^{20}$ cm^{-2}, which is more important at low energies.

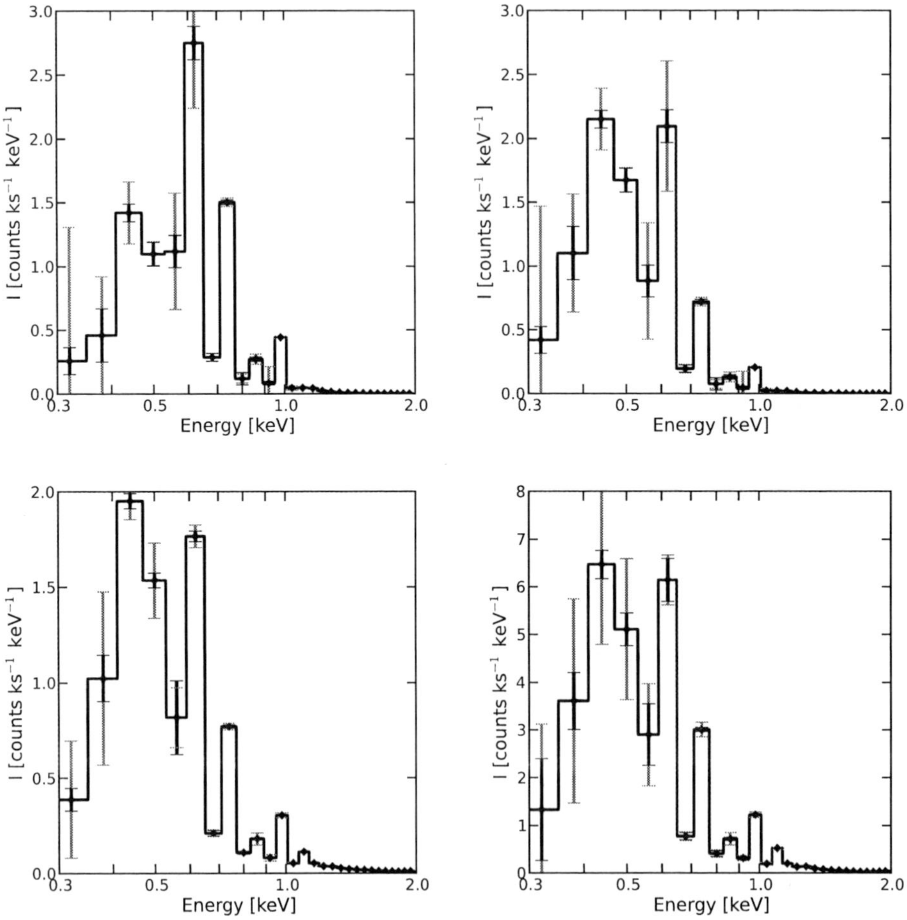

Figure 11. Synthetic absorbed, convolved and binned spectra for the $1.5M_\odot$ model: the error bars correspond to samples with 50 counts (grey) and 200 counts (black). Top row: early time; bottom row: late time. Left-hand panels: without conduction; right-hand panels: with thermal conduction. The absorbing neutral column density is $N_H = 8 \times 10^{20}$ cm^{-2} in all cases.

4.3. X-ray Spectra

The reported X-ray temperatures are derived from X-ray spectra obtained from the *Chandra* and *XMM-Newton* telescopes. We can process our numerical simulation results using the DEM profiles and the CHIANTI code to obtain synthetic spectra. Carbon and nitrogen lines are important at low energies, while oxygen and neon lines are important contributors at higher energies. However, the intrinsic spectrum needs to be corrected for interstellar absorption by neutral H along the line of sight, which affects the lowest energy X-rays $E < 0.5$ keV (see Figure 10). The resultant absorbed spectrum then needs to be convolved with the instrumental response matrices and binned to the typical bin resolution of the reported PN observations. The final spectrum has lost much of the detail of the spectral lines (see Figure 11). The changing spectral shape reflects the evolution of the DEM profile and is more marked for models without conduction. The models with thermal conduction show little variation over time, in common with the almost constant mean temperature shown in Figure 9.

These absorbed, convolved, synthetic spectra represent ideal, infinite-exposure-time spectra.

Real observations last a finite amount of time (for PN typically 30,000 seconds) and the photons received in this time register as counts. Several PNe in the CHANPLANS program have fewer than 50 counts due to reduced exposure times. To mimic the observations, we sample the cumulative absorbed spectra a finite number, $N_{\rm counts}$, times and repeat this operation a large number (10,000) times to obtain good statistics. The error bars shown in Figure 11 represent the standard deviations obtained for $N_{\rm counts} = 50$ and $N_{\rm counts} = 200$ sample sizes. We conclude that it is only possible to give a reasonable description of the spectral shape, and hence obtain a reliable X-ray temperature for the gas, with ≥ 200 counts.

5. Conclusions

Radiation-hydrodynamic simulations reproduce the general properties of Wolf-Rayet and planetary nebulae. 2D simulations allow for the development of instabilities in the swept-up shell of circumstellar material and the formation of clumps and filaments, which promote the turbulent mixing of nebular and fast-wind material. Interactions of non-radial flows result in a full range of plasma temperatures, from the immediate post-shock temperature in the hot wind bubble ($> 10^7$ K) down to the photoionized nebular gas ($\sim 10^4$ K) even when there is no heat diffusion (i.e., thermal conduction).

The emission coefficient integrated over the *Chandra* and *XMM Newton* soft X-ray energy bands is sharply peaked at $T \sim 2 \times 10^6$ K, when typical chemical abundances are taken into account. It acts as a filter for the Differential Emission Measure distribution and emphasizes the contribution from gas with temperature close to this peak. This is a natural explanation for the apparent ubiquity of the $T_{\rm X} \sim 1$–3×10^6 K temperature range derived from observations of the diffuse X-ray emission from planetary nebulae and Wolf-Rayet wind bubbles, without the need to invoke additional physical processes.

Acknowledgments

SJA is grateful for financial support from UNAM DGAPA PAPIIT project IN112816. This work was largely carried out in collaboration with Jesús Toalá.

References

[1] Kastner J H, Montez R Jr, Balick B, et al. 2012 *Astron. J.* **144** 58
[2] Freeman M, Montez R Jr, Kastner J H, et al. 2014 *Astrophys. J.* **794** 99
[3] Chu Y-H 2008 *Massive Stars as Cosmic Engines* (IAU Symposium vol 250) ed F Bresolin, P A Crowther and J Puls (Cambridge: CUP) p 341
[4] Toalá J A, Guerrero M A, Chu Y-H, et al. 2016 *Mon. Not. R. Astron. Soc.* **456** 4305
[5] Toalá J A, Marston A P, Guerrero M A, Chu Y-H and Gruendl R A 2017 *Astrophys. J.* **846** 76
[6] Meynet G and Maeder A 2003 *Astron. & Astrophys.* **404** 975
[7] Toalá J A and Arthur S J 2011 *Astrophys. J.* **737** 100
[8] Toalá J A and Arthur S J 2014 *Mon. Not. R. Astron. Soc.* **443** 3486
[9] Toalá J A and Arthur S J 2016 *Mon. Not. R. Astron. Soc.* **463** 4438
[10] Toalá J A and Arthur S J 2017 *Mon. Not. R. Astron. Soc.* **464** 178
[11] Weaver R, McCray R, Castor J, Shapiro P and Moore R 1977 *Astrophys. J.* **218** 377
[12] Steffen M, Schönberner D and Warmuth A 2008 *Astron. & Astrophys.* **489** 173
[13] Dere K P, Landi E, Mason H E, Monsignori Fossi B C and Young P R 1997 *Astron. &Astrophys. Suppl.* **125** 149
[14] Dere K P, Landi E, Young P R, et al. 2009 *Astron. & Astrophys.* **498** 915
[15] Landi E, Young P R, Dere K P, Del Zanna G and Mason H E 2013 *Astrophys. J.* **763** 86

IOP Conf. Series: Journal of Physics: Conf. Series **1031** (2018) 012004 doi:10.1088/1742-6596/1031/1/012004

Spiral Shocks in Young Circumbinary Disks

Tomoyuki Hanawa[1], **Keijiro Miyazawa**[2], **Tomoaki Matsumoto**[3]

[1] Center for Frontier Science, Chiba University, Inage-ku, Chiba 263-8522, Japan
[2] Graduate School of Science, Chiba University, Inage-ku, Chiba 263-8522, Japan
[3] Faculty of Humanity and Environment, Hosei University, Chiyoda-ku, Tokyo 102-8160, Japan

E-mail: `hanawa@faculty.chiba-u.jp`

Abstract. Numerical simulations of binary star formation suffer from serious mismatch with observations. Equal mass binaries are abundant in numerical simulations while also unequal mass binaries are commonly observed. The discrepancy should be due to errors in the numerical simulations. In this paper we discuss the evaluation of the Coriolis force as a source of errors in the numerical simulations. Simulations of an accreting young binary is often performed in the frame co-rotating with the binary. We demonstrate that the specific angular momentum changes spuriously at a shock front, if it is evaluated either solely with the density and velocity at the cell center. We show that the spurious change is erased out if a half of it is evaluated from the numerical flux on the cell surface. We name this method of evaluating the Coriolis force HH type since a half is evaluated from the numerical mass flux and the other half is from the momentum density in the cell. We prove that simulations conserve the momentum measured in the rest frame only when HH type Coriolis force is adopted. The numerical error is serious around shock waves since the difference between the numerical mass flux and momentum density is large there. The shock waves drive gas accretion through angular momentum transfer and should be simulated accurately.

1. Introduction

A young binary accretes gas from the circumbinary disk and each component star grows in mass by the gas accretion. Thus the accretion is important to the primary and secondary masses at the zero age main sequence. Artymowicz and Lubow [1] studied this stage based on their SPH simulations. Their and most of the simulations thus far show that the secondary accretes more gas. It implies that nearly equal mass binaries should be abundant. However, this theoretical prediction is not consistent with the observations that secondaries have considerably lower masses than primaries in many binaries (see, e.g., Reipurth et al. [2] and the references therin). The discrepancy suggests some unknown numerical errors in our simulations.

A clue to solve the discrepancy can be obtained from direct comparison with observations. Takakuwa et al. [3] have discovered spiral waves in the circumbinary disk of L1551 NE from high resolution imaging observation with ALMA in its cycle 0. The spiral waves are also seen in the simulations and drive accretion through angular momentum transfer. Since the resolution and sensitivity of ALMA have been improved, more detailed comparison will be possible in near future.

In this paper we examine the Coriolis force taken into account in the simulations of an accreting young binary as a source of numerical errors. The simulations thus far have employed the frame co-rotating with the binary in order to reduce numerical errors due to advection of the circumstellar disks. In compensation, the hydrodynamical equations contain an extra source term, the Coriolis force, and are no more in the conservation form in a strict sense. As shown in the subsequent sections, this can be a source of serious error around strong shock waves. We propose the best numerical recipe to evaluate the Coriolis force in this paper.

This paper is organized as follows. We propose our method of evaluating the Coriolis force after inspecting the hydrodynamical equations in §2. We confirm the validity of our method by comparing types of the methods of evaluating the Coriolis force in §3. Implications of our results are given in §4.

2. Methods

We solve the hydrodynamical equations in the co-rotating frame to simulate accretion onto a young binary. For simplicity, the binary is assumed to rotate in the $z = 0$ plane in the Cartesian coordinates with a constant angular velocity, Ω. The primary and secondary of the binary are assumed to have the masses, $M/(1+q)$ and $Mq/(1+q)$, respectively, where M and q denote the total mass and the mass ratio, respectively. We neglect the mass of the accreting gas for simplicity. Then the gravitational potential of the binary is expressed as

$$\Phi = -\frac{GM}{(1+q)\,|\boldsymbol{r}-\boldsymbol{r}_1|} - \frac{GMq}{(1+q)\,|\boldsymbol{r}-\boldsymbol{r}_2|}, \tag{1}$$

$$\boldsymbol{r}_1 = \frac{aq}{1+q}\boldsymbol{e}_x, \tag{2}$$

$$\boldsymbol{r}_2 = -\frac{a}{1+q}\boldsymbol{e}_x, \tag{3}$$

$$a = \left(\frac{GM}{\Omega^2}\right)^{1/3}, \tag{4}$$

where G, a, and $\boldsymbol{e}_x$ denote the gravitational constant, binary separation, and the unit vector in the x-direction. We use the unit system in which a and Ω are unity.

In the co-rotating frame the hydrodynamical equations are expressed as

$$\frac{\partial \rho}{\partial t} + \boldsymbol{\nabla}\cdot(\rho\boldsymbol{v}) = 0, \tag{5}$$

$$\frac{\partial}{\partial t}(\rho\boldsymbol{v}) + \boldsymbol{\nabla}\cdot(\rho\boldsymbol{v}\boldsymbol{v}) + \boldsymbol{\nabla}\left(\rho c_s^2\right) = -\rho\left[\boldsymbol{\nabla}\Phi + 2\boldsymbol{\Omega}\times\boldsymbol{v} + \boldsymbol{\Omega}\times(\boldsymbol{\Omega}\times\boldsymbol{r})\right], \tag{6}$$

where the symbols, ρ, $\boldsymbol{v}$, $\boldsymbol{r}$, c_s, and Φ denote the density, velocity, position, the sound speed, and the gravitational potential, respectively. The gas is assumed to be isothermal for simplicity. The vector, $\boldsymbol{\Omega} = \Omega\boldsymbol{e}_z$, denotes the angular velocity of the co-rotating frame.

In numerical simulations we employ the conservation form of the hydrodynamical equations to ensure the conservation of mass and momentum of the flow. Equation of continuity has no source term, while equation of momentum conservation has some external source term such as gravity. In addition to the gravity, the Coriolis force and centrifugal force appear as source terms, i.e., in the right hand side [cf. Eq. (6)] in the co-rotating frame. We argue that a half of the Coriolis force should be evaluated by the cell center value and the other half from the numerical flux. As shown later, this evaluation ensures momentum conservation in the rest frame.

Our argument is based on the following theoretical consideration. Eq. (6) is derived from equation of momentum conservation in the rest frame,

$$\frac{\partial}{\partial t}(\rho\boldsymbol{u}) + \boldsymbol{\nabla}\cdot(\rho\boldsymbol{u}\boldsymbol{u}) + \boldsymbol{\nabla}\left(\rho c_s^2\right) = -\rho\boldsymbol{\nabla}\Phi, \tag{7}$$

where $\boldsymbol{u}$ denotes the velocity in the rest frame. The x-component of Eq. (6) is obtained by the inner product of the unit vector in the x direction and Eq. (7),

$$\boldsymbol{e}_x \cdot \left[\frac{\partial}{\partial t} (\rho \boldsymbol{u}) + \boldsymbol{\nabla} \cdot (\rho \boldsymbol{u} \boldsymbol{u}) + \boldsymbol{\nabla} \left(\rho c_s^2 \right) \right] = -\rho \boldsymbol{e}_x \cdot \boldsymbol{\nabla} \Phi. \tag{8}$$

It is rewritten as

$$\frac{\partial}{\partial t} \left[\rho \left(\boldsymbol{e}_x \cdot \boldsymbol{u} \right) \right] + \boldsymbol{\nabla} \cdot \left[\rho \boldsymbol{u} \left(\boldsymbol{e}_x \cdot \boldsymbol{u} \right) \right] + \left(\boldsymbol{e}_x \cdot \boldsymbol{\nabla} \right) \left(\rho c_s^2 \right) = -\rho \left(\boldsymbol{e}_x \cdot \boldsymbol{\nabla} \right) \Phi + \Omega \rho \left(\boldsymbol{e}_y \cdot \boldsymbol{u} \right), \tag{9}$$

since the time derivative of the unit vector is expressed

$$\frac{\partial}{\partial t} \boldsymbol{e}_x = \boldsymbol{\Omega} \times \boldsymbol{e}_x = \Omega \boldsymbol{e}_y. \tag{10}$$

Here the symbol, $\boldsymbol{e}_y$, denotes the unit vector in the y-direction. Similarly we obtain

$$\frac{\partial}{\partial t} \left[\rho \left(\boldsymbol{e}_y \cdot \boldsymbol{u} \right) \right] + \boldsymbol{\nabla} \cdot \left[\rho \boldsymbol{u} \left(\boldsymbol{e}_y \cdot \boldsymbol{u} \right) \right] + \left(\boldsymbol{e}_y \cdot \boldsymbol{\nabla} \right) \left(\rho c_s^2 \right) = -\rho \left(\boldsymbol{e}_y \cdot \boldsymbol{\nabla} \right) \Phi - \Omega \rho \left(\boldsymbol{e}_x \cdot \boldsymbol{u} \right). \tag{11}$$

The last term in Eq. (9) and that in Eq (11) yield a half of the Coriolis force.

The rest half of the Coriolis force is derived as follows. The rest frame velocity, $\boldsymbol{u}$, is related with the co-rotating frame velocity, $\boldsymbol{v}$, via

$$\boldsymbol{u} = \boldsymbol{v} + \boldsymbol{\Omega} \times \boldsymbol{r}. \tag{12}$$

Thus we obtain

$$\boldsymbol{e}_x \cdot \boldsymbol{u} = \boldsymbol{e}_x \cdot \boldsymbol{v} - \Omega y = v_x - \Omega y, \tag{13}$$
$$\boldsymbol{e}_y \cdot \boldsymbol{u} = \boldsymbol{e}_y \cdot \boldsymbol{v} + \Omega x = v_y + \Omega x. \tag{14}$$

Substituting Eqs. (13) and (14) into Eqs. (9) and (11) we obtain

$$\frac{\partial}{\partial t} \left[\rho \left(v_x - \Omega y \right) \right] + \frac{\partial}{\partial x} \left[\rho v_x \left(v_x - \Omega y \right) \right] + \frac{\partial}{\partial y} \left[\rho v_y \left(v_x - \Omega y \right) \right] + \frac{\partial}{\partial z} \left[\rho v_z \left(v_x - \Omega y \right) \right] + \frac{\partial}{\partial x} \left(\rho c_s^2 \right) = -\rho \frac{\partial \Phi}{\partial x} + \rho \Omega \left(v_y + \Omega x \right), \tag{15}$$

$$\frac{\partial}{\partial t} \left[\rho \left(v_y + \Omega x \right) \right] + \frac{\partial}{\partial x} \left[\rho v_x \left(v_y + \Omega x \right) \right] + \frac{\partial}{\partial y} \left[\rho v_y \left(v_y + \Omega x \right) \right] + \frac{\partial}{\partial z} \left[\rho v_z \left(v_y + \Omega x \right) \right] + \frac{\partial}{\partial y} \left(\rho c_s^2 \right) = -\rho \frac{\partial \Phi}{\partial y} - \rho \Omega \left(v_x - \Omega x \right), \tag{16}$$

We obtain Eq. (6) by substituting the mass conservation,

$$\frac{\partial \rho}{\partial t} + \frac{\partial}{\partial x} \left(\rho v_x \right) + \frac{\partial}{\partial y} \left(\rho v_y \right) + \frac{\partial}{\partial z} \left(\rho v_z \right) = 0, \tag{17}$$

into Eq. (15) and (16). Note that the second half of the Coriolis force is introduced by substituting the mass conservation. It implies that this half of the Coriolis force should be evaluated consistently with the evaluation of the mass conservation. It means that we should use the mass flux across the numerical cell surface to evaluate the Coriolis force. Otherwise the momentum of the flow is not conserved in the rest frame. On the other hand, the first half of

IOP Conf. Series: Journal of Physics: Conf. Series **1031** (2018) 012004 doi:10.1088/1742-6596/1031/1/012004

the Coriolis force is due to rotation of the unit vectors in the co-rotating frame. Thus it should be evaluated by using the velocity and density in the numerical cell considered.

We examine three types of methods of evaluating the Coriolis force in order to assess the argument mentioned above.

In the first type, the source term, $2\rho\boldsymbol{\Omega} \times \boldsymbol{v}$, is evaluated by the density and velocity in the cell. Then the x and y-component of the source term due to the Coriolis force is evaluated to be

$$F_{\mathrm{CC},x}(x_i, y_j, z_k) = 2\Omega\rho(x_i, y_j, z_k)v_y(x_i, y_j, z_k), \tag{18}$$

$$F_{\mathrm{CC},y}(x_i, y_j, z_k) = -2\Omega\rho(x_i, y_j, z_k)v_x(x_i, y_j, z_k), \tag{19}$$

where (x_i, y_j, z_k) denotes the central position of the numerical cell considered. We call this cell center (CC) type in the following. CC type is likely to be used in most of numerical simulations, although it is not written explicitly in the literature.

In the second type, the source term is evaluated from the mass flux across the cell surface, i.e.,

$$F_{\mathrm{NF},x}(x_i, y_j, z_k) = \Omega\left[F_{\rho,y}\left(x_i, y_j + \frac{\Delta y}{2}, z_k\right) + F_{\rho,y}\left(x_i, y_j - \frac{\Delta y}{2}, z_k\right)\right], \tag{20}$$

$$F_{\mathrm{NF},y}(x_i, y_j, z_k) = -\Omega\left[F_{\rho,x}\left(x_i + \frac{\Delta x}{2}, y_j, z_k\right) + F_{\rho,y}\left(x_i - \frac{\Delta x}{2}, y_j, z_k\right)\right], \tag{21}$$

where Δx and Δy denote the cell width in the x- and y-directions, respectively. The symbols, $F_{\rho,x}$ and $F_{\rho,y}$ denote the x- and y-components of the numerical mass flux used in the equation of mass conservation [Eq. (5)]. We call this numerical flux (NF) type in the following. NF type is employed in Hanawa et al.[4] and Takakuwa et al. [3].

We propose to evaluate the source term by half & half mixture of CC and NF types,

$$F_{\mathrm{HH},x} = \frac{1}{2}\left(F_{\mathrm{CC},x} + F_{\mathrm{NF},x}\right), \tag{22}$$

$$F_{\mathrm{HH},y} = \frac{1}{2}\left(F_{\mathrm{CC},y} + F_{\mathrm{NF},y}\right). \tag{23}$$

This is called half & half (HH) in the following.

We use the numerical simulation code SFUMATO (Matsumoto [5]) to solve the hydrodynamical equations. The temporal and spatial accuracies are set to be of the second order. We use the nested grids of 5 levels in which the coarsest grid covers the region of $|x| \leq 12.8a$, $|y| < 12.8a$, and $|z| < 6.4a$. Finer grids cover smaller spaces around the origin with higher spatial resolution. Each grid contains $128 \times 128 \times 64$ cubic cells.

The fixed boundary is set on the cylindrical surface of $\sqrt{x^2 + y^2} = 12.8a$ and the planes of $|z| = 6.4a$. The specific angular momentum is fixed to be $j_{\mathrm{inf}} = 1.2\Omega a^2$. The radial infall velocity is set to be

$$v_r = \begin{cases} -\left(\dfrac{2GM}{\sqrt{x^2+y^2+z^2}} - \dfrac{j_{\mathrm{inf}}^2}{x^2+y^2}\right)^{1/2} & \left(\dfrac{2GM}{\sqrt{x^2+y^2+z^2}} > \dfrac{j_{\mathrm{inf}}^2}{x^2+y^2}\right) \\ 0 & \text{(otherwise)} \end{cases}. \tag{24}$$

Note that the specific angular momentum is defined in the rest frame. The density is fixed to be $\rho_{\mathrm{inf}} = 1.0$ outside the boundary.

The initial density inside the boundary is set to be uniform at ρ_0 and the initial velocity is set to vanish in the rest frame, i.e., $\boldsymbol{v} = -\boldsymbol{\Omega} \times \boldsymbol{r}$. The CFL number is taken to be 0.7 unless otherwise noted.

3. Results

In this section we compare the three types of evaluating the Coriolis force (CC, NF, and HH) and demonstrate how numerical simulations depend on them. First we compare the numerical simulations in which the mass ratio and sound speed are set to be $q = 0.2$ and $c_s = 0.1$, respectively. The initial density is $\rho_0 = 0.1$ inside the boundary and $\rho_{\rm inf} = 1.0$ for the gas infalling from the boundary. The infalling gas has a constant specific angular momentum, $j_{\rm inf} = 1.2$.

Figure 1 shows the density and velocity distributions at $t = 9.43$. The top, middle and bottom panels denote the results obtained with CC, NF, and HH types of the Coriolis force. The left panels denote the density on the mid plane ($z = 0$) by color and the velocity in the co-rotating frame, $\boldsymbol{v}$, on the mid plane by arrows. The right panels denote the specific angular momentum, j_z, on the midplane by color and the velocity in the rest frame, $\boldsymbol{u}$, by the arrows. The color bars shown in the right of the panels. Some gas elements have specific angular momentum outside the range shown in the color bar.

In the period $t < 9.43$ the infalling gas remain in the outer region of $\sqrt{x^2 + y^2} > 6$ on the mid plane. The density and velocity are both symmetric around the z-axis. The gravity is almost symmetric around the axis since the quadruple moment of the gravity is very small there. Thus the specific angular momentum of each gas element should be conserved. However, it is largely changed from its initial value, $j_{\rm inf} = 1.2$, when either CC type or NF type of the Coriolis force is employed. Note that the gas element rotate clockwise in the ring like region of $\sqrt{x^2 + y^2} \simeq 8$ in the top right panel (CC type). The specific angular momentum is larger than the initial value outside the ring. On the other hand, the specific angular momentum is enhanced in the ring of $\sqrt{x^2 + y^2} \simeq 8$ and lowered outside the ring when NF type Coriolis force is employed. Both CC and NF types of the Coriolis force induce unphysical change in the specific angular momentum but in the opposite directions each other.

When HH type Coriolis force is employed, the specific angular momentum remains nearly constant although it is slightly enhanced in the ring of $\sqrt{x^2 + y^2} \simeq 8.5$. This is reasonable since HH type is an intermediate of CC and NF types. The unphysical changes seen in CC and NF types are canceled out each other in HH type.

The unphysical change in the specific angular momentum may have been missed in simulations thus far. The velocity distributions ($\boldsymbol{v}$) shown in the left panels of Fig. 1 look quite normal, although the top right and middle right panels show apparently abnormal rotation. This serious error is detectable only when the rest frame velocity is displayed. It also should be noted that the error is emphasized in the specific angular momentum, j. The corresponding error in the rotation velocity is lower since $\Delta v_{\rm rot} = \Delta j / \sqrt{x^2 + y^2}$.

The unphysical change in the specific angular momentum is more serious, when the density contrast is higher between the infalling gas ($\rho_{\rm inf}$) and pre-existing gas (ρ_0). We have confirmed this dependence by lowering the density of pre-existing gas to $\rho_0 = 0.01$ and 10^{-3} for NF type Coriolis force. When the density contrast is higher, a stronger shock wave arises between the pre-existing and newly infalling gases. Remember that the numerical mass flux differs very much from the product of the density and velocity at the cell center. Thus the results depend seriously on the evaluation of the Coriolis force.

Even after 10 revolution of the binary the numerical simulation depend considerably on the evaluation of the Coriolis force. Figure 2 is the same as Figure 1 but for the stage of $t \simeq 71.73$. The color scales are modified to cover a wider range in the density and specific angular momentum. The specific angular momentum is unnaturally high in the model obtained with CC type Coriolis force. Hence the circumbinary disk is more extended in the model. In the model obtained with NF type Coriolis force, the specific angular momentum is artificially decreased in the model computed and accordingly more gas is accreted onto the binary.

Figure 3 shows the structure of the circumbinary disk at the stage shown in Figure 2. The

IOP Conf. Series: Journal of Physics: Conf. Series **1031** (2018) 012004 doi:10.1088/1742-6596/1031/1/012004

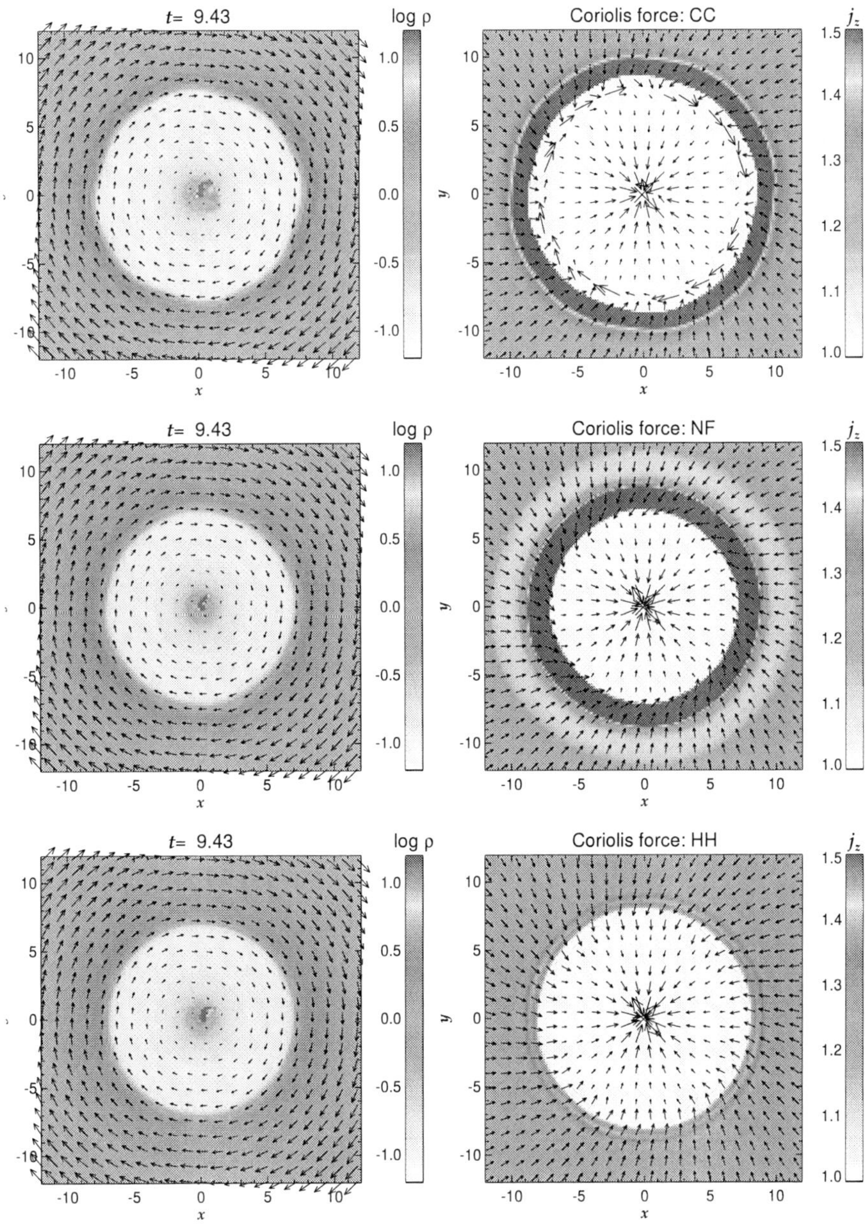

Figure 1. The density and velocity distributions at $t = 9.43$ for the model of $q = 0.2$, $c_s = 0.1$ and $j_{\rm inf} = 1.2$. Color denote the density on the mid plane ($z = 0$) in the left panels and the specific angular momentum in the right panels. The arrows denote $\boldsymbol{v}$ in the left panels and $\boldsymbol{u}$ in the right panels. CC, NF and HH types Coriolis force is employed in the models shown in the top, middle and bottom panels, respectively.

left, middle and right panels denote the models obtained with CC, NF and HH types Coriolis force, respectively. The upper panels show the cross section in the xz plane ($y = 0$) while the lower panels do the distribution in the xy plane ($z = 0$). The density is denoted by color and the velocity is shown by arrows.

IOP Conf. Series: Journal of Physics: Conf. Series **1031** (2018) 012004 doi:10.1088/1742-6596/1031/1/012004

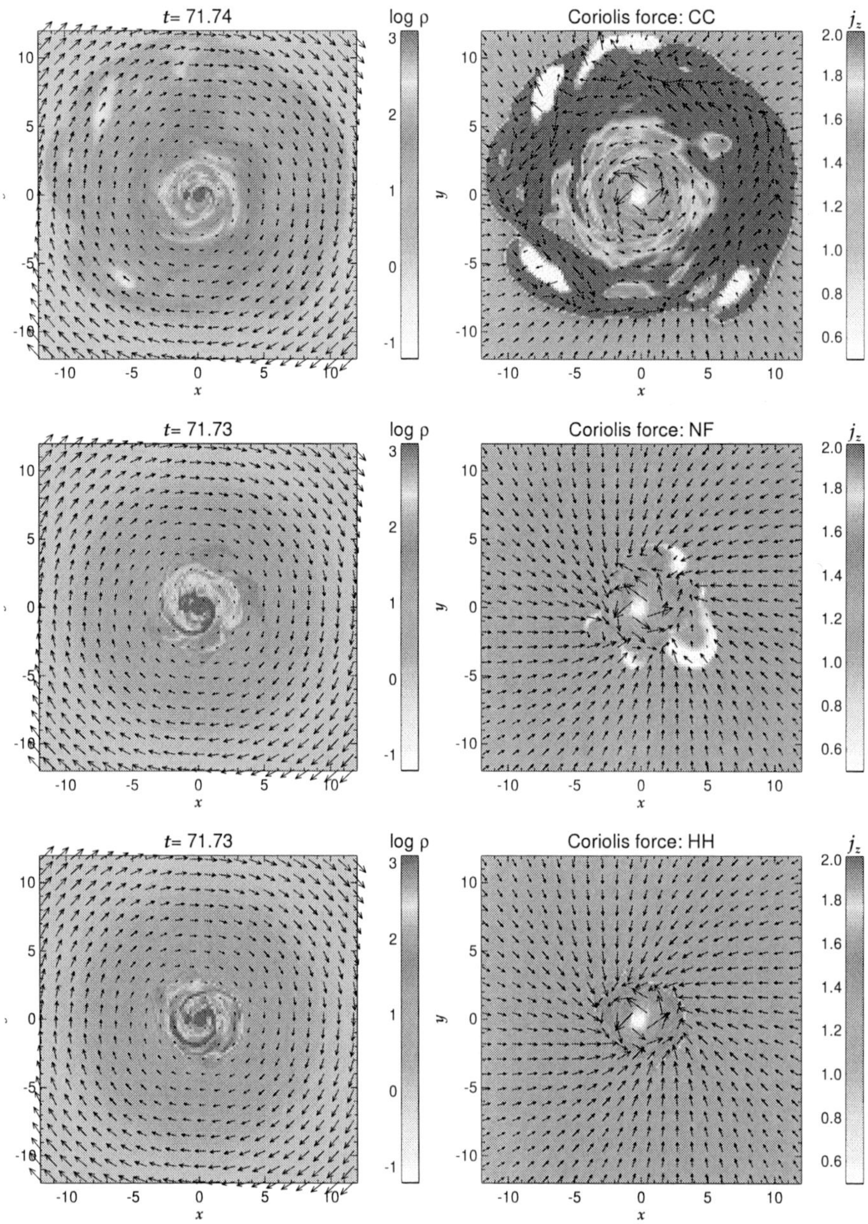

Figure 2. The same as Fig. 1 but for $t \simeq 71.73$ with different color scales.

All the models show a geometrically thin circumbinary disk, S-shaped spiral shock waves, and vertically expanded infalling envelope. However, the difference is not negligibly small. The difference is in part due to the specific angular momentum of infalling gas. The specific angular momentum is higher in the model obtained with CC type Coriolis force (see Fig. 2). Thus the circumbinary disk is extended more in the radial direction in CC model. Some part of the difference is due to spiral shock waves seen in the circumbinary disk. This possibility should be examined more carefully.

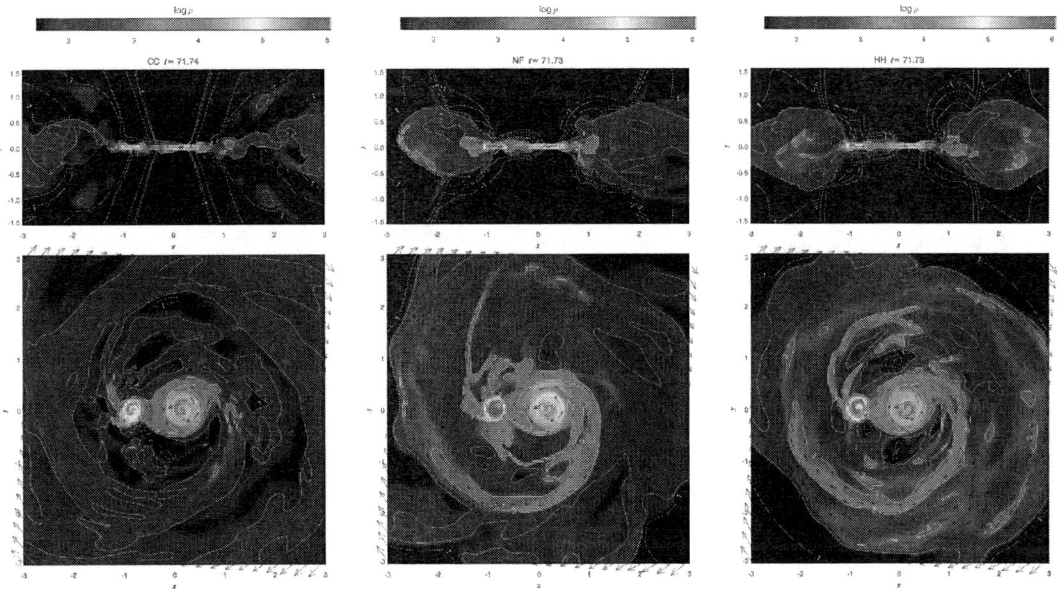

Figure 3. The structure of the circumbinary disk is shown for the models shown in Fig. 2. The left, middle, and right panels show the models obtained with CC, NF, and HH type Coriolis force, respectively. The upper panels denote the cross sections in the xz plane while the lower panels do those in the xy plane.

4. Discussions

As shown in the previous section, numerical simulations depend on the method of evaluating Coriolis force. A serious numerical error arises when the Coriolis force evaluated either by CC type or by NF type. On the other hand, the numerical error is much smaller when the Coriolis force is evaluated by HH type, i.e., the average of CC and NF types. We have also found that the structure of circumbinary disk and the S-shaped spiral shock waves are also affected by the methods of evaluating the Coriolis force. Strong spiral waves are also excited near the primary and secondary. They are expected to suffer from similar numerical errors and may be a source of discrepancy.

The numerical error is ascribed to the fact that the Coriolis force appears as a source term in the hydrodynamical equations. In other words, the hydrodynamical equations are not in the conservation form in a strict sense. The source term should vanish when the equations are transformed into those in the rest frame. However, it may not be true when the hydrodynamical equations are discretized. The mass flux and momentum density are evaluated differently in numerical simulations, although both of them are the product of the density and velocity, $\rho \boldsymbol{v}$. The former is evaluated on the cell surface and named numerical flux, while the latter is regarded as a cell average. The numerical flux is intentionally taken to be different from the simple average in modern numerical hydrodynamics. This difference is a key to avoid numerical oscillation around a shock wave. We need to think whether the Coriolis force is proportional to either the mass flux or the momentum density.

In §2 we have demonstrated that the half of the Coriolis force is due to the mass conservation and accordingly proportional to the mass flux. The other half is ascribed to the rotation of unit vector in the co-rotating frame and should be proportional to the momentum density. We have confirmed this theoretical argument in §3.

Our method of evaluating the Coriolis force is simple and easy to implement. Still the advantage is large. The conservation of specific angular momentum is greatly improved. This recipe can be applied to any numerical simulations performed in a rotating frame. It could be applied to rotational frame dragging (Lense-Thirring) effect in general relativistic hydrodynamics since this effect corresponds to rotation of the frame.

We also argue that the rest frame velocity should be examined when the numerical simulation is performed in a co-rotating frame. As demonstrated in Figure 1, a serious numerical error can be hidden and missed easily when the velocity is shown in the co-rotating frame.

Acknowledgments

This work was supported by KAKENHI, Grant-in-Aid for Scientific Research (15K1507).

References

[1] Artymowicz, P., Lubow, S.H. 1996, Astrophys. J. 467, L77

[2] Reipurth, B., Clarke, C.J., Boss, A.P., Goodwin, S.P., Rodríguez, L.F., Stassun, K.G., Tokovinin, A., Zinnecker, H. 2014, in Protostars and Planets VI ed. H. Beuther, R.S. Klessen, C.P. Dullemond, T. Henning (Univ. Arizona, Tucson) p. 267

[3] Takakuwa, S.. Saito, M., Saigo, K., Matsumoto, T., Lim, J., Hanawa, T., Ho, P.T.P. 2014 Astrophys. J., 796, 1

[4] Hanawa, T., Ochi, Y., Ando, K. 2010 Astrophys. J., 708, 485

[5] Matsumoto, T. 2007 Publ. Astron. Soc. Japan, 59, 905

IOP Conf. Series: Journal of Physics: Conf. Series **1031** (2018) 012005 doi:10.1088/1742-6596/1031/1/012005

Nuclear Rings, Nuclear Spirals, and Mass Accretion to Black Holes in Disk Galaxies

Woong-Tae Kim

Department of Physics & Astronomy, Seoul National University, Seoul 151-742, Republic of Korea

E-mail: wkim@astro.snu.ac.kr

Abstract. We use two-dimensional hydrodynamic simulations to study the formation of nuclear rings and nuclear spirals and the associated mass inflow rates at the centers of barred galaxies. We find nuclear rings form by the centrifugal barrier that the inflowing gas cannot overcome. The size of nuclear rings depend on various galaxy properties such as the bar strength, the bar pattern speed, and the bulge central density: they are smaller in galaxies with a stronger or slower bar, and with a more centrally concentrated bulge. Even a very weak bar potential can induce nuclear spirals that eventually develop into shocks. In galaxies with high shear, nuclear spirals are tightly wound and the shocks are inclined, forming a circumnuclear disk. On the other hand, galaxies with low shear produce loosely wound spirals and perpendicular shocks, without forming a circumnuclear disk. The mass inflow rates driven by the nuclear spiral shocks are enough to account for the observed level of AGN activities in Seyfert galaxies.

1. Introduction

One of the characteristic features of barred-spiral galaxies is the presence of nuclear rings and nuclear spirals at their centers (e.g., [1]). Nuclear rings are usually bright in Hα, indicating that they are actively forming new stars [2]. Observations suggest that about 20% of local spirals host star-forming nuclear rings [3]. The star formation rate (SFR) in nuclear rings appears to depend on the bar strength such that the SFR is quite small ($\sim 0.1\,\mathrm{M_\odot\,yr^{-1}}$) in strongly-barred galaxies, while it varies widely ($\sim 0.1 - 10\,\mathrm{M_\odot\,yr^{-1}}$) in weakly-barred galaxies [4], [5], [6]. The ring size also appears to depend on the bar strength, with a strongly-barred galaxy tending to have a smaller ring [5]. However, it is not well understood how the bar strength affects the ring formation and the associated SFR.

Nuclear spirals found at the very centers of barred galaxies [7], [8], [9] are thought to be a channel for gas inflows to feed supermassive black holes (e.g., [10], [11]. They even exist in galaxies with weak bar-like or oval potentials (e.g., [12], [13]). Statistically, weakly barred galaxies tend to harbor tightly wound nuclear spirals, while they are preferentially loosely wound in galaxies with a strong bar [9]. The linear theory on a curvature instability suggests that the shape of nuclear spirals should also depend on shear in the background rotation [14]. Since observed nuclear spirals are nonlinear, the linear prediction has to be checked by direct numerical simulations.

In order to study how nuclear rings and spirals form and evolve, we have been running a series of two-dimensional numerical simulations in which a bar is modeled by a fixed gravitational potential. These studies extend the previous works [15], [16], [17], [18] to explore the effects of the sound speed [19], magnetic fields [20], bar strength [21], star formation [22], [23], bulge density and pattern speed [24], and background shear [25]. Here, we present the highlights of these studies, focusing on the properties of nuclear rings and spirals formed in our simulations. We refer the reader to the papers mentioned above for technical details as well as more quantitative results.

2. Nuclear Rings

To study how the ring size depends on various galaxy parameters such as the bar strength, pattern speed, and bulge density, we consider an initially-uniform gaseous disk subject to a fixed bar potential that rotates rigidly about the galaxy center at a pattern speed $\mathbf{\Omega}_b = \Omega_b\hat{\mathbf{z}}$. The gas is taken to be isothermal with a sound speed of $c_s = 10\,\mathrm{km\,s^{-1}}$. All the simulations are run in a frame corotating with the bar. The basic hydrodynamic equations we solve read

$$\left(\frac{\partial}{\partial t} + \mathbf{v}\cdot\nabla\right)\Sigma = -\Sigma\nabla\cdot\mathbf{v}, \tag{1}$$

$$\left(\frac{\partial}{\partial t} + \mathbf{v}\cdot\nabla\right)\mathbf{v} = -c_s^2\frac{\nabla\Sigma}{\Sigma} - \nabla\Phi_{\rm ext} + \Omega_b^2\mathbf{R} - 2\mathbf{\Omega}_b\times\mathbf{v}, \tag{2}$$

where $\Phi_{\rm ext}$ is the external gravitational potential.

In our models, $\Phi_{\rm ext}$ comes from four components: a stellar disk modeled by a Kuzmin-Toomre disk, a stellar bulge, a stellar bar, and a central black hole represented by the Plummer sphere with mass $M_{\rm BH}$. For the bulge, we take a Hubble profile

$$\rho(r) = \rho_{\rm bul}\left(1+\frac{r^2}{r_b^2}\right)^{-3/2}, \tag{3}$$

where $\rho_{\rm bul}$ and r_b is the central density and scale length of the bar, respectively. In our models, the bar is modeled by Ferrers prolate spheroids with density distribution

$$\rho(r,\phi,z) = \begin{cases} \rho_{\rm bar}\left(1-g^2\right)^n, & \text{for } g<1, \\ 0, & \text{elsewhere}, \end{cases} \tag{4}$$

where $\rho_{\rm bar}$ is the central density and $n = 1$ measures the degree of central density concentration. In Equation (4), g is defined by

$$g^2(x,y,z) = \frac{y^2}{a^2} + \frac{x^2+z^2}{b^2}, \tag{5}$$

where $a = 5\,\mathrm{kpc}$ and b $(\leq a)$ denote the semimajor and semiminor axes of the bar, respectively. The bar aspect ratio is $\mathcal{R} = a/b$, which we control as a free parameter.

We run the following three series of models:

(1) In the first series of models, we vary the aspect ratio $\mathcal{R}$ in the range between 1.5 and 3.5 and the bar mass fraction $f_{\rm bar} = M_{\rm bar}/(M_{\rm bar}+M_{\rm bul})$ in the range between 0.08 and 0.6, where $M_{\rm bar}$ and $M_{\rm bar}$ denote the mass of the bar and bulge inside $r = 10\,\mathrm{kpc}$, respectively. We fix the bar pattern speed to $\Omega_b = 33\,\mathrm{km\,s^{-1}\,kpc^{-1}}$ and the bulge central density to $\rho_{\rm bul} = 10^{10}\,\mathrm{M_\odot\,kpc^{-3}}$. The variations of the bar mass and the aspect ratio change the bar strength defined by

$$Q_b = \max_{r,\phi}\frac{F_T(r,\phi)}{F_R(r)}, \tag{6}$$

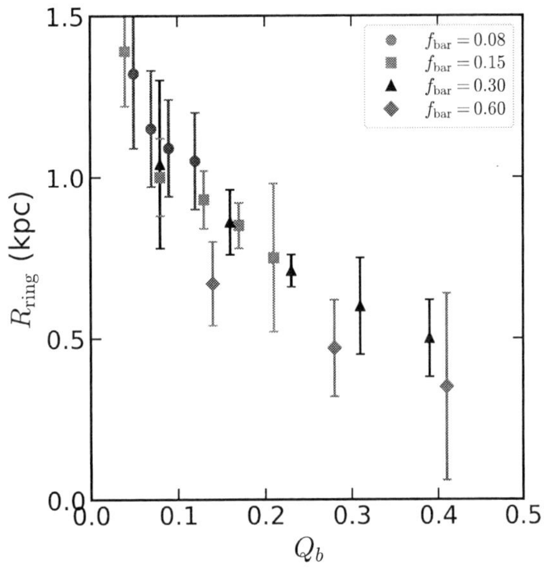

Figure 1. Azimuthally- and temporally-averaged ring radius $R_{\rm ring}$ as a function of the bar strength Q_b. The symbols and errorbars give the mean values and the standard deviations.

where F_T is the tangential force due to the non-axisymmetric bar potential and F_R is the centrifugal force of the galaxy rotation (e.g., [26], [27], [28], [29]). Numerically, our galaxy models have

$$Q_b = \begin{cases} 0.58 f_{\rm bar}^{0.89}(\mathcal{R}-1), & \text{for } n=0, \\ 0.44 f_{\rm bar}^{0.87}(\mathcal{R}-1), & \text{for } n=1, \\ 0.38 f_{\rm bar}^{0.70}(\mathcal{R}-1), & \text{for } n=2, \end{cases} \tag{7}$$

[21]. These models allow to study how the bar strength affects the size of nuclear rings that form.

(2) In the second series of models, we vary Ω_b between 21 and 49 km s^{-1} kpc^{-1}, while fixing $\rho_{\rm bul} = 10^{10}\,{\rm M_\odot\,kpc^{-3}}$, and $\mathcal{R}=2.5$, $M_{\rm bar} = 1.5\times10^{10}\,{\rm M_\odot}$ corresponding to $f_{\rm bar}=0.3$. This series of models is useful to isolate the effect of the bar pattern speed on the ring formation.

(3) In the third series of models, we vary $\rho_{\rm bul}$ from 1.2 to $4.0\times10^{10}\,{\rm M_\odot\,kpc^{-3}}$ by fixing $M_{\rm bul} = 3.37\times10^{10}\,{\rm M_\odot}$ and $\Omega_b = 33\,{\rm km\,s^{-1}\,kpc^{-1}}$ to study the effect of the bulge central density (or compactness) on the ring size.

We use the `CMHOG` code to run the first series of models [21], while the simulations for the second and third series of models [24] utilize the `Athena` code. `CMHOG` is third-order accurate in space, has very little numerical diffusion, and solves the basic hydrodynamic equations in cylindrical geometry [18]. On the other hand, `Athena` is based on a higher-order Godunov scheme in Cartesian geometry that conserves mass and momentum within machine precisions [30], [31], [32]. In all models, we turn on the bar potential slowly over one bar revolution time $2\pi/\Omega_b$ in order to minimize transients in the flows caused by a sudden introduction of the bar potential.

2.1. Effects of the Bar Strength

The imposed non-axisymmetric bar potential provides perturbations for the gas that would otherwise follow circular orbits. The perturbations are strong enough to induce shocks, called

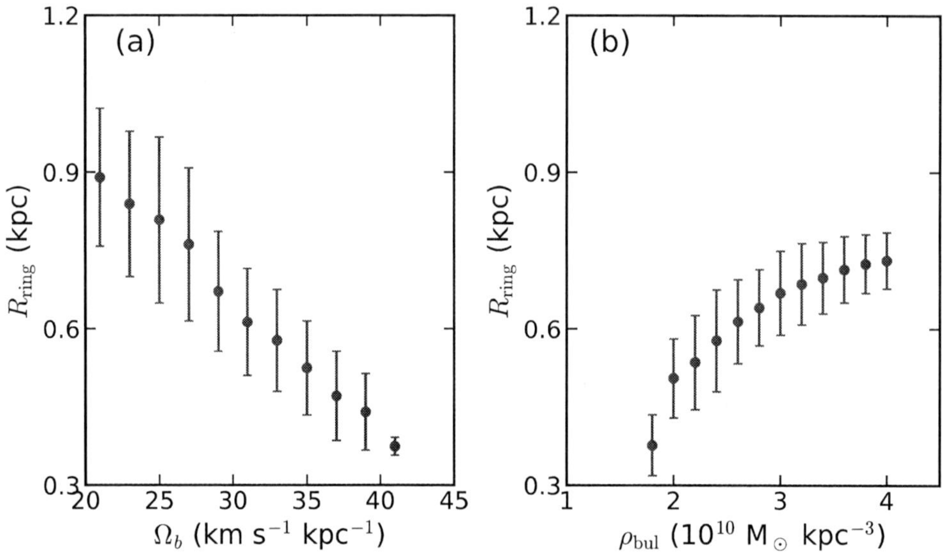

Figure 2. Azimuthally- and temporally-averaged ring radius $R_{\rm ring}$ as functions of (a) the bar pattern speed Ω_b and (b) the bulge central density $\rho_{\rm bul}$. The filled circles mark the mean values, with the errorbars corresponding to ring thickness.

dust lanes, located at the downstream side of galaxy rotation from the bar semimajor axis. Gas passing through the shocks loses angular momentum and thus experiences radial infalls. The inflowing gas tends to rotate faster as it moves radially inward, eventually forming a nuclear ring at the position where the inflowing gas achieves the initial velocity of galaxy rotation. The shape of the dust-lane shocks is well described by x_1 orbits, while nuclear rings closely follow x_2 orbits. This implies that the formation of x_2-type rings is caused by the centrifugal barrier that the inflowing gas cannot overcome [19], [20], [21].

Figure 1 plots the the azimuthally- and temporally-averaged ring radius $R_{\rm ring}$ as a function of Q_b for the first series of models [21]. It is apparent that $R_{\rm ring}$ becomes smaller with increasing Q_b, entirely consistent with the observational results that a stronger bar hosts a smaller nuclear ring [5]. This is because a stronger bar induces stronger shocks and can thus take away a larger amount of angular momentum, making the gas move radially further in and form a ring closer to the center. In general, $R_{\rm ring}$ in our models is smaller than the location of the inner Lindblad resonance (ILR), suggesting that the ring position is not determined by resonances with the bar potential but by the amount of angular momentum loss at dust-lane shocks.

2.2. Effects of the Pattern Speed

Figure 2(a) plots the azimuthally- and temporally-averaged ring radius $R_{\rm ring}$ as a function of Ω_b for x_2-type rings obtained from the second series of models [24]. Models with $\Omega_b > 41\,{\rm km\,s^{-1}\,kpc^{-1}}$ form x_1-type rings elongated along the bar semimajor axis. Apparently, the ring size decreases with increasing Ω_b almost linearly. This is presumably because the only the gas inside the corotation resonance loses angular momentum to move in, and corotation radius $R_{\rm CO}$ is a decreasing function of Ω_b. In addition, gas in models with smaller Ω_b takes longer time to be exposed to the full bar strength, so that it experiences a weaker bar torque at early time when a ring is beginning to form. The change of $R_{\rm ring}$ over $\Omega_b = 21 - 41\,{\rm km\,s^{-1}\,kpc^{-1}}$ is by less than a factor of 3.

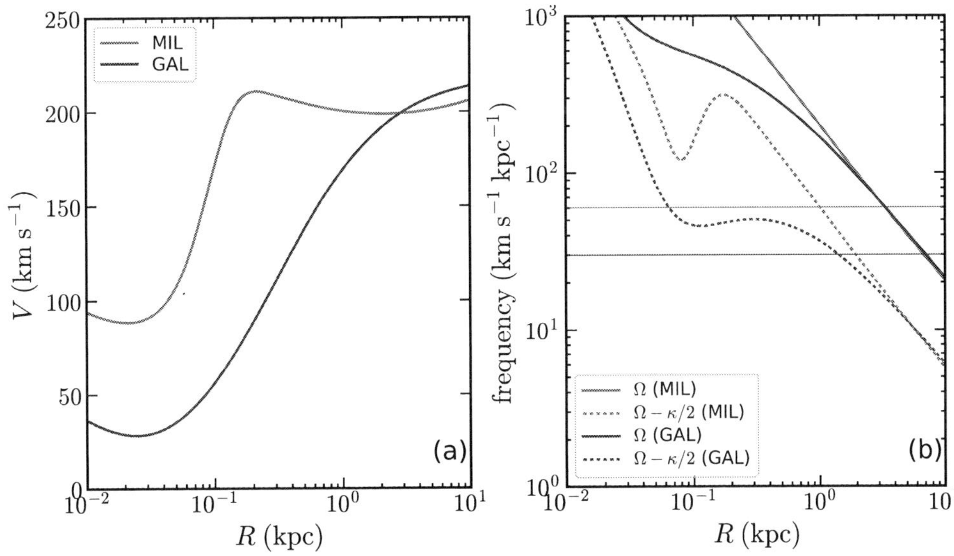

Figure 3. Radial profiles of (a) the rotational velocity V and (b) the angular frequencies Ω (solid) and $\Omega - \kappa/2$ (dotted) of the MIL (red) and GAL (blue) models. The thin horizontal lines in (b) indicate the bar pattern speed $\Omega_b = 60$ and $30\,\mathrm{km\,s^{-1}\,kpc^{-1}}$ for the MIL and GAL models, respectively.

2.3. Effects of the Bulge Central Density

Figure 2(b) plots the variation of $R_{\rm ring}$ as a function of $\rho_{\rm bul}$ for x_2-type rings obtained from the third series of models [24]. Models with $\rho_{\rm bul} < 1.8 \times 10^{10}\,\mathrm{M_\odot\,kpc^{-3}}$ produce x_1-type rings. Note that $R_{\rm ring}$ is an increasing function of $\rho_{\rm bul}$. Physically, this is because models with larger $\rho_{\rm bul}$ rotates faster and can thus provide the required centrifugal barrier for the inflowing gas at larger radii. While $R_{\rm ring}$ varies by a factor of about 2 over the variation of $\rho_{\rm bul} = 1.8$ to $4.0 \times 10^{10}\,\mathrm{M_\odot\,kpc^{-3}}$, $R_{\rm ring}$ is insensitive to $\rho_{\rm bul} \gtrsim 3 \times 10^{10}\,\mathrm{M_\odot\,kpc^{-3}}$.

3. Nuclear Spirals and Gas Accretion

We now turn to nuclear spirals formed at the central parts of weakly barred galaxies. To study the effect of shear on the shape of nuclear spirals, we consider two galaxy models with rotational velocities

$$V = \begin{cases} 65 + 95\tanh\left(\dfrac{R-0.07}{0.06}\right) - 50\log R + 1.5(\log R + 3)^3, & \text{for MIL model,} \\ \left[\dfrac{GM_{\rm BH}}{R} + \left(\dfrac{V_0 R}{R_0 + R}\right)^2\right]^{1/2}, & \text{for GAL model,} \end{cases} \tag{8}$$

where V and R for the MIL model are in units of $\mathrm{km\,s^{-1}}$ and kpc, respectively, and $V_0 = 220$ km s^{-1}, $R_0 = 0.3\,$kpc, and $M_{\rm BH} = 3 \times 10^6\ \mathrm{M_\odot}$ for the GAL model. The MIL model is designed to simulate galaxies like the Milky Way with strong shear at the centers, while the GAL model is for galaxies with low shear, like NGC 3041 [33].

Figure 3 plots the radial distributions of V as well as the angular frequencies $\Omega = V/R$ and $\Omega - \kappa/2$, where $\kappa^2 = R^{-3}d(\Omega^2 R^4)/dR$ is the epicycle frequency. The MIL model has strong shear near $R \sim 0.1\,$kpc, while shear is overall weak in the GAL model. The gas is taken to be initially uniform with surface density Σ_0 and isothermal with a sound speed of $c_s = 10\,\mathrm{km\,s^{-1}}$.

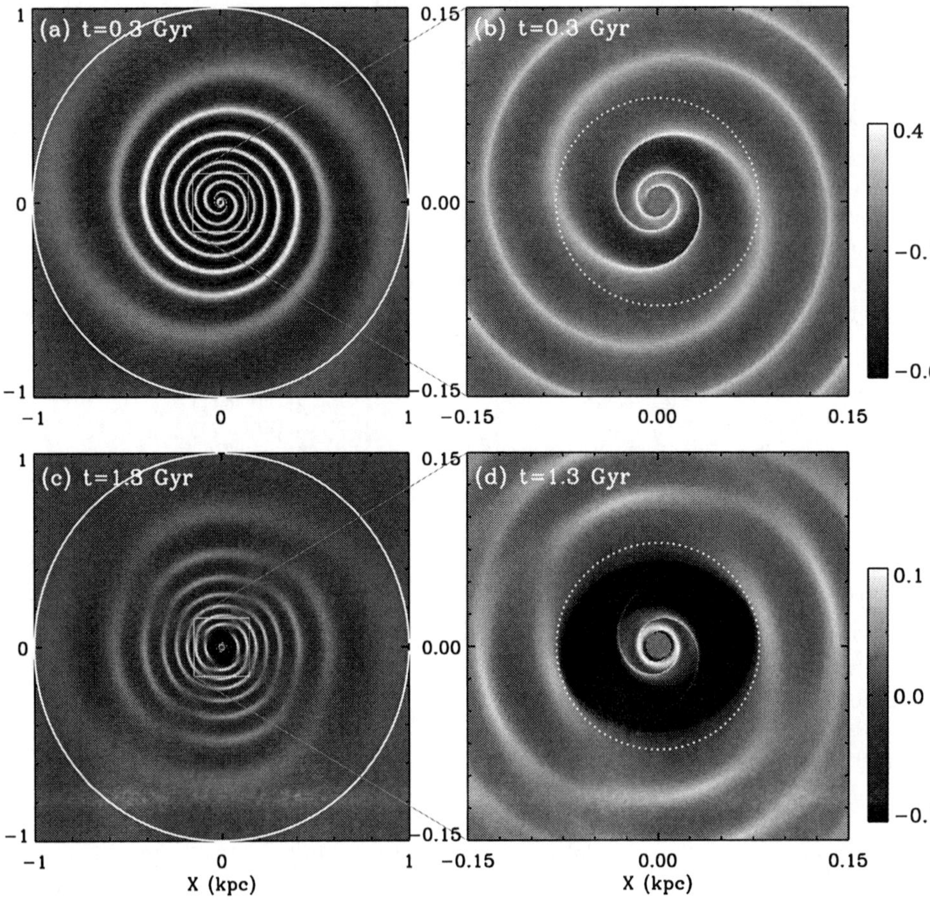

Figure 4. Snapshots of the gas surface density Σ at $t = 0.3\,\mathrm{Gyr}$ (upper panels) and $t = 1.3\,\mathrm{Gyr}$ (lower panels) of the high-shear MIL model with $\Phi_0/c_s^2 = 0.1$. The circles in the left panels mark the ILR. The right panels zoom in the central 0.15 kpc regions. The dotted circles with radius 80 pc draw the regions influenced by shocks. Colorbars label $\log(\Sigma/\Sigma_0)$.

For the perturbing potential representing a weak bar or an oval distortion, we take a simple bi-symmetric form

$$\Phi_b(R, \phi, t) = \Phi_0 \cos(2\phi - 2\Omega_b t), \tag{9}$$

with amplitude Φ_0 and pattern speed Ω_b. To study the effect of the bar strength, we consider two models with $\Phi_0/c_s^2 = 0.1$ and 0.01. The bar pattern speed is taken to be $\Omega_b = 60$ and $30\,\mathrm{km\,s^{-1}\,kpc^{-1}}$ for the MIL and GAL models, respectively, which are indicated as thin horizontal lines in Figure 3(b). The corresponding ILR and corotation resonance are at $R_{\mathrm{ILR}} = 1.00\,\mathrm{kpc}$ and $R_{\mathrm{CO}} = 3.32\,\mathrm{kpc}$ in the MIL model, and at $R_{\mathrm{ILR}} = 1.42\,\mathrm{kpc}$ and $R_{\mathrm{CO}} = 7.03\,\mathrm{kpc}$ in the GAL model, respectively. We use `Athena++`, a newly developed grid-based code based on a higher-order Godunov scheme to evolve the gas in two-dimensional cylindrical geometry. The simulation domain extends from $R_{\mathrm{in}} = 10\,\mathrm{pc}$ to $R_{\mathrm{out}} = R_{\mathrm{CO}}$.

3.1. Density Structure

The imposed gravitational potential induces waves in the otherwise uniform disk that are gradually organized into a piecewise logarithmic spiral shape. The spiral waves grow as thermal

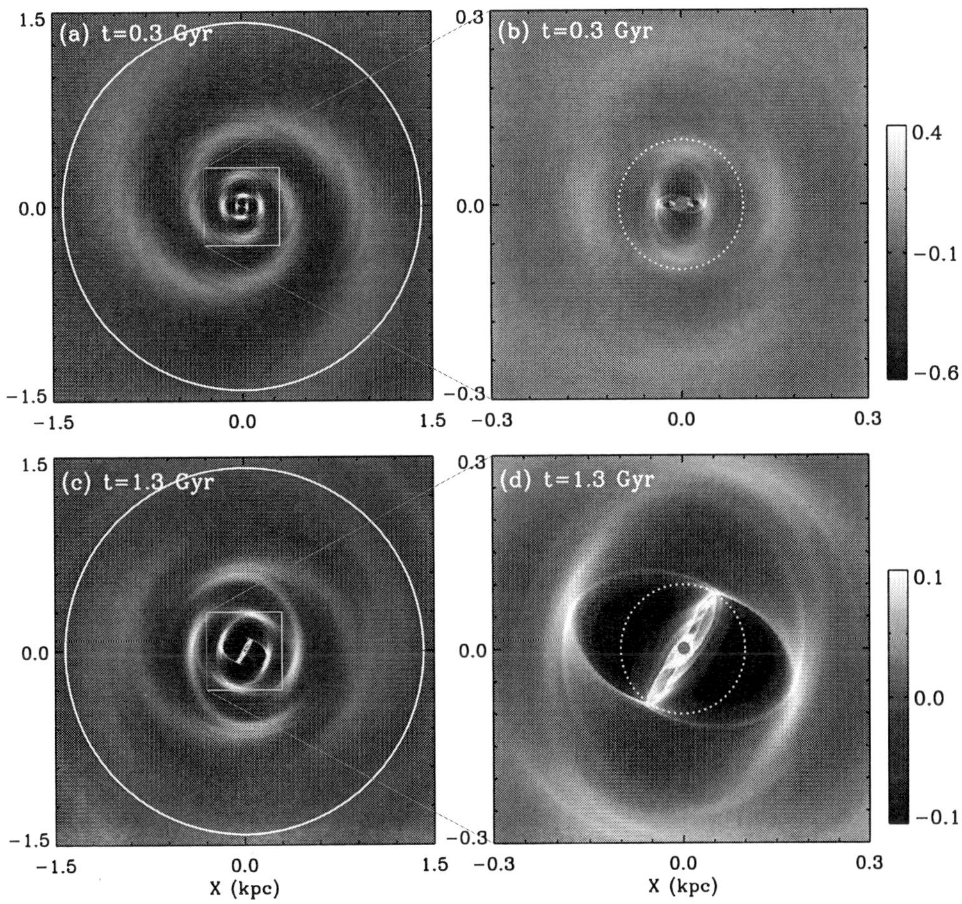

Figure 5. Same as Figure 4, but for the low-shear GAL model with $\Phi_0/c_s^2 = 0.1$. The radius of the dotted circles is 100 pc.

pressure align the apocenters of perturbed gas orbits inside the ILR [14]. They amplify further to become nonlinear as they propagate inward due to a geometric effect, eventually developing into shocks even for very weak potentials. Figures 4 and 5 plot the distributions of the gas surface density at $t = 0.3$ Gyr (upper panels) and 1.3 Gyr (lower panels) of the MIL and GAL models with $\Phi_0/c_s^2 = 0.1$, respectively. Clearly, the spirals exist only inside the ILR, denoted by the solid circles in both models, while the regions outside the ILR are almost featureless.

Comparison of Figures 4 and 5 reveals the following notable differences. First, the shape of the spirals depends sensitively on the background shear: spirals are tightly wound (with a pitch angle of $i_p \sim 10°$) in the high-shear MIL model, while they are loosely wound (with $i_p \sim 35°$) in the low-shear GAL model. Second, shocks are inclined in the MIL model, resulting in relatively small perpendicular Mach numbers of $\mathcal{M}_\perp \sim 1.5$. On the other hand, the shocks in the GAL model are less inclined and unwind over time to be perpendicular, presumably due to an increase of the angular momentum flux [34]. The resulting perpendicular Mach numbers are $\mathcal{M}_\perp \sim 2.5$. Third, the shocks take away angular momentum from the gas that encounters them, causing mass inflows. In the MIL model, the inflowing gas moves on more-or-less circular orbits, piles up near the center due to the geometric convergence effect, and forms a circumnuclear disk with radius of $\sim$ 20–30 pc. In the GAL model, on the other hand, the inflowing gas after

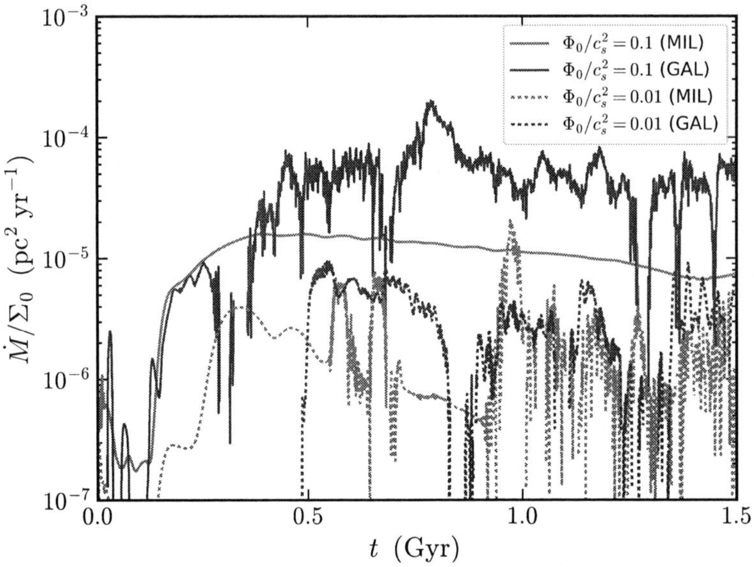

Figure 6. Mass inflow rates $\dot{M}$ measured at the inner radial boundary $R_{\rm in} = 10\,{\rm pc}$ for the MIL (red) and GAL (blue) models. The solid and dotted lines correspond to the models with $\Phi_0/c_s^2 = 0.1$ and 0.01, respectively.

the perpendicular shocks moves on almost radial orbits, directly plugging to the inner radial boundary of the simulation domain.

3.2. Mass Inflows

Figure 6 plots the mass inflow rates $\dot{M}$, normalized by Σ_0, measured at the inner radial boundary $R_{\rm in} = 10\,{\rm pc}$ of the simulation domain. The solid and dotted lines correspond to the models with $\Phi_0/c_s^2 = 0.1$ and 0.01, respectively. Overall, the low-shear GAL models have larger $\dot{M}$ than the high-shear MIL models since the shocks are stronger and perpendicular in the former. The mass inflow rate is approximately proportional to the strength of the imposed bar potential. The time-averaged mass inflow rates in our simulations can be written as

$$\langle \dot{M} \rangle = 10^{-5} f \left(\frac{\Sigma_0}{{\rm M_\odot\,yr^{-1}}} \right) \left(\frac{\Phi_0/c_s^2}{0.1} \right) {\rm M_\odot\,yr^{-1}}, \tag{10}$$

where f is a factor, of order unity, responsible for the effect of background shear [25]. Assuming that all the inflowing gas is accreted to a central black hole with mass $M_{\rm BH}$, the corresponding Eddington ratio is

$$\lambda = 1.5 \times 10^{-2} f \left(\frac{\epsilon}{0.1} \right) \left(\frac{\Sigma_0}{10^2\,{\rm M_\odot\,yr^{-1}}} \right) \left(\frac{\Phi_0/c_s^2}{0.1} \right) \left(\frac{M_{\rm BH}}{3 \times 10^6\,{\rm M_\odot}} \right), \tag{11}$$

where $\epsilon \sim 0.1$ is the efficiency of an active galactic nucleus (AGN). Since Seyfert galaxies have $\lambda \sim 10^{-3}$–10^{-1} [35], [36], [37], this suggests that gas inflows driven by nonlinear nuclear spirals can account for observed levels of AGN activity in Seyfert galaxies.

4. Summary

Using two-dimensional hydrodynamic simulations, we have investigated the formation of nuclear rings and spirals and the associated mass inflow rates at galaxy centers. Our results can be summarized as follows.

(1) Nuclear rings form not by resonances but by the centrifugal barrier that the inflowing gas driven by the bar potential cannot overcome. This predicts that nuclear rings in more strongly barred galaxies are smaller in size, consistent with the results of our simulations and also with observations.
(2) The size of nuclear rings depend not only on the bar strength but also on the bar pattern speed and the bulge central density. Nuclear rings are larger in models with smaller pattern speed and/or larger bulge central density. This suggests that one should be careful in inferring the galaxy properties from the sizes of observed nuclear rings.
(3) Nuclear spirals exist even in very weekly barred galaxies. They amply due to the geometric convergence effect and eventually evolve into shocks, causing radial mass inflows. In high-shear models, nuclear spirals tend to be tightly wound and shocks are inclined. This results in relatively small rates of the mass inflows and the formation of a circumnuclear disk. In low-shear models, on the other hand, nuclear spirals are loosely wound and shocks are perpendicular, resulting in a comparatively large rate of the mass inflows and no formation of a circumnuclear disk.
(4) Even under a very-weak external potential, the mass inflow rates driven by nonlinear nuclear spirals are sufficient to power the observed level of AGN activity in diverse Seyfert galaxies.

Acknowledgments
I am grateful to my collaborators including Bruce Elmegreen, Zhi Li, Woo-Young Seo, and Juntai Shen for discussions and helps. This work was supported by the National Research Foundation of Korea (NRF) grant funded by the Korean government (MEST; No. 3348-20160021). The computation of this work was supported by the Supercomputing Center/Korea Institute of Science and Technology Information with supercomputing resources including technical support (KSC-2016-C3-0044).

References

[1] Buta R and Combes F 1996 *Fund. Cosmic Phys.* **17** 95
[2] Kormendy J and Kennicutt R C Jr 2004 *ARA&A* **42** 603
[3] Knapen J H 2005 *A&A* **429** 141
[4] Mazzuca L M, Knapen J H, Veilleux S, and Regan M W 2008 *ApJ* **174** 337
[5] Comerón S, Knapen J H, Beckman J E, Laurikainen E, Salo H, Martínez-Valpuesta I, and Buta R J 2010 *MNRAS* **402** 2462
[6] Mazzuca L M, Swaters R A, Knapen J H, and Veilleux S 2011 *ApJ* **739** 104
[7] Martini P, Regan M R, Mulchaey J S, and Pogge R W 2003a *ApJS* **146** 353
[8] Martini P, Regan M R, Mulchaey J S, and Pogge R W 2003b *ApJ* **589** 774
[9] Peeples M S and Martini P 2006 *ApJ* **652** 1097
[10] Shlosman I, Begelman M C, and Frank J 1990 *Nature* **345** 679
[11] van de Ven G and Fathi K 2010 *ApJ* **723** 767
[12] Elmegreen B G, Elmegreen D M, Brinks E, Yuan C, Kaufman M, Klarić M, Montenegro L, Struck C, and Thomasson M 1998 *ApJL* **503** L119
[13] Elmegreen D M, Elmegreen B G, and Eberwein K S 2002 *ApJ* **564** 234
[14] Montenegro L E, Yuan C, and Elmegreen B G 1999 *ApJ* **520** 592
[15] Sanders R H and Huntley J M 1976 *ApJ* **209** 53
[16] Roberts W, Huntley M, and Van Albada G D 1979 *ApJ* **233** 67
[17] Athanassoula E 1992 *MNRAS* **259** 345
[18] Piner B G, Stone J M, and Teuben P J 1995 *ApJ* **449** 508
[19] Kim W-T, Seo W-Y, Stone J M, Yoon D, and Teuben P J 2012 *ApJ* **747** 60
[20] Kim W-T and Stone J M 2012 *ApJ* **751** 124
[21] Kim W-T, Seo W-Y, and Kim Y 2012 *ApJ* **758** 14
[22] Seo W-Y and Kim W-T 2013 *ApJ* **769** 100
[23] Seo W-Y and Kim W-T 2014 *ApJ* **792** 47
[24] Li Z, Shen J, and Kim W-T 2015 *ApJ* **806** 150

[25] Kim W-T and Elmegreen B G 2017 *ApJ* **841** L4
[26] Combes F and Sanders R H 1981 *A&A* **96** 164
[27] Laurikainen E and Salo H 2002 *MNRAS* **337** 1118
[28] Block D L, Buta R, Knapen J H, Elmegreen D M, Elmegreen B G, and Puerari I 2004 *AJ* **128** 183
[29] Laurikainen E, Salo H, Buta R, Knapen J, Speltincx T, and Block D 2006 *AJ* **132** 2634
[30] Gardiner T A and Stone J M 2005 *J. Comp. Phys.* **205** 509
[31] Stone J M and Gardiner T A 2009 *New Astr.* **14** 139
[32] Stone J M, Gardiner T A, Teuben P, Hawley J F, and Simon J B 2008 *ApJS* **178** 137
[33] Erroz-Ferrer S 2016 *MNRAS* **458** 1199
[34] Lee E and Goodman J 1999 *MNRAS* **308** 984
[35] Bian W and Gu Q 2007 *ApJ* **657** 159
[36] Ho L C 2008 *ARA&A* **46** 475
[37] Meyer-Hofmeister E and Meyer F 2011 *A&A* **527** A127

Dynamical friction on hot bodies in gaseous, opaque media, and application to embedded protoplanets

F S Masset[1], D A Velasco Romero[1] and H Eklund[2]

[1] Instituto de Ciencias Físicas, Universidad Nacional Autónoma de México, Av. Universidad s/n, Col. Chamilpa, 62210 Cuernavaca, Mor., Mexico

[2] Institute of Theoretical Astrophysics, University of Oslo, Postboks 1029 Blindern, N-0315 Oslo, Norway

E-mail: masset@icf.unam.mx

Abstract. A massive and luminous perturber moving across an opaque gas is subjected to a force different from the gravitational friction that it would experience if it were cold. The heat released by the perturber diffuses in the surrounding gas, where it gives rise to a low density region behind the perturber that exerts a force (that we call heating force) in the direction of motion, thus opposed to the standard dynamical friction. We present numerical simulations with nested meshes that confirm the analytical expression of the heating force in the limits of a low and high Mach number, respectively, and we present simulations that show that the dynamical friction exerted on a cold perturber in a gas with thermal diffusion is markedly different from that in an adiabatic gas. We then present numerical simulations of low-mass protoplanets embedded in opaque, viscous discs, that show that when these bodies have a sufficiently large luminosity their eccentricity and inclination can be excited to values comparable to the aspect ratio of the disc. We finally present numerical experiments with very high resolution that try to resolve the flow within the Bondi sphere, in an attempt to study the dependence of the heating force as a function of the ratio of the diffusive to acoustic times across the Bondi radius.

1. Introduction

While an important part of the interactions between a protoplanetary disc and a forming planet is of tidal nature (i.e. the planet's response to the disturbance excited in the disc by its gravitational field), the planet is also subjected to a non-tidal force that arises from the disturbance created by the heat that it releases in the surrounding gas. The first investigation on this topic was about the influence of the planet's luminosity on its own migration [1]. It was found that the disturbance created by the planet's luminosity yields in general a positive extra torque, which may be able to counteract or reverse the inward migration of small mass planetary embryos (below a few Earth masses). The impact of heat release was subsequently investigated in the context of dynamical friction by [2]. The low-density, heated region found downstream of a hot perturber yields an extra force, called heating force, that is in the direction of the motion, and is therefore opposed to the dynamical friction. This analysis suggested that hot embryos in protoplanetary discs could have their eccentricity and inclination excited by the release of heat (a prospect that was not investigated by [1], who limited themselves to planets on fixed, circular orbits). This expectation was later confirmed by the numerical simulations of [3]. Finally, the impact of thermal diffusion, even in the absence of heat release, was investigated by [4]. It was found, in the context of Keplerian protoplanetary discs that, for largely subsonic point-like massive perturbers, the diffusion of heat induces a perturbation of the flow with respect to the adiabatic response that is equivalent to that triggered by a massless heat sink. These proceedings are organised as follows: in section 2 we recall the basic equations of the problem at hand, in section 3 we briefly present the numerical code used in this work, in section 4 we present the analytical and numerical results about the heating force and in section 5 the analytical and numerical results about the dynamical friction on a cold body. In section 6 we summarise our results on dynamical friction, and turn to the excitation of eccentricity and inclination of embedded protoplanets in section 7. We finally present a high resolution study aimed at resolving the flow at the sub-Bondi scale in section 8, and give our conclusions in section 9.

2. Physical setup and basic equations

We consider a point-like body of mass M and luminosity L moving across a homogeneous, three-dimensional gas of density ρ, thermal diffusivity χ and sound speed c_s. We denote $(u, v, w)^T$ the velocity of the gas, $\vec{V}$ the velocity vector of the body, p and e respectively the pressure and internal energy density of the gas. As we assume it to be ideal, we have:

$$p = (\gamma - 1)e, \tag{1}$$

where γ is the adiabatic index. The equations that govern the response of the gas to the perturber are the continuity equation

$$\partial_t \rho + \vec{\nabla} \cdot (\rho \vec{v}) = 0, \tag{2}$$

the Euler equation

$$\partial_t \vec{v} + \vec{v} \cdot \vec{\nabla} \vec{v} = -\frac{\vec{\nabla} p}{\rho} - \vec{\nabla} \Phi_p, \tag{3}$$

where Φ_p is the gravitational potential of the perturber and the energy equation

$$\partial_t e + \vec{\nabla} \cdot (e\vec{v}) = -\vec{\nabla} \cdot \vec{F}_H + L\delta(\vec{r} - \vec{r}_p), \tag{4}$$

where $\vec{r}_p$ is the position of the body and $\vec{F}_H$ is the heat flux given by:

$$F_H = -\rho\chi\vec{\nabla}\left(\frac{e}{\rho}\right). \tag{5}$$

The linearised version of Eqs. (1)-(5) have been studied by [2]. We do not reproduce here the derivation of [2], but underline its main features. We denote with a 0 subscript the unperturbed quantities and with a prime the perturbations of the quantities. We therefore have:

$$\rho = \rho_0 + \rho' \tag{6}$$

and similar relations on the velocity components, pressure and internal energy. Upon linearisation, the governing equations can be cast as:

$$S(\vec{Q}) = M\vec{R}_1 + L\vec{R}_2, \tag{7}$$

where $\vec{Q} = (\rho', u', v', w', e')^T$ is the vector of the perturbations, S stands for the linearized system of equations and $\vec{R}_1$ and $\vec{R}_2$ are vectors of the right-hand-side that correspond to the forcing terms respectively due to the gravity and luminosity of the body:

$$\vec{R}_1 = \frac{G}{|\vec{r} - \vec{r}_p|^3}(0, x_p - x, y_p - y, z_p - z, 0)^T \tag{8}$$

and

$$\vec{R}_2 = [0, 0, 0, 0, \delta(\vec{r} - \vec{r}_p)]^T. \tag{9}$$

The linearity of the operator S implies that the solution $\vec{Q}$ can be written $\vec{Q} = \vec{Q}_1 + \vec{Q}_2$, where $\vec{Q}_1$ and $\vec{Q}_2$ verify

$$S(\vec{Q}_1) = M\vec{R}_1, \tag{10}$$
$$S(\vec{Q}_2) = L\vec{R}_2 \tag{11}$$

and correspond respectively to the disturbance excited in the gas by a non-luminous perturber of mass M and a massless perturber of luminosity L. In the following we will consider separately the responses $\vec{Q}_1$ and $\vec{Q}_2$.

3. Numerical setup

We have performed numerical simulations to verify the analytic dependencies of [2]. We use the FARGO3D code [5], that we have modified to deal with an arbitrary number of nested meshes with a resolution doubled in each direction for each new level of refinement. This is required to make the calculations tractable: as can be seen in Fig. 1 of [2], obtaining the heating force in the subsonic regime with an accuracy of 10 % requires to have a large range of distances to the perturber covered by the mesh (approximately two orders of magnitude). Obtaining this force with an even better accuracy ($\sim$ 5 %) requires to cover three orders of magnitude of distance to the perturber. We use a Cartesian mesh and up to 9 levels of refinement, depending on our parameters. We use a sub-cycling technique to alleviate the computational cost: for each time step performed at a given level, two smaller time steps are performed on the level with immediately finer resolution, in a recursive manner. The mesh is fixed with respect to the body, and the gas is injected at a constant speed $-V$ at the upper $z-$boundary of the coarsest mesh.

4. Heating force

4.1. Analytical considerations

The force induced by the density disturbance of $\vec{Q}_2$ has been dubbed *heating force* by [2], who have worked out its analytic expression in the limits of a small and large Mach number. In the low Mach number limit, they find that the perturbation of density ρ' has the following form:

$$\rho' = -\frac{\gamma(\gamma-1)L}{4\pi\chi c_s^2 r}\exp\left(-\frac{\gamma V r}{\chi}\cos^2\frac{\theta}{2}\right). \tag{12}$$

This form is fundamentally the r^{-1} law that corresponds to the diffusion from a point-like source in a medium at rest, with a cut-off that depends on the angle θ with respect to the direction of the velocity $\vec{V}$. In the particular case $\theta = \pi$, which corresponds to the downstream flow in the body's frame, there is no cut-off. This perturbation of density exerts the force

$$\vec{F}_H = \frac{\gamma(\gamma-1)GML}{2\chi c_s^2}\cdot\frac{\vec{V}}{V}. \tag{13}$$

This force has the same direction as the velocity $\vec{V}$ of the perturber. It is therefore opposed to the dynamical friction, and has a motor role on the perturber's motion. Remarkably, its magnitude $\gamma(\gamma-1)GML/2\chi c_s^2$ does not depend on the perturber's velocity. This can be understood from the form of ρ' given at Eq. (12). Close to the perturber, the distribution of density is nearly symmetric. The asymmetry between the front and rear distribution of material occurs at a length scale $\lambda \sim \chi/V$. The volume of the corresponding shell is $V \sim \lambda^3$, and it corresponds to front and rear masses that are proportional to V/λ, since the front coefficient of Eq. (12) is inversely proportional to the radius. The masses involved are thus proportional to λ^2, and the forces they exert on the body do not depend on λ, hence on V.

In the large Mach number limit, [2] find that the heating force has the approximate following form:

$$\vec{F}_H = \frac{(\gamma-1)GML}{\chi V^3}\left[-1.96 - \log\left(\frac{r_{\min}V}{4\chi}\right)\right]\vec{V}, \tag{14}$$

where $r_{\min}$ is a minimal distance of integration of the force, which can amount to the body's physical radius, or to the mean free path of photons (whichever is largest). The magnitude of the force scales as V^{-2}, as does the standard dynamical friction. As in the subsonic regime, the force is along the direction of motion and has a motor role on the body.

4.2. Numerical results

We have performed various series of runs covering a range of Mach numbers from 0.02 to 10, in geometric sequence. The coarsest mesh has $x, y, z_{\min,\max} = \pm 408.13$ and 32^3 cells. The parameters are: $c_s = 1$, $\chi = 0.5$, $L = 10^{-5}$, $G = 1$ and $\gamma = 1.4$. We take $\rho_0 = 1$ (although the value of ρ_0 does not feature in the heating force expression, if it is too small, the perturbation can become non-linear even for the low luminosity considered here). We have two series of runs: one with 8 and another one with 6 levels of refinement. We plot in Fig. 1 the results of these runs. We find that the asymptotic value of the force at low Mach number is ~ 10 % below the analytic expectation for the low resolution case, and ~ 5 % below for the high resolution case. We also find that the force does exhibit a V^{-2} dependence in the supersonic case, and that the force depends strongly on the resolution in this regime, as expected from the logarithmic divergence of the heating force on $r_{\min}$. We can further investigate the supersonic regime by evaluating the heating force for different resolutions. We neither apply a truncation of the force summation nor soften the body's potential. The effective minimal radius of Eq. (14) therefore scales with the resolution, and is divided by a factor of 2 every time a new level of refinement is added. We therefore expect the specific force to increase by the following amount every time the resolution is doubled:

$$\Delta F = \frac{(\gamma-1)GL}{\chi V^2}\log 2 \tag{15}$$

Fig. 2 shows that this is indeed the case. We note that Eq. (14) is only an approximation, which is valid in the regime where $r_{\min}V \ll 4\chi$. When this relation breaks down (which occurs for the lowest resolutions in Fig. 2), the spacing between successive values of the force is no longer ΔF.

5. Dynamical friction on a cold body

5.1. Analytic considerations

We now turn to the force exerted by the perturbation corresponding to $\vec{Q}_1$ (see section 2). This is the perturbation imparted by a massive, non-luminous body on the gas. This perturbation and the resulting force have been worked out by [6] for an adiabatic gas. Here the situation is different, as thermal diffusion is present and the gas is not adiabatic. [2] have speculated that the force would correspond to the force expression worked out by [6], in which the sound speed to be used should have a value intermediate between the adiabatic and isothermal sound speeds, depending on the amount of thermal diffusion. The situation is actually different from this early expectation. In the low Mach number limit, the effect of thermal diffusion is adding to the perturbation of the adiabatic case the perturbation that would be imparted by a point-like heat sink with luminosity $-L_c$ with $L_c = 4\pi GM\chi\rho_0/\gamma$. The derivation will be presented elsewhere (Velasco Romero and Masset, *in prep.*, see also [4]). This implies that the dynamical friction has the following form, using the expression of L_c, Eq. (13) and the expression given by [6] for

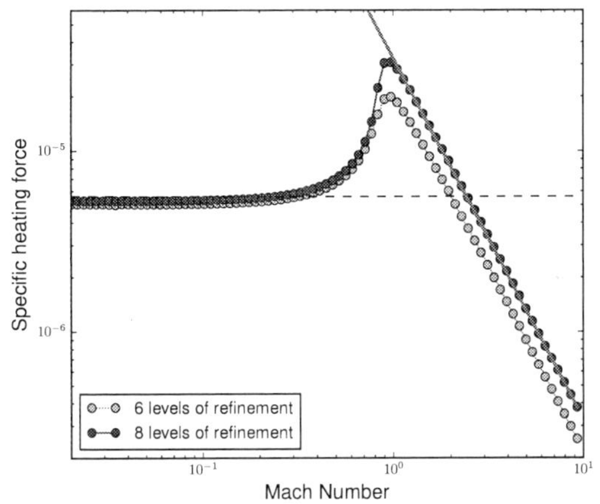

Figure 1. Specific heating force as a function of the Mach number for the parameters given in the main text, for two different resolutions. The horizontal dashed line represents the result of Eq. (13), while the red line shows a V^{-2} dependence as expected in the supersonic regime. The force is measured once a steady state is reached.

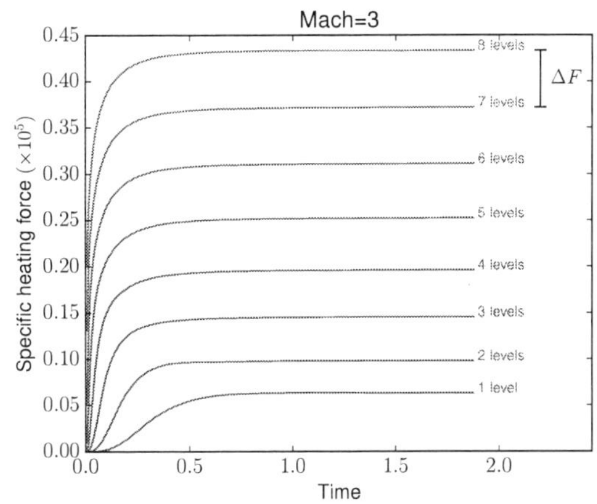

Figure 2. Specific heating force in a supersonic case as a function of time, for different resolutions. The vertical bar at the top right part of the plot depicts the expected force difference between two successive resolution levels, given by Eq. (15). The physical parameters for this series of runs are the same as those of Fig. 1.

the subsonic case:

$$\vec{F} = -\frac{2\pi(GM)^2\rho_0}{c_s^2}\left(\gamma - 1 + \frac{2}{3}\mathcal{M}\right)\frac{\vec{V}}{V}, \tag{16}$$

where $\mathcal{M} = V/c_s$ is the Mach number. Eq. (16) is an expansion of the dynamical friction to first order in $\mathcal{M}$. There is no first order term in $\mathcal{M}$ in the expansion of the new force component arising from the finite thermal diffusivity (Velasco Romero and Masset, *in prep.*). This new component therefore accounts for the constant term in Eq. (16), while the first order term in $\mathcal{M}$ comes from the expansion of Ostriker's formula.

The dynamical friction in the supersonic regime with thermal diffusion is presently under study and will be presented elsewhere.

5.2. Numerical results

We show in Fig. 3 the results of preliminary numerical simulations of the dynamical friction in a medium with thermal diffusion, in the subsonic regime.

The blue curve (in the electronic version) represents the results obtained in an adiabatic gas. They match satisfactorily the analytic expectation of [6] (see also [7]).

The set of the other three solid curves are obtained with three different values of the thermal diffusivity. They clearly confirm that the force does not tend to zero when the body's velocity does, and that the value of the friction is nearly independent of the value of the thermal diffusivity in the limit of a low Mach number. We also see that the friction measured in the simulations falls short of the analytic expectation by $\sim 10-20$ %. This is along the lines of what we noted for the heating force in section 4.2, and can be traced back to the restricted range of length scales accessible to the numerical simulations. Higher resolution simulations with more levels of refinement and larger coarsest boxes are under way and will be presented elsewhere.

IOP Conf. Series: Journal of Physics: Conf. Series **1031** (2018) 012006 doi:10.1088/1742-6596/1031/1/012006

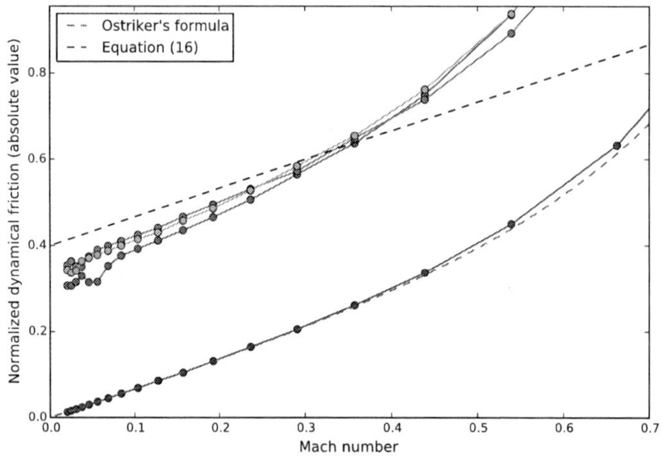

Figure 3. Dynamical friction in an adiabatic disc (blue dots) and in discs with thermal diffusion [respectively $\chi = 0.05$ (green), $\chi = 0.1$ (red) and $\chi = 0.2$ (cyan)]. The dynamical friction is normalized to $2\pi(GM)^2\rho_0/c_s^2$. We use seven levels of refinement (including the coarsest mesh). Each refined level has 16^3 cells.

As the dynamical friction is independent of the velocity in the low Mach number limit, it may be regarded as "dry" or "solid" friction in that limit. For this statement to be true, however, the stopping time of the body MV/F should be much shorter than the response time of the force (which in this limit merely amounts to the response time of the cold, dense trail χ/V^2, see [2]), or equivalently the stopping distance $MV^2/(2F)$ should be much shorter than the trail's size χ/V. This property is verified in the subsonic regime if $L_{\text{grav}} \ll L_c$, the gravitational luminosity being $L_{\text{grav}} = c_s^5/G$.

6. Putting together the results on dynamical friction

The net force experienced by the perturber is the sum of that exerted by the perturbation of density associated to $\vec{Q}_1$ and that associated to $\vec{Q}_2$ (see section 2). In the subsonic regime, a first-order expansion of the force in the Mach number, using Eqs. (13) and (16), is:

$$\vec{F} = \left[\frac{\gamma(\gamma-1)GML}{2\chi c_s^2} - \frac{2\pi(GM)^2\rho_0}{c_s^2}\left(\gamma - 1 + \frac{2}{3}\mathcal{M}\right)\right]\frac{\vec{V}}{V} \tag{17}$$

$$= \left[\frac{\gamma(\gamma-1)GM(L-L_c)}{2\chi c_s^2} - \frac{4\pi(GM)^2\rho_0}{3c_s^2}\mathcal{M}\right]\frac{\vec{V}}{V}. \tag{18}$$

This expression slightly differs from that of [2], in which the term in L_c is absent. As a consequence, a body at rest is unstable and will eventually acquire a finite velocity with respect to the medium if and only if $L > L_c$. When its terminal speed is largely subsonic, it scales with $L - L_c$. We note that the luminosity above which the gas response becomes non-linear over the extent of the hot region, given by [2] as

$$L_{\text{NL}} = \frac{4\pi\chi^2 c_s^2\rho_0}{\gamma^2(\gamma-1)|V|} \tag{19}$$

is much larger than L_c when $GM/(\chi c_s) \ll \mathcal{M}^{-1}$. The ratio of the left-hand-side appears as the ratio of the mass to critical mass $\chi c_s/G$ above which one expects a cut-off of the heating force (see [2, 3, 4] and section 8 below). One therefore expects sub-critical mass bodies with $L > L_c$ to experience a net positive force and therefore to be accelerated.

7. Application to planets embedded in protoplanetary discs

The above analysis suggests that planetary embryos heated by planetesimal or pebble accretion in opaque, gaseous protoplanetary discs may be subjected to a force directed along their velocity with respect to the gas. This can have important consequences on the evolution of their eccentricity and inclination. However, the results obtained in the context of dynamical friction can only serve as a proxy for the more complex case of hot embryos in Keplerian disc, where the shear leads to a more complex response than in a uniform medium at rest. The shear can only be neglected when the response time of the wake is shorter than the time scale of the shear Ω_p^{-1}, where Ω_p is the planet's orbital frequency. For the tidally excited wake, this condition is equivalent to $e \gg h$, where e is the eccentricity and h the disc's aspect ratio [8, 9]. For the response excited by the heat release, the condition on time scales is equivalent to $e \gg \sqrt{\chi/\Omega_p}/a$ where a is the planet's semi-major axis [3]. This last condition is less stringent than that on the tidal wake [3]. In order to study the time behaviour of the evolution of embedded, hot embryos, we have resorted to numerical simulations using the public hydrocode FARGO3D which we have coupled to a module of radiative transfer in the flux limited diffusion limit (see [1] for details). We find that hot embryos, with accretion rates typical of those found in the literature, experience a growth of eccentricity and inclination [3]. We present in Fig. 4 the time behaviour of the eccentricity e and inclination i of an Earth-mass embryo which has initially either $e = 0.01$ or $i = 0.01$. The embryo has a mass doubling time $\tau = 10^5$ yr due to accretion of solids, which corresponds to a luminosity:

$$L = \frac{GM^2}{R_p \tau}, \tag{20}$$

where R_p, the embryo's physical radius, is estimated assuming the embryo to have a mean density $\rho_p = 3$ g cm^{-3}. It is located at 5.2 au from the central star, which has a solar mass. The disc in which it is embedded has a surface density $\Sigma = 200$ g cm^{-2}, an effective viscosity $\nu = 10^{15}$ cm^2 s^{-1}, and a Rosseland opacity $\kappa = 1$ cm^2 g^{-1}. It is relaxed, prior to the simulations, toward hydrostatic and thermal equilibrium, and tends to $H/r = 0.043$, where H is the pressure length scale and r the radius. We note that the critical luminosity L_c is $\sim L/5$, so that it is nearly negligible compared to L.

We find that the eccentricity and inclination grow respectively to $e_c = 0.031$ and $i_c = 0.028$ for these fiducial calculations. These values are comparable in order of magnitude to the aspect ratio of the disc. Similar results have been obtained for the eccentricity in two-dimensional calculations by [10]. These results are in sharp contrast with the standard view that the interaction with the disc damps the eccentricity and inclination to negligible values over time scales much shorter than the migration time scale.

8. Cut-off at larger mass

The linear analysis performed by [2] requires that the heat released by the central object diffuses to the region where the flow's perturbation is linear. A ratio that features prominently in that analysis is $GM/\chi c_s$. This ratio is also the ratio of the diffusion to acoustic time scales across the Bondi radius $r_B = GM/c_s^2$. This radius corresponds to the distance from the perturber below which the flow is non-linear when $L < L_c$. When the heat diffusion across the Bondi radius is shorter than the acoustic time (i.e. $M < \chi c_s/G$), most of the heat released by the central body diffuses to the region where the perturbation is linear, and the perturbation of density

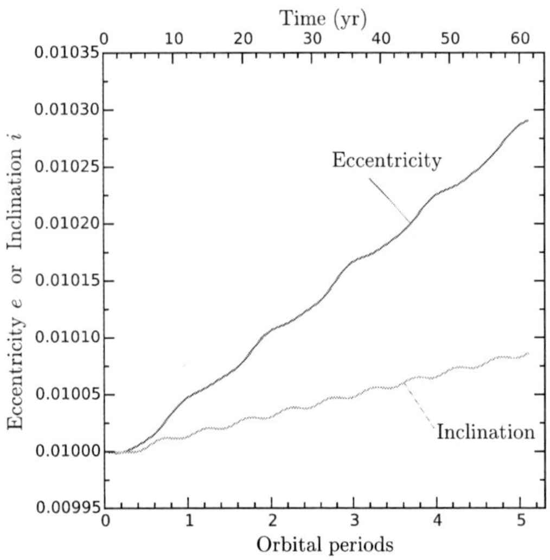

Figure 4. Time behaviour of the eccentricity (red curve in the electronic version) and inclination (orange curve in the electronic version). These curves represent the results of two different calculations: one in which the planet is coplanar with the disc ($i = 0$) and has a finite eccentricity, and another one in which the planet has a circular orbit ($e = 0$) and a finite inclination. The details of the setup are given in the text.

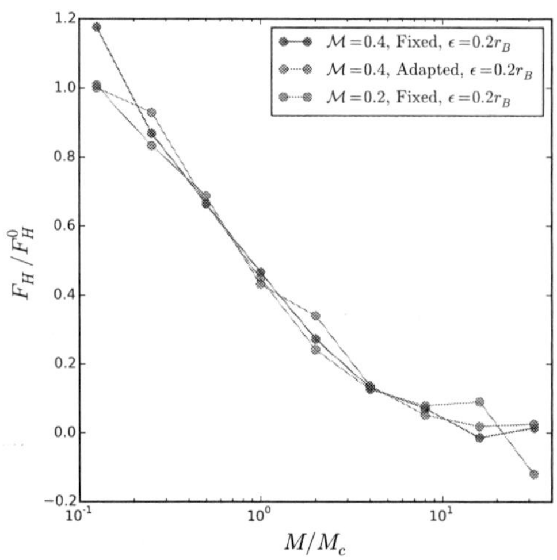

Figure 5. Heating force F_H measured in numerical simulations (obtained by subtracting the force measured in a run with a luminous body and a run with a non-luminous body), normalised to the heating force expected from the linear analysis in the limit of a vanishing Mach number [given by Eq. (13)], as a function of the body's mass, normalised to the critical mass $M_c = \chi c_s/G$.

in that region takes therefore the form given by Eq. (12). In the opposite situation (that of a mass larger than the critical mass $\chi c_s/G$) the energy deposited near the body may not emerge out of the Bondi sphere as an excess of temperature but can also be carried away by acoustic waves and the density perturbation may differ from that given by Eq. (12). As an investigation of this effect requires to resolve the Bondi sphere, where the flow is highly non-linear, we have undertaken numerical simulations with nested meshes, using FARGO3D, to study the behaviour of the heating force when the body's mass varies over a range including the critical mass $\chi c_s/G$.

We have run for this purpose two series of runs: one in which the resolution is fixed, so that the Bondi sphere of large mass bodies is well resolved while it is barely resolved for the lowest mass considered, and one in which the resolution adapts to the setup and is one sixth of the Bondi radius (on the finest level of refinement), regardless of the mass. These calculations have been run with $\mathcal{M} = 0.4$, and the size of the mesh at each level was 48^3. In addition, we have also performed a series of runs with a fixed resolution and $\mathcal{M} = 0.2$ (such calculations are more expensive as the size of the hot trail is twice larger than in the former, and the time required for convergence is a factor of four larger). The calculations becomes increasingly expensive at smaller mass for the adapted series because of the higher resolution requirements and because by definition, for sub-critical masses, the timestep is not limited by the sound speed but by the thermal diffusion. The results are presented in Fig. 5. We clearly see that the force approximately tends toward the analytic expectation when the mass is small compared to the critical mass, whereas it decays to zero when the mass is much larger. The dispersion of the measurements at larger mass is due to the fact that the force measured in that case is not stationary and may oscillate. The ratio at smaller mass can be larger than one: this is because the force is measured at $\mathcal{M} = 0.4$ and not for $\mathcal{M} \to 0$, which results in a heating force larger than its asymptotic value at low Mach number (see Fig. 1).

9. Conclusions

Thermal diffusion strongly alters the response of the gas in the vicinity of a massive perturber. The dynamical friction is enhanced, and behaves much differently than in the adiabatic case at low Mach number. If, in addition, the body releases heat in the surrounding gas, it triggers a response that has a form similar to that of the perturbation due to thermal diffusion, but of opposite sign. In particular, the extra force arising from the heat release is along the motion and may result, if the luminosity is large enough, in a net force that accelerates the body, rather than slowing it down. In protoplanetary discs, the release of heat can lead to an excitation of the eccentricity and inclination. This has potentially a very important effect on scenarios of planetary growth. However, there is a cut-off of these effects that occurs for masses larger than the critical mass $\chi c_s/G$. In the planet forming regions of protoplanetary discs, this mass is of the order of an Earth mass. We have presented here a study of the heat release for masses in excess of the critical mass by means of high resolution calculations with nested meshes that resolve the flow within the Bondi sphere (such as those of [11]). As the sub-Bondi flow is only marginally resolved in this study, its results should be regarded as preliminary. A complete understanding of the high mass regime is required to get reliable estimates of the values of eccentricity and inclination that a hot protoplanet can ultimately attain.

References

[1] Benítez-Llambay P, Masset F, Koenigsberger G and Szulágyi J 2015 *Nature* **520** 63–65
[2] Masset F S and Velasco Romero D A 2017 *Mon. Not. R. Astron. Soc.* **465** 3175–3184 (*Preprint* 1611.05684)
[3] Eklund H and Masset F S 2017 *Mon. Not. R. Astron. Soc.* **469** 206–217 (*Preprint* 1704.01931)
[4] Masset F S 2017 *Mon. Not. R. Astron. Soc.* **472** 4204–4219 (*Preprint* 1708.09807)
[5] Benítez-Llambay P and Masset F S 2016 *Astrophys. J. Suppl. Ser.* **223** 11 (*Preprint* 1602.02359)
[6] Ostriker E C 1999 *Astrophys. J.* **513** 252–258 (*Preprint* astro-ph/9810324)
[7] Sánchez-Salcedo F J and Brandenburg A 1999 *Astrophys. J. Lett.* **522** L35–L38
[8] Papaloizou J C B and Larwood J D 2000 *Mon. Not. R. Astron. Soc.* **315** 823–833 (*Preprint* astro-ph/9911431)
[9] Muto T, Takeuchi T and Ida S 2011 *Astrophys. J.* **737** 37–+ (*Preprint* 1106.0417)
[10] Chrenko O, Brož M and Lambrechts M 2017 *ArXiv e-prints* (*Preprint* 1706.06329)
[11] Fung J, Masset F, Lega E and Velasco D 2017 *Astron. J.* **153** 124 (*Preprint* 1610.09375)

IOP Conf. Series: Journal of Physics: Conf. Series **1031** (2018) 012007 doi:10.1088/1742-6596/1031/1/012007

Low mass planet migration in Hall-affected disks

Colin P McNally,[1] Richard P Nelson,[1] Sijme-Jan Paardekooper,[1,3] Oliver Gressel [4] and Wladimir Lyra [5,6]

[1] Astronomy Unit, School of Physics and Astronomy, Queen Mary University of London, London E1 4NS, UK

[3] DAMTP, University of Cambridge, Wilberforce Road, Cambridge CB3 0WA, UK

[4] Niels Bohr International Academy, The Niels Bohr Institute, Blegdamsvej 17, DK-2100, Copenhagen Ø, Denmark

[5] Department of Physics and Astronomy, California State University Northridge, 18111 Nordhoff St, Northridge, CA 91330, USA

[6] Jet Propulsion Laboratory, California Institute of Technology, 4800 Oak Grove Drive, Pasadena, CA 91109, USA

E-mail: c.mcnally@qmul.ac.uk

Abstract. Recent developments in non-ideal magnetohydrodynamic simulations of protoplanetary disks suggest that instead of being traditional turbulent (viscous) accretion disks, they have a largely laminar flow with accretion driven by large-scale wind torques. These disks are possibly threaded by Hall-effect generated large-scale horizontal magnetic fields. We have examined the dynamics of the corotation region of a low mass planet embedded in such a disk and the evolution of the associated migration torque. These disks lack strong turbulence and associated turbulent diffusion, and the presence of a magnetic field and radial gas flow presents a situation outside the applicability of previous corotation torque theory. We summarize the analytical analysis of the corotation torque, give details on the numerical methods used, and in particular the relative merits of different numerical schemes for the inviscid problem.

1. Introduction

Protoplanetary disks, the formation sites of solar systems, are comprised of a very low ionization gas and a mixture of dust, boulders, and protoplanets. Simulations have shown that if the disk is threaded by a weak net vertical field, the disk develops an essentially laminar flow, as the magnetorotational instability is quenched by large Ohmic and Ambipolar diffusion. The spatial arrangement of these regimes is sketched in Figure 1. The accretion is driven by a wind from the most ionized region near the disk surfaces [1, 2]. Near the disk midplane, where Ohmic diffusion dominates, a 'dead zone' forms ($0.5 \lesssim R \lesssim 10\,\mathrm{au}$ from the star), but at the same time the large Hall effect can give rise to strong fields when the background field is aligned with the disk rotation [3, 4]. These laminar magnetic fields can give rise to significant accretion at the midplane [1, 5, 6]. Recently in [7] we considered the problem of low mass planet migration in such a wind-driven protoplanetary disk that support accretion at the midplane [1, 5, 6, 8–14]. In this proceeding we summarize the problem and discuss the numerical issues in greater detail.

To study the flow near a low mass planet, we adopt a common two-dimensional formulation of the disk-planet interaction, and reduce the Hall-generated field to this dimensionality. In [7]

we find a equilibrium field for the unperturbed disk:

$$B_r = B_0 \left(\frac{r}{r_0}\right)^{-1} \quad B_\phi = -2B_0\Omega_0 r_0^2 \frac{\mu_0}{\eta} \left(\frac{r}{r_0}\right)^{-1/2} \tag{1}$$

where B_r and B_ϕ are the radial and azimuthal magnetic fields, r is the cylindrical radius, r_0 is a fiducial radius, Ω_0 is the Keplerian orbital frequency at r_0, η is the Ohmic resistivity, and μ_0 the vacuum permeability. This magnetic field drives radial motion of the disk gas at a velocity denoted v_r. Importantly, the resistivity of the dead zone region we study is very high, such that the timescale for magnetic field perturbations to diffuse across the length scale of interest (the planet's corotation region) is very small compared to the timescale for fluid elements to make the horseshoe turn at the end of a librating orbit in that corotation region. So, although the disk is well enough ionized so that the magnetic field drives the radial motion of the disk gas, the magnetic field response to the presence of the planet may be very small. Indeed, upon investigation we found it to to be negligible.

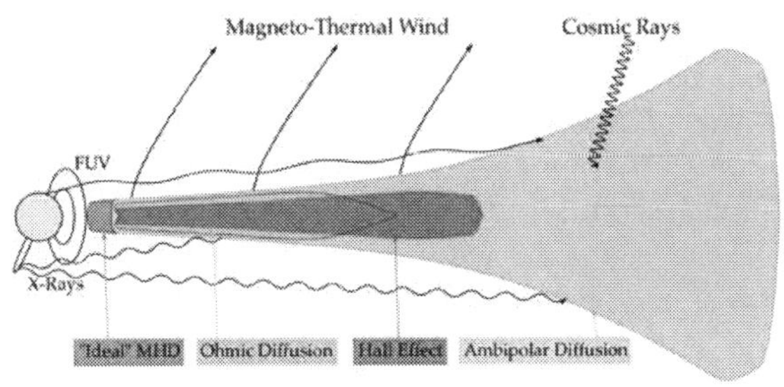

Figure 1. Schematic of the structure of a wind-driven protoplanetary disk in terms of ionization sources and non-ideal MHD terms.

2. Corotation Torques

When a low mass planet is placed in a disk, the response of the disk to this perturbation creates under- and over-densities. These density perturbations gravitationally tug on the planet in the form of a torque which modifies its orbit. Here, we focus exclusively on the corotation torque, arising from perturbations in the coorbital region, and for brevity neglect discussion of the spiral wave and associated Lindblad torque as they are not modified by the magnetic field in this case. In this section we summarize the theory for the corotation torque arising in this inviscid flow by considering the magnetic field as a fixed body force acting on the gas. The presentation is a brief summary of the analytical results of [7].

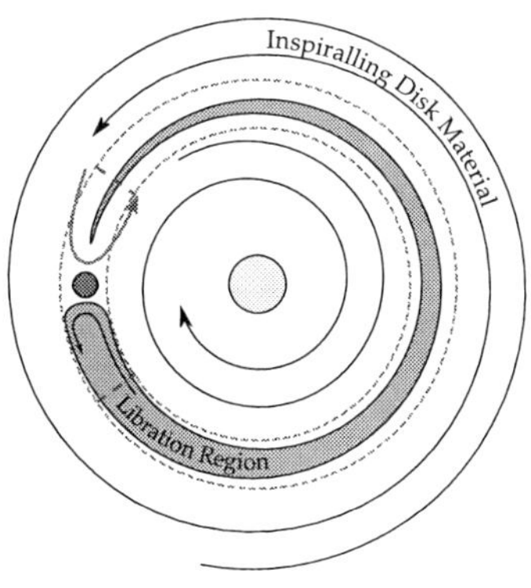

Figure 2. Cartoon of the flow in and around the corotation region of a low mass planet in a magnetically torqued dead zone. The magnetic torque drives disk gas inwards, and distorts the island of librating material (green) into an asymmetrical tadpole-like shape. This asymmetry of the librating material, and the introduction of the flow-through stream (red) lead to the corotation torque.

In our disk with a laminar magnetic field driving accretion, the laminar flow past the planet is illustrated in Figure 2. Disk gas spiral inwards as it loses angular momentum to the magnetic field. When it encounters the planet from the outside it executes a turn, giving angular momentum to the planet and making a quick jump inwards toward the star. The width of the turn is $2x_s$, where x_s is the half-width of the corotation region. Within the corotation region, there is gas that is trapped on librating orbits. These orbits execute two horseshoe turns, in front and behind of the planet. Due to the action of the magnetic field these orbits are also asymmetrical, as the gas drifts radially inwards between turns. Thus, on the turn in front of the planet, both gas flowing through the coorbital region and gas on librating orbits makes the turn. In the turn behind the planet, only gas on librating orbits makes the turn. The difference between the rate of angular momentum exchange with the planet by the gas on these two turns gives rise to the corotation torque.

To isolate the corotation torque due to the radial gas flow, and remove thermal effects (the entropy-related torque), we employ a globally isothermal disk with power law surface density $\Sigma = \Sigma_0 \left(r/r_0\right)^{-\alpha}$. In [7] we show that the corotation torque can be calculated in terms of the inverse vortensity $w = \Sigma/\nabla \times \mathbf{v}$ of the material undergoing the two horseshoe turns. Due to the steady-state magnetic accretion stress in the disk, as gas spirals inwards, the value of the vortensity at any given radial location in the flow is constant, even though the vortensity of a given fluid element increases at it moves inwards. At the same time, the gas in the libration region has constantly increasing vortensity because the same Lorentz force acts on it but it is not free to fall inwards towards the star. The analysis in [7] yields the expression for the torque associated with these two horseshoe turns

$$\Gamma_{\rm hs} = 2\pi\left(1 - \frac{w_{\rm c}(t)}{w(r_{\rm p})}\right)\Sigma_{\rm p} r_{\rm p}^2 x_s \Omega_{\rm p}(-v_r)\,, \tag{2}$$

where $w(r_{\rm p})$ is the inverse vortensity of the unperturbed disk at the planet location, $w_{\rm c}(t)$ is the time-evolving characteristic value of the inverse vortensity in the libration region, Ω is the orbital frequency of the disk, x_s is the corotation region half width, and quantities at the planet position are denoted with a subscript p. The corotation region half width for a low mass planet is approximately $x_s = 1.2 r_{\rm p}\sqrt{(q/h)}$ where q is the planet-star mass ratio, and $h = H/r$ is the non-dimensional pressure scale height. For the purposes of this discussion, we give the evolution of $w_{\rm c}$ in a simple form as

$$w_{\rm c}(t) = w(r_{\rm p})\left[1 + t/\tau_w\right]^{-1}, \quad \tau_w = \left[\left(\frac{3}{2} - \alpha\right)\frac{(-v_r)}{r_{\rm p}}\left(\frac{r_{\rm p}}{r_0}\right)^{\alpha - \frac{5}{2}}\right]^{-1}. \tag{3}$$

The full development of this formula, analysis of relevant timescales and its applicability are discussed in [7].

From this discussion of the analytical approach to the problem, we can see that to arrive at good numerically approximated solutions to this inviscid problem we require that the vortensity of the material trapped in the libration region should not artificially mix with the flow-through material, so that the evolution of $w_{\rm c}(t)$ is correctly obtained. This becomes the main criteria for successfully running a fluid simulation.

3. Methods for full magnetohydrodynamics

The first code the problem was attacked with was NIRVANA3.5 [15, 16]. As employed here, it is a second order accurate Godunov method, with an HLLD Riemann solver and a CTU scheme for the magnetic field evolution and second-order Runge-Kutta time integration. Because of the very large Ohmic diffusion, the parabolic term is treated with a Runge-Kutta-Legendre

IOP Conf. Series: Journal of Physics: Conf. Series **1031** (2018) 012007 doi:10.1088/1742-6596/1031/1/012007

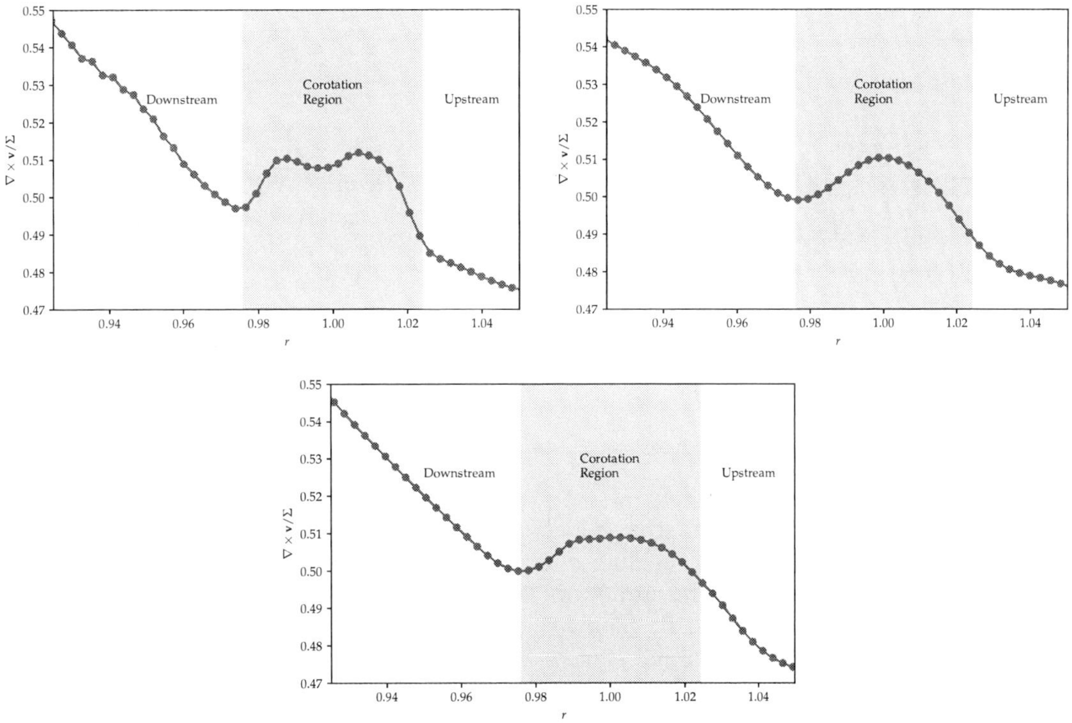

Figure 3. Illustration of the difference between simulations in three codes. Examples of the vortensity perturbation in the corotation region, in a radial cut opposite to the planet after 240 orbits at a low resolution as specified in the text. Upper Left: FARGO3D, Upper Right: The PENCIL CODE, Lower Center: NIRVANA3.5

polynomial based second-order accurate super-time-stepping scheme (RKL2) [17], as previously employed in [2]. In addition, we have newly implemented a conservative orbital advection scheme following [18], employing a FFT-based spectral interpolation for the azimuthal advection.

Typically, we find that the super-time-stepping allows a time step $\sim$ 2000 times greater than the stability limit for a basic explicit finite difference integration of the diffusion operator, by employing a super time stepping solver with $\sim$ 65 stages on each stage of the two-stage Runge-Kutta hydrodynamics time step with a wall clock speed-up of $\sim$ 17 times over a simple explicit integration. The RKL2 super time stepping scheme displayed no instability despite these extreme time step ratios. Moreover, through these experiments we find that the problem can be described well by fixing the magnetic field and neglecting the induction equation. This allowed the problem to be described as one in pure isothermal hydrodynamics, with a simple body force added to describe the Lorentz force from the magnetic field [7]. Thus we were able to progress with a simplified problem, and search for even more efficient codes for this problem.

4. The fixed magnetic field problem and vortensity evolution

The PENCIL CODE [19, 20] is a high-order finite difference method based on point collocation values and sixth-order centered differences in space to construct a method-of-lines approach to the MHD equations and applying a third-order time integration scheme. In addition, for disk problems like the one here, we have implemented an orbital advection technique by splitting of the time integration [21], without decomposing the velocity field as in NIRVANA.

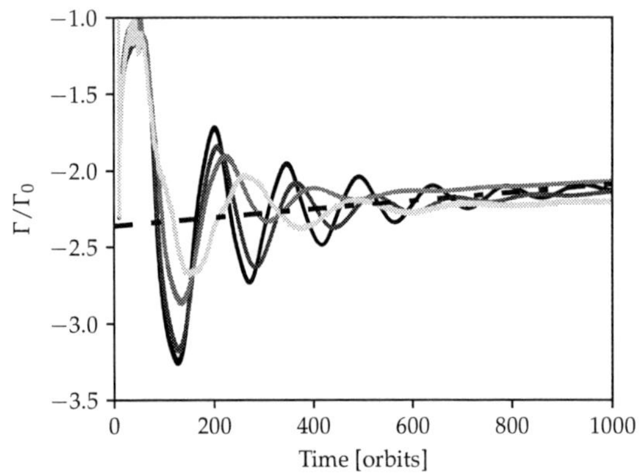

Figure 4. Convergence of torque histories from FARGO3D simulations and the analytical prediction. Grid resolutions are doubled between each run, with the sequence light green, red, blue, black, and the analytical formula equation (2) is plotted with a dashed line. The total torque Γ is scaled by $\Gamma_0 = (q/h)^2 \Sigma_{\rm p} r_{\rm p}^4 \Omega_{\rm p}^2$.

With this technique, the bulk azimuthal advection due to Keplerian motion is achieved with a FFT-based spectral interpolation, yielding a significant speedup and lower numerical diffusion while preserving the third order accuracy in time on the integration scheme. Stability for the integration scheme is achieved by modifying the governing equations with the addition of diffusion operators. The usual minimal choices in the PENCIL CODE for a problem like the one at hand are a hyperdiffusion of the form $\nabla^6 = \nabla^2(\nabla^2(\nabla^2))$ applied to all fields to stabilize the grid scale, and a shock-viscosity. The full form of the hyperdiffusion operator and the scaling of the hyperdiffusion coefficient as applied in cylindrical coordinates is given in the appendix of [21]. Importantly, this hyperdiffusion has shear components which can mix vortensity into the libration region at the grid scale.

FARGO3D [22–24] is based on an operator splitting technique with a second-order conservative finite differences in space. However, to gain stability the scheme employs only a bulk shock-viscosity with no shear components. Additionally, the method is known to have good properties for the conservation and advection of vortensity $(\nabla \times \mathbf{v}/\Sigma)$ [23]. Although the PENCIL CODE has a high formal order of accuracy and has excellent fidelity for quickly evolving shear flow problems [25], for this problem the slow formation (hundreds of orbits) of the vortensity structure in the corotation region is washed out by the shear component of the stabilizing hyperdiffusion. Thus, although it has a lower formal order of accuracy, we found FARGO3D to be a efficient in practice for achieving solutions with approximately the correct physical properties, particularly the sharp, relatively undiffused vortensity profile shown in Figure 3. This comparison uses a cylindrical grid of 512 zone in radius and 1536 in azimuth, with a radial range $[0.3, 1.7]$ and boundary conditions and planet potential as employed in [7]. The planet/star mass ratio was 2×10^{-5}. In the PENCIL CODE the `hyper3-cyl` [21] diffusions of velocity and density were employed with the coefficient set to 10^{-4}, the smallest value found to yield stability. In addition, the shock-capturing artificial viscosity `nu-shock` was employed on the velocity field with the coefficient set to 1. In this comparison, NIRVANA3.5 solved only hydrodynamics with the magnetic force replaced by a body force as in the other codes, and the monotonized central slope limiter was selected. In agreement with previous commentary on Godunov methods [23] for similar problems, we found that the algorithm we applied in NIRVANA3.5 diffused the sharp vortensity profile in the corotation region more than FARGO3D. We note that the choice of limiter does have a significant effect on the numerical diffusion of vortensity in that method. With higher resolution versions of these FARGO3D simulations, we were able to verify our analytical prediction for the corotation torque, as shown by the example in Figure 4.

5. Conclusions

In summary, we have:

- Implemented orbital advection schemes appropriate for disk problems in NIRVANA3.5 and the PENCIL CODE
- Demonstrated the practical application of RKL2 super time stepping for Ohmic resistivity to very diffusive setups with no numerical stability issues arising
- Described the relative strengths of the main methods used in NIRVANA3.5, the PENCIL CODE, and FARGO3D, all well-proven and high-quality codes in their own rights, for an inviscid disk problem where evolution of vortensity is paramount, finding the best results with FARGO3D
- Developed a torque formula for the corotation torque on a low mass planet embedded in a magnetically torqued dead zone

Acknowledgments

This research was supported by STFC Consolidated grants awarded to the QMUL Astronomy Unit 2015-2018 ST/M001202/1 and 2017-2020 ST/P000592/1. The simulations presented in this paper utilized Queen Mary's MidPlus computational facilities, supported by QMUL Research-IT and funded by EPSRC grant EP/K000128/1; the DiRAC Complexity system, operated by the University of Leicester IT Services, and the DiRAC Data Centric system at Durham University, operated by the Institute for Computational Cosmology which form parts of the STFC DiRAC HPC Facility (www.dirac.ac.uk). This equipment is funded by BIS National E-Infrastructure capital grants ST/K000373/1, ST/K00042X/1; STFC capital grant ST/K00087X/1, and STFC DiRAC Operations grants ST/K0003259/1 and ST/K003267/1. DiRAC is part of the National E-Infrastructure. This work used the NIRVANA3.5 code developed by Udo Ziegler at the Leibniz Institute for Astrophysics (AIP). This research was supported in part by the National Science Foundation under Grant No. NSF PHY11- 25915. The research leading to these results has received funding from the European Research Council (ERC) under the European Unions Horizon 2020 research and innovation programme (grant agreement No 638596). SJP is supported by a Royal Society University Research Fellowship.

References

[1] Bai X N and Stone J M 2013 ApJ **769** 76 (*Preprint* 1301.0318)
[2] Gressel O, Turner N J, Nelson R P and McNally C P 2015 ApJ **801** 84 (*Preprint* 1501.05431)
[3] Wardle M and Ng C 1999 MNRAS **303** 239–246 (*Preprint* astro-ph/9810468)
[4] Pandey B P and Wardle M 2008 MNRAS **385** 2269–2278 (*Preprint* 0707.2688)
[5] Lesur G, Kunz M W and Fromang S 2014 A&A **566** A56 (*Preprint* 1402.4133)
[6] Béthune W, Lesur G and Ferreira J 2017 A&A **600** A75 (*Preprint* 1612.00883)
[7] McNally C P, Nelson R P, Paardekooper S J, Gressel O and Lyra W 2017 MNRAS **472** 1565–1575 (*Preprint* 1708.05721)
[8] Bai X N 2013 ApJ **772** 96 (*Preprint* 1305.7232)
[9] Bai X N 2014 ApJ **791** 73
[10] Bai X N 2014 ApJ **791** 137 (*Preprint* 1402.7102)
[11] Bai X N 2015 ApJ **798** 84 (*Preprint* 1409.2511)
[12] Simon J B, Lesur G, Kunz M W and Armitage P J 2015 MNRAS **454** 1117–1131 (*Preprint* 1508.00904)

IOP Conf. Series: Journal of Physics: Conf. Series **1031** (2018) 012007 doi:10.1088/1742-6596/1031/1/012007

[13] Bai X N, Ye J, Goodman J and Yuan F 2016 ApJ **818** 152 (*Preprint* 1511.06769)

[14] Bai X N 2016 ApJ **821** 80 (*Preprint* 1603.00484)

[15] Ziegler U 2004 *Journal of Computational Physics* **196** 393–416

[16] Gressel O 2017 Toward realistic simulations of magneto-thermal winds from weakly-ionized protoplanetary disks *Journal of Physics Conference Series* (*Journal of Physics Conference Series* vol 837) p 012008 (*Preprint* 1611.09533)

[17] Meyer C D, Balsara D S and Aslam T D 2012 MNRAS **422** 2102–2115

[18] Mignone A, Flock M, Stute M, Kolb S M and Muscianisi G 2012 A&A **545** A152 (*Preprint* 1207.2955)

[19] Brandenburg A and Dobler W 2002 *Computer Physics Communications* **147** 471–475 (*Preprint* astro-ph/0111569)

[20] Brandenburg A and Dobler W 2010 Pencil: Finite-difference Code for Compressible Hydrodynamic Flows Astrophysics Source Code Library (*Preprint* 1010.060)

[21] Lyra W, McNally C P, Heinemann T and Masset F 2017 AJ **154** 146 (*Preprint* 1708.05057)

[22] Benítez-Llambay P and Masset F S 2016 ApJS **223** 11 (*Preprint* 1602.02359)

[23] Masset F S and Benítez-Llambay P 2015 Planet-Disk Interaction on the GPU: The FARGO3D code *Numerical Modeling of Space Plasma Flows ASTRONUM-2014* (*Astronomical Society of the Pacific Conference Series* vol 498) ed Pogorelov N V, Audit E and Zank G P p 216

[24] Masset F 2000 A&AS **141** 165–173 (*Preprint* astro-ph/9910390)

[25] McNally C P, Lyra W and Passy J C 2012 ApJS **201** 18 (*Preprint* 1111.1764)

IOP Conf. Series: Journal of Physics: Conf. Series **1031** (2018) 012008 doi:10.1088/1742-6596/1031/1/012008

The Black Hole Accretion Code: adaptive mesh refinement and constrained transport

HR Olivares Sánchez, O Porth, Y Mizuno

Institute for Theoretical Physics, Max-von-Laue-Str. 1, 60438 Frankfurt, Germany.

E-mail: `olivares@th.physik.uni-frankfurt.de`

Abstract. With the forthcoming VLBI images of Sgr A* and M87, simulations of accretion flows onto black holes acquire a special importance to aid with the interpretation of the observations and to test the predictions of different accretion scenarios, including those coming from alternative theories of gravity. The Black Hole Accretion Code (`BHAC`) is a new multidimensional general-relativistic magnetohydrondynamics (GRMHD) module for the MPI-AMRVAC framework. It exploits its adaptive mesh refinement techniques (AMR) to solve the equations of ideal magnetohydrodynamics in arbitrary curved spacetimes with a significant speedup and saving in computational cost. In a previous work, this was shown using a Generalized Lagrange Multiplier (GLM) to enforce the solenoidal constraint of the magnetic field. While GLM is fully compatible with `MPI-AMRVAC` 's AMR infrastructure, we found that simulations were sensible to the divergence control technique employed, resulting in an improved behavior for those using Constrained Transport (CT). However, cell-centered CT is incompatible with AMR, and several modifications were required to make AMR compatible with staggered CT. We present here preliminary results of these new additions, which achieved machine precision fulfillment of the solenoidal constraint and a significant speedup in a problem close to the intended scientific application.

1. Introduction

The Black Hole Accretion Code `BHAC` is an extension of the `MPI-AMRVAC` framework to perform General Relativistic Magnetohydrodynamics (GRMHD) simulations in 1, 2, and 3 dimensions using finite volume methods and a variety of modern numerical methods, described more in detail in [1]. It exploits `MPI-AMRVAC` 's infrastructure for parallelization and block-based automated Adaptive Mesh Refinement (AMR), resulting in a significant saving in computational time and resources.

In fact, despite the variety of General Relativistic Hydrodynamics and Magnetohydrodynamics codes currently available [2, 3, 4, 5, 6, 7, 8, 9, 10, 11, 12, 13, 14, 15, 16, 17, 18, 19] aside from some exceptions as [8, 20, 17] AMR is still not a commonly exploited tool.

However, AMR capabilities can be extremely useful for some problems that are currently computationally prohibitive for most codes. These include resolving simultaneously the formation and propagation of relativistic jets from black holes, due to the interaction between very different physical scales (see e.g., [21]), or tilted accretion disks, where the highly asymmetric evolution prevents the use of the static stretched grid commonly employed to increase resolution at the equator.

The most immediate application envisaged for BHAC is the simulation of Sgr A* and M87, the two primary targets of the Event Horizon Telescope (EHT). Both these objects belong to the class of advection dominated accretion flows (ADAFs), for which ideal magnetohydrodynamics without radiation feedback constitutes a reasonable approximation of the plasma properties. In order to properly study the impact of plasma and gravitational conditions on the EHT images, BHAC is coupled to the General Relativistic Radiative Transfer (GRRT) codes BHOSS [22] and RAPTOR [23].

An important motivation for our research is the possibility to distinguish departures from General Relativity in the images obtained by the EHT. For this reason, BHAC is designed with a modular structure that can handle arbitrary spacetimes, including numerical ones as well as those coming from alternative theories of gravity. For instance, [24] successfully performed GRMHD simulations of accretion flows onto a dilaton black hole in the Einstein-Maxwell-Axion-Dilaton theory of gravity. As an example of another application, the code has recently been used to study quasi-periodic-oscillations (QPOs) in accretion discs around neutron stars [25].

In a previous work [1], we tested BHAC in several standard problems and validated it by comparing the results of some of them to those obtained in control simulations performed using the well known code HARM3D [5, 26].

The new additions presented in this work concern changes in the AMR infrastructure, necessary to allow AMR to operate simultaneously with Constrained Transport (CT), a divergence control scheme which already showed considerable advantages with respect to GLM, the technique used in [1] (see section 3.2 for more details). One of these advantages is the ability to keep a discretization of $\nabla \cdot \mathbf{B}$ equal to zero to machine precision. In fact, since no cell-centered divergence-free discretization is currently known to be compatible with AMR [27], and staggered versions require special divergence-free prolongation and restriction operators for face- and edge-allocated quantities, GLM was the only such technique available for AMR simulations in BHAC .

The paper is organized as follows: section 2 summarizes the equations and the formulation of GRMHD used in the code; section 3 briefly describes the numerical methods employed in the code focusing on the newly implemented divergence control strategies; and section 4 reports on two numerical tests performed. Throughout this work, we employ geometrized units ($G = c = 1$) and use the Einstein summation convention. Greek indices run from 0 to 3, while Latin indices run from 1 to 3.

2. Equations of GRMHD

The equations of ideal GRMHD are those of particle conservation, local conservation of energy-momentum and the homogeneous Maxwell equations

$$\nabla_\mu(\rho u^\mu) = 0\,,\ \ \nabla_\mu T^{\mu\nu} = 0\,,\ \text{and}\ \nabla_\mu {}^*F^{\mu\nu} = 0\,, \tag{1}$$

where ∇_μ denotes the covariant derivative, ρ is the particle number in the fluid frame, u^μ the fluid 4-velocity, $T^{\mu\nu}$ the energy-momentum tensor and ${}^*F^{\mu\nu}$ the dual of the Faraday tensor $F^{\alpha\beta}$.

The Faraday tensor and its dual are such that, for a frame moving at 4-velocity n^ν, the electric and magnetic fields are given by

$$E^\mu = F^{\mu\nu} n_\nu \ \ \text{and}\ \ B^\mu = {}^* F^{\mu\nu} n_\nu\,. \tag{2}$$

In ideal MHD, only the magnetic field is evolved (using the homogeneous Maxwell equations), since the electric field is determined by the ideal MHD condition, which requires that the electric field in the frame co-moving with the fluid is $e^\mu = F^{\mu\nu} u_\nu = 0$.

To formulate system (1) as a set of evolution equations, we use the 3+1 decomposition of spacetime (see for example [28] and [29]). The spacetime is sliced into spacelike 3-dimensional hypersurfaces with metric γ_{ij}. The 4-dimensional line element is expressed as

$$ds^2 = -\alpha^2 dt^2 + \gamma_{ij}(dx^i + \beta^i dt)(dx^j + \beta^j dt)\,, \tag{3}$$

where α and β^i are called the lapse function and the shift vector. The 4-velocity of the Eulerian observers is just the vector normal to each hypersurface, $n_\mu = -\alpha\nabla_\mu t$. The 4-metric can then be decomposed as $g_{\mu\nu} = \gamma_{\mu\nu} - n_\mu n_\nu$; therefore, $\gamma^\mu{}_\nu$ acts as a projection operator on the hypersurface.

When projecting the equations of system (1) along n_μ and $\gamma^\mu{}_\nu$, the result is a set of conservation equations with geometry-dependent sources

$$\partial_t(\sqrt{\gamma}\mathbf{U}) + \partial_i(\sqrt{\gamma}\mathbf{F}^i) = \sqrt{\gamma}\mathbf{S} \tag{4}$$

and the solenoidal constraint for the magnetic field $\partial_i\sqrt{\gamma}B^i = 0$, which results from the projection $n_\mu\nabla_\nu {}^*F^{\mu\nu}$. Here, γ is the metric determinant, and the vectors of conserved quantities $\mathbf{U}$, fluxes $\mathbf{F}^i$, and sources $\mathbf{S}$ are given by

$$\boldsymbol{U} = \begin{bmatrix} D \\ S_j \\ \tau \\ B^j \end{bmatrix}, \quad \boldsymbol{F}^i = \begin{bmatrix} \mathcal{V}^i D \\ \alpha W^i_j - \beta^i S_j \\ \alpha(S^i - v^i D) - \beta^i \tau \\ \mathcal{V}^i B^j - B^i \mathcal{V}^j \end{bmatrix} \text{ and } \boldsymbol{S} = \begin{bmatrix} 0 \\ \frac{1}{2}\alpha W^{ik}\partial_j\gamma_{ik} + S_i\partial_j\beta^i - U\partial_j\alpha \\ \frac{1}{2}W^{ik}\beta^j\partial_j\gamma_{ik} + W^j_i\partial_j\beta^i - S^j\partial_j\alpha \\ 0 \end{bmatrix}, \tag{5}$$

where $\mathcal{V}^i := \alpha v^i - \beta^i$ are the *transport velocities*, and the others variables are quantities in the Eulerian frame: $D = -\rho u^\nu n_\nu$ is the number density, $S_i = n_\mu\gamma_{\nu i}T^{\mu\nu}$ the covariant 3-momentum, $U = n_\mu n_\nu T^{\mu\nu}$ the total energy and $W^{ij} = \gamma_{\mu i}\gamma_{\nu j}T^{\mu\nu}$ the spatial stress tensor. Evolving $\tau = U - D$ instead of U makes the evolution more accurate in regions of low energy and allows to recover the Newtonian limit.

Evolution cannot be carried out using only conservative variables, since the computation of some quantities in the expressions for the fluxes requires the knowledge of the primitive variables $\mathbf{P} = [\rho, \Gamma v^i, p, B^i]$. Here, Γ is the Lorentz factor, $v^i = u^i/\Gamma - \beta^i/\alpha$ and p is the pressure in the fluid frame. While it is straightforward to find $\mathbf{U}(\mathbf{P})$, $\mathbf{P}(\mathbf{U})$ requires numerical inversion. To this end, `BHAC` extends the vector $\mathbf{U}(\mathbf{P})$ by the auxiliary variables $\mathbf{A} = [\Gamma, \xi]$, where $\xi := \Gamma^2\rho h$ and h is the specific enthalpy. The inversion process then consists of finding $\mathbf{A}$ compatible with $\mathbf{U}$ and $\mathbf{P}$.

3. Numerical methods and implementation

In this section we will describe briefly the numerical methods employed in `BHAC` , focusing on generalities of the finite volume implementation and the new features of staggered-mesh based divergence control methods and adaptive mesh refinement for the staggered variables.

For an in depth description of the methods available in the code, including equations of state, coordinates and handling of the metric data structure, reconstruction schemes, Riemann solvers and procedures for primitive variable recovery, we refer the reader to [1].

3.1. Finite volume scheme

To obtain the finite-volume scheme used by `BHAC` , we discretize the domain into control volumes $\Delta V_{i,j,k}$ and integrate equation (4) over each of them. This leads to the equations of evolution for the average of the conserved quantities inside each cell,

$$\frac{d\bar{\boldsymbol{U}}_{i,j,k}}{dt} = -\frac{1}{\Delta V_{i,j,k}}\Big[\boldsymbol{F}^1\Delta S^1\big|_{i+1/2,j,k} - \boldsymbol{F}^1\Delta S^1\big|_{i-1/2,j,k} + \boldsymbol{F}^2\Delta S^2\big|_{i,j+1/2,k} - \boldsymbol{F}^2\Delta S^2\big|_{i,j-1/2,k} + \boldsymbol{F}^3\Delta S^3\big|_{i,j,k+1/2} - \boldsymbol{F}^3\Delta S^3\big|_{i,j,k-1/2}\Big] + \bar{\boldsymbol{S}}_{i,j,k}\,. \tag{6}$$

The quantities as $\boldsymbol{F}^1\Delta S^1\big|_{i+1/2,j,k}$ are integrals of the fluxes over the surfaces $\Delta S^1\big|_{i+1/2,j,k}$ bounding the control volume and $\bar{\boldsymbol{S}}_{i,j,k}$ is the volume average of the sources. Both kinds of integrals are approximated to second order, by assigning to $\boldsymbol{F}^n$ ($n = 1, 2, 3$) the point value of the flux at the interface center and to $\bar{\boldsymbol{S}}_{i,j,k}$ the point value at the cell barycenter. $\boldsymbol{F}^n$ is obtained through the approximate solution of a Riemann problem at the interface, and static integrals such as cell volumes, interface areas and barycenter positions are calculated at initialization using fourth-order Simpson's rule and stored in memory. Equation 6 can then be solved using the integrators present in the `MPI-AMRVAC` toolkit. These include the simple predictor-corrector, the third order Runge-Kutta RK3 [30] and the strong-stability preserving s-step, pth-order RK schemes SSPRK(s,p) schemes: SSPRK(4,3), SSPRK(5,4) due to [31]. (For implementation details, see [32].)

3.2. Divergence control

Although the induction equation can also be expressed in the form of equation (4), using the finite volume scheme of equation (6) alone to evolve the magnetic field usually results in the creation and rapid growth of numerical magnetic monopoles, driving the evolution towards flagrantly unphysical states. In order to keep violations to $\nabla \cdot \mathbf{B} = 0$ small, three schemes are available in `BHAC` that can be used together with AMR. The first one is the scheme known as the *Generalized Lagrange Multiplier* (GLM) of the Maxwell equations, a generalization of the Dedner scheme [33] used in Newtonian MHD. This method consists in solving an additional evolution equation that has the effect of damping and advecting away the violations to $\nabla \cdot \mathbf{B} = 0$. GLM has already been applied to GRMHD by e.g., [34]. Though this technique can be straightforwardly included in `BHAC` 's algorithm, some of its Newtonian versions have been shown to suffer from spurious oscillations in the magnetic energy and from an artificial growth of the magnetic fields, effects attributed, respectively, to the loss of locality due to the parabolic nature of the additional equation and to the resulting scheme being non-conservative [35, 36]. The other two available

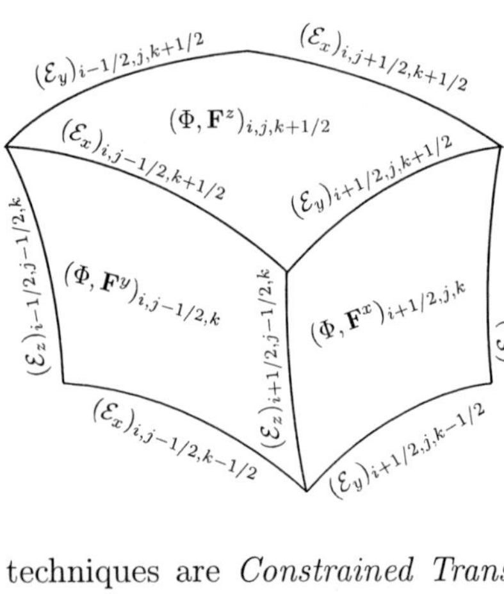

Figure 1. Spatial location of variables for a cell with indices (i, j, k). Line integrals of the electric field $\mathcal{E}$ are located at its edges, and magnetic and numerical fluxes Φ and $\mathbf{F}^i$ (the latter used for the BS algorithm) are located at its faces. The rest of variables (not shown) are located at cell centers.

techniques are *Constrained Transport* (CT) schemes. These are obtained by integrating the induction equation ($\gamma^i{}_j \nabla_\mu {}^*F^{\mu j} = 0$) on the *boundary* of the control volume, and their central feature is that they fulfill to machine precision a discretized version of the solenoidal constraint when constraint-satisfying initial data is supplied (it can be easily obtained by setting the initial magnetic field as the curl of a potential). CT was first devised for ideal GRMHD by [37], although a similar idea was exploited before in the Yee algorithm [38]. In these algorithms, the electromagnetic variables are given a special space location (see Figure 1): on each face of the cell resides a magnetic flux calculated, e.g., as

$$\Phi_{i+1/2,j,k} = \int_{\partial V(x^1_{i+1/2})} \gamma^{1/2} B^1 dx^2 dx^3 \, , \tag{7}$$

and on each edge resides a line integral of the electric field, e.g.,

$$\mathcal{E}_{i+1/2,j+1/2,k} = -\int_{x^3_{k-1/2}}^{x^3_{k+1/2}} E_3|_{x^1_{i+1/2}, x^2_{j+1/2}} dx^3 \, . \tag{8}$$

The magnetic flux at each face is updated using the integral form of Faraday's law:

$$\frac{d}{dt}\Phi_{i+1/2,j,k} = \mathcal{E}_{i+1/2,j+1/2,k} - \mathcal{E}_{i+1/2,j-1/2,k} - \mathcal{E}_{i+1/2,j,k+1/2} + \mathcal{E}_{i+1/2,j,k-1/2} \, . \tag{9}$$

Since each of the line integrals is shared by two faces, but appears with opposite sign in the time update formula for each of them, the rate of change of $(\nabla \cdot \mathbf{B})$, i.e., the sum of the rate of change of the outgoing flux through all faces, vanishes. So far, equation (9) is exact. Each variant of CT arises from different ways of approximating the line integrals of the electric field.

The two variants available in our code are the method of Balsara & Spicer (BS) [39] and *Upwind Constrained Transport* (UCT) [40]. In BS, $\mathcal{E}_{i+1/2,j+1/2,k}$ is calculated simply as the arithmetic average of the fluxes obtained by the Riemann solver that correspond to the electric component E_3 at the faces $\Delta S_{i+1/2,j,k}$, $\Delta S_{i+1/2,j+1,k}$, $\Delta S_{i,j+1/2,k}$ and $\Delta S_{i+1,j+1/2,k}$.

A cell-centered version of the BS algorithm, known as *Flux-interpolated Constrained Transport* (flux-CT), was found in [41] and is widely used in the literature (see e.g., [5, 26]). Unlike the staggered version, the cell-centered scheme is not compatible with AMR; however, it has been reported that in otherwise identical GRMHD simulations performed using GLM and flux-CT, the latter produced less spurious structures in the magnetic field, and was able to preserve for a longer time an exact stationary solution [1]. This provided a strong motivation for us to implement the staggered algorithm in `BHAC` to gain the advantages of AMR. Nevertheless, BS has also known deficiencies [42] that can be overcome by upwinding the electromotive force, as it is done in the the algorithm by Gardiner & Stone [43] and in UCT. The latter, also implemented in `BHAC` , is another staggered algorithm, devised to incorporate the correct continuity and upwind properties of the magnetic field by using limited reconstructions and by taking into account the transport velocities. In contrast to BS, UCT has the additional property that it reduces to the correct 1-dimensional limit when the correspondent symmetry is assumed. For details on the specific UCT implementation in `BHAC` , we refer the reader to a more complete work [44], currently in preparation.

3.3. Adaptive mesh refinement

Most of the infrastructure for AMR is inherited from the `MPI-AMRVAC` toolkit. The grid is a fully adaptive block-based octree (in 3D) with a fixed refinement factor of two between successive levels. Operations on the grid as time update, IO and problem initialization are performed on a loop over a Morton Z-order curve. The time step is calculated globally and is the same for all levels, thus load-balancing is simply done by cutting the space-filling curve in equal parts and distributing them among the MPI-processes. This strategy is applied in various astrophysical codes, for example in those employing the PARAMESH library [45, 46, 47] , or the recent Athena++ framework [48]. Refinement can be triggered in a completely automated way either using the Löhner scheme [49] or user defined prescriptions. Details on the prolongation and restriction operations and the ghost cell exchange can be found in [50]. To ensure machine precision conservation of $\boldsymbol{U}$, re-fluxing is performed every (partial) time step, i.e., the fluxes on the coarse side of coarse/fine interfaces are replaced by the sum of the co-spatial fluxes on the fine side.

New additions specific to BHAC are divergence-free restriction and prolongation operators for the staggered variables and an electric field fixing step to avoid producing numerical monopoles across resolution jumps, and which also consists on replacing the electric fields on the coarse side with their co-spatial fine representation. Details about the prolongation operator and the electric field fixing formulas will be documented in a forthcoming work [44].

4. Numerical tests

4.1. Validation of the code

BHAC has been thoroughly validated using the GLM and the FCT schemes for the magnetic field evolution. This was done by performing several test problems in 1, 2 and 3 dimensions, as well as comparisons with simulations performed using the code HARM3D , as is reported in [1]. The results obtained using the newly available features here described were verified against the validated results to ensure that the implementation was correct. This will be documented in detail in [44]. In the next sections, we will describe the results of applying AMR and the staggered FCT algorithm in two test problems.

4.2. Relativistic Orszag-Tang vortex

The Orszag-Tang vortex [51] is a common setup to highlight the impact of the violations to the solenoidal constraint in the numerical solution. Starting from a configuration in which $\nabla \cdot \mathbf{B} = 0$ to machine precision, the problem quickly develops magnetic shocks and turbulence, both challenging conditions for the preservation of the constraint. In this 2-dimensional, special-relativistic realization of the test, we set $\rho = 1$, $p = 10$, $v^x = \frac{0.99}{\sqrt{2}} \sin y$, $v^y = \frac{0.99}{\sqrt{2}} \sin x$, $B^x = -\sin y$ and $B^y = \sin 2x$. The equation of state is that of an ideal fluid with $\gamma = 4/3$. The domain is the square $x, y \in [0, 2\pi]$ with periodic boundary conditions. We adopt three AMR levels, where the lowest resolution is equivalent to resolve the whole domain with 64×64 cells. The numerical methods to evolve the system are an RK3 integrator with HLL fluxes and the Koren reconstruction (third order accuracy in smooth parts of the solution [52]). The divergence control method is the staggered CT algorithm with arithmetic averaging and the CFL factor is set to 0.4.

Figure 2 shows two snapshots of the evolution. On the left panel is the divergence of the magnetic field at $t = 2$, near the time when the strongest shocks form. At that moment, all the three AMR levels are present in the simulation, and it can be seen that the algorithm is able to keep the largest violations to the level of 10^{-13}, in contrast to the $\sim 10^{-1} - 10^{0}$ that are produced in a similar set up with GLM (see [1]). The right panel displays the magnetic field intensity and the magnetic field lines at the same time of the simulation, showing the formation of current sheets at the same location of the maximum creation of divergence.

In order to perform a more quantitative comparison of how well AMR performs, Figure 3 displays the density and magnetic field strength profiles along a cut at $y = 0.5$ and at the same time, for the simulation described above and for another one identical except for being run at a uniform high resolution correspondent to that of the highest AMR level. A very good agreement can be observed between both simulations. When evolving up to $t = 10$ the AMR case obtained a modest speedup factor of 1.35. This is due to the fact that at later times most of the domain shows large variations in the quantities that trigger refinement, thus most of the simulation reaches the highest AMR level.

This is, however, not expected to be the case in the intended astrophysical applications of the code, where likely large parts of the domain are emulating vacuum, as will be seen in the next section.

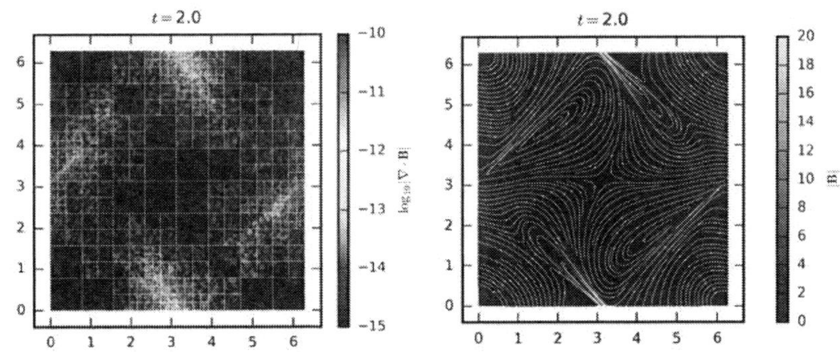

Figure 2. Snapshots of the relativistic Orszag-Tang problem at $t = 2$. *Left:* Divergence of the magnetic field and AMR blocks. *Right:* Magnetic field intensity and magnetic field lines.

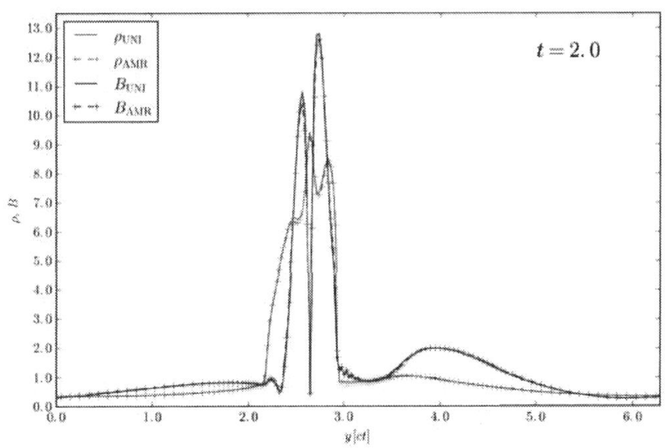

Figure 3. Density and magnetic field strength profiles at $y = 0.5$ and $t = 2$ for uniform resolution and 3-level AMR. The AMR structure is represented by the symbols '+'.

4.3. Magnetized accretion onto Kerr black hole

To study the speedup and accuracy achievable using AMR in a case closer to the intended application of the code, we perform two simulations of the same 2D problem of accretion from a magnetized torus onto a Kerr black hole, one using a grid with uniformly high resolution and the other one using AMR, with the maximum resolution corresponding to that of the first simulation.

The spacetime is described using logarithmic Kerr-Schild coordinates, correspondent to the standard Kerr-Schild coordinates $r \in [1.213, 2500\,M]$ and $\theta \in [0, \pi]$. This allows the propagation of the jet over a long distance and prevents signals from the boundaries to affect the inner region when evolving until $t = 5000\,M$. The spin parameter of the black hole is $a = 0.9375$, and the event horizon is located at $r = 1.348\,M$.

The fluid obeys an ideal equation of state with $\gamma = 4/3$. As initial condition, we set up an equilibrium torus with inner radius at $r_{\rm in} = 6\,M$, and density maximum at $r_{\rm max} = 12\,M$ (orbital period of $247\,M$ at the density maximum). A single-loop poloidal magnetic field is built from the vector potential $A_\phi \propto \max(\rho/\rho_{\rm max} - 0.2, 0)$ and is normalized in such a way that the minimum

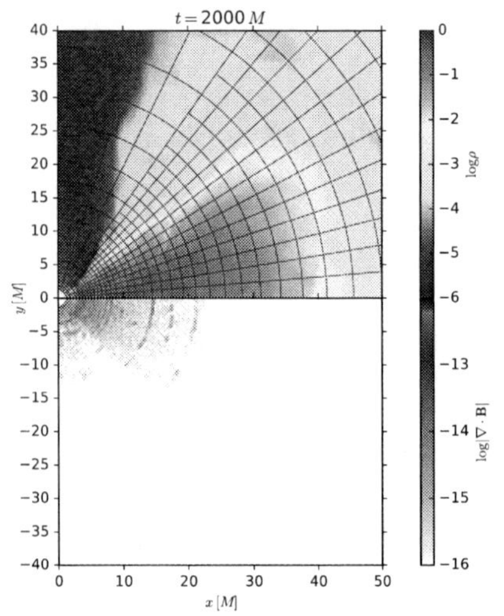

Figure 4. *Upper half*: Logarithmic density and AMR blocks at time $t = 2000\,M$.
Lower half: Divergence of the magnetic field.

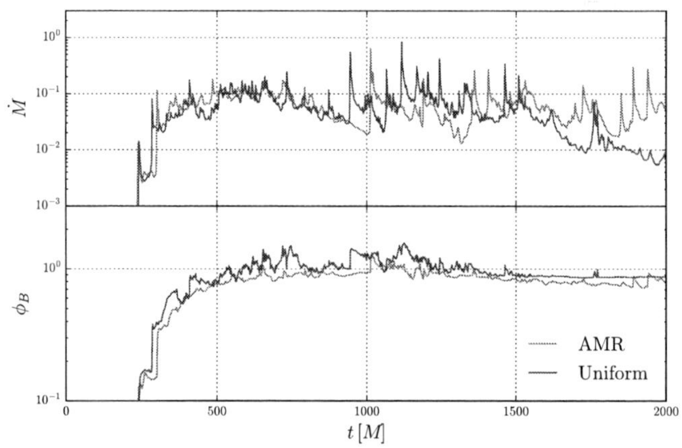

Figure 5. Mass accretion rate and absolute magnetic flux through the horizon for uniform resolution and AMR

plasma $\beta = p_{\rm fluid}/p_{\rm mag} = 100$. To avoid vacuum regions, the rest of the simulation is filled with a tenuous atmosphere with density $\rho_{\rm fl} = 10^{-4} r^{-3/2}$ and fluid pressure $p_{\rm fl} = 1/3 \times 10^{-6}\ r^{-5/2}$. We reset the density or the pressure whenever they fall below these floor values. To perturb this equilibrium state, random perturbations of 4% are added to the pressure. This eventually triggers the magneto-rotational instability, allowing the plasma to accrete.

The simulations are evolved using a two-step predictor-corrector method, TVDLF fluxes, and PPM reconstruction. The CFL number is set to 0.35. For evolving the magnetic field we use staggered CT with arithmetic averaging. The boundary conditions at the inner and outer radial boundaries are set to zero gradient in the primitive variables, i.e., their values at the last cell of the physical domain are copied to fill the ghost cells, except for the ingoing component of the velocity, which is set to zero. The numerical fluxes and line integrals of the azimuthal electric field in contact with the polar axis are also set to zero, since they correspond to integrals over zero-area surfaces and zero-length paths. The number of cells in the simulation with uniform high resolution is $N_r \times N_\theta = 800 \times 400$. The simulation with AMR has three levels, the highest with the same resolution as the one with the uniform mesh. AMR is triggered automatically using the Löhner scheme.

Figure 4 displays the grid blocks of the AMR simulation at time $t = 2000\,M$, showing how resolution increases in regions with large variations in density. To perform a more quantitative comparison, Figure 5 shows that the mass accretion rate and absolute magnetic flux through the horizon for both simulations have comparable magnitude and variability. While the simulation at uniform resolution required 2324 cpu-hours to reach $t = 5000\,M$, the simulation using three AMR levels required only 327, yielding a significant speedup factor of 7.1.

5. Conclusion

BHAC is a new versatile tool to study magneto-hydrodynamic flows in arbitrary spacetimes in General Relativity and other metric theories of gravity, which incorporates modern numerical methods and an efficient AMR infrastructure inherited from `MPI-AMRVAC` .

We have made new additions to this infrastructure in order to allow staggered-mesh constrained transport to run with the advantages of AMR.

In this work, we have presented the first results of simulations performed with `BHAC` using AMR and a CT together. These indicate that the code is now capable of evolving efficiently the GRMHD equations in multi-scale problems, with the solenoidal constraint fulfilled to machine precision.

As a matter of fact, for a problem close to its intended application, `BHAC` was able to attain a very significant speedup of 7.1, which can be of crucial importance, also for e.g., performing parameter studies to contrast with the EHT data.

In a forthcoming work, we will describe in greater detail such modifications as well as other additions to the numerical methods besides those described in [1].

Acknowledgments

We would like to express our gratitude to Luciano Rezzolla, Elias Most, Christian Fromm, Ziri Younsi, Alejandro Cruz Osorio, David Kling, Jonas Köhler and Mariafelicia de Laurentis for useful discussions. This research is supported by the ERC synergy grant "BlackHoleCam: Imaging the Event Horizon of Black Holes" (Grant No. 610058), by "NewCompStar", COST Action MP1304, by the LOEWE-Program in HIC for FAIR, and by the European Union's Horizon 2020 Research and Innovation Programme (Grant 671698) (call FETHPC-1-2014, project ExaHyPE). HO is supported in part by a CONACYT-DAAD scholarship. The simulations were performed on LOEWE at the CSC-Frankfurt and Iboga at ITP Frankfurt. We acknowledge technical support from Thomas Coelho.

References

[1] Porth O, Olivares H, Mizuno Y, Younsi Z, Rezzolla L, Moscibrodzka M, Falcke H and Kramer M 2017 *Computational Astrophysics and Cosmology* **4** 1 (*Preprint* 1611.09720)
[2] Hawley J F, Smarr L L and Wilson J R 1984 *Astrophys. J.* **277** 296–311
[3] Kudoh S K D M K S T 2000 *Astrophys. J* **536** 668–674
[4] De Villiers J P and Hawley J F 2003 *Astrophys. J.* **589** 458–480 (*Preprint* arXiv:astro-ph/0210518)
[5] Gammie C F, McKinney J C and Tóth G 2003 *Astrophys. J.* **589** 458 (*Preprint* astro-ph/0301509)
[6] Baiotti L, Hawke I, Montero P J, Löffler F, Rezzolla L, Stergioulas N, Font J A and Seidel E 2005 *Phys. Rev. D* **71** 024035 (*Preprint* gr-qc/0403029)
[7] Duez M D, Liu Y T, Shapiro S L and Stephens B C 2005 *Phys. Rev. D* **72** 024028 (*Preprint* astro-ph/0503420)
[8] Anninos P, Fragile P C and Salmonson J D 2005 *Astrophys. J.* **635** 723
[9] Antón L, Zanotti O, Miralles J A, Martí J M, Ibáñez J M, Font J A and Pons J A 2006 *Astrophys. J.* **637** 296 (*Preprint* astro-ph/0506063)
[10] Mizuno Y, Nishikawa K I, Koide S, Hardee P and Fishman G J 2006 *ArXiv Astrophysics e-prints* (*Preprint* astro-ph/0609004)
[11] Del Zanna L, Zanotti O, Bucciantini N and Londrillo P 2007 *Astron. Astrophys.* **473** 11–30 (*Preprint* 0704.3206)
[12] Giacomazzo B and Rezzolla L 2007 *Class. Quantum Grav.* **24** 235 (*Preprint* gr-qc/0701109)
[13] Radice D and Rezzolla L 2012 *Astron. Astrophys.* **547** A26 (*Preprint* 1206.6502)
[14] Radice D, Rezzolla L and Galeazzi F 2014 *Mon. Not. R. Astron. Soc. L.* **437** L46–L50 (*Preprint* 1306.6052)
[15] McKinney J C, Tchekhovskoy A, Sadowski A and Narayan R 2014 *Mon. Not. R. Astron. Soc.* **441** 3177–3208 (*Preprint* 1312.6127)
[16] Etienne Z B, Paschalidis V, Haas R, Mösta P and Shapiro S L 2015 *Class. Quantum Grav.* **32** 175009 (*Preprint* 1501.07276)
[17] White C J, Stone J M and Gammie C F 2016 *Astrophys. J., Supp.* **225** 22 (*Preprint* 1511.00943)
[18] Zanotti O and Dumbser M 2015 *Computer Physics Communications* **188** 110–127 (*Preprint* 1312.7784)
[19] Meliani Z, Mizuno Y, Olivares H, Porth O, Rezzolla L and Younsi Z 2017 *Astron. Astrophys.* **598** A38 (*Preprint* 1606.08192)
[20] Zanotti O, Fambri F and Dumbser M 2015 *Mon. Not. R. Astron. Soc.* **452** 3010–3029 (*Preprint* 1504.07458)

[21] Mizuno Y, Gómez J L, Nishikawa K I, Meli A, Hardee P E and Rezzolla L 2015 *Astrophys. J.* **809** 38 (*Preprint* 1505.00933)
[22] Younsi Z and *et al* 2017 In preparation
[23] Bronzwaer, T and *et al* 2017 In preparation
[24] Mizuno Y and *et al* 2017 In preparation
[25] de Avellar M and *et al* 2017 In preparation
[26] Noble S C, Krolik J H and Hawley J F 2009 *Astrophys. J.* **692** 411–421 (*Preprint* 0808.3140)
[27] Tóth G and Roe P 2002 *Journal of Computational Physics* **180** 736 – 750 ISSN 0021-9991 URL http://www.sciencedirect.com/science/article/pii/S0021999102971202
[28] Alcubierre M 2008 *Introduction to* $3+1$ *Numerical Relativity* (Oxford, UK: Oxford University Press)
[29] Rezzolla L and Zanotti O 2013 *Relativistic Hydrodynamics* (Oxford, UK: Oxford University Press) ISBN 9780198528906
[30] Gottlieb S and Shu C 1998 *Math. Comp.* **67** 73–85
[31] Spiteri R J and Ruuth S J 2002 *SIAM J. Numerical Analysis* **40** 469–491
[32] Porth O, Xia C, Hendrix T, Moschou S P and Keppens R 2014 *Astrophys. J., Supp.* **214** 4 (*Preprint* 1407.2052)
[33] Dedner A, Kemm F, Kröner D, Munz C D, Schnitzer T and Wesenberg M 2002 *Journal of Computational Physics* **175** 645–673
[34] Palenzuela C, Lehner L, Reula O and Rezzolla L 2009 *Mon. Not. R. Astron. Soc.* **394** 1727–1740 (*Preprint* 0810.1838)
[35] Balsara D S and Kim J 2004 *Astrophys. J.* **602** 1079–1090 (*Preprint* astro-ph/0310728)
[36] Mocz P, Pakmor R, Springel V, Vogelsberger M, Marinacci F and Hernquist L 2016 *Mon. Not. R. Astron. Soc.* **463** 477–488 (*Preprint* 1606.02310)
[37] Evans C R and Hawley J F 1988 *Astrophys. J.* **332** 659–677
[38] Yee K 1966 *IEEE Transactions on Antennas and Propagation* **14** 302–307 ISSN 0018-926X
[39] Balsara D S and Spicer D S 1999 *J. Comput. Phys.* **149** 270–292
[40] Londrillo P and Del Zanna L 2004 *Journal of Computational Physics* **195** 17–48
[41] Toth G 2000 *J. Comput. Phys.* **161** 605–652
[42] Flock M, Dzyurkevich N, Klahr H and Mignone A 2010 *Astron. Astrophys.* **516** A26 (*Preprint* 0906.5516)
[43] Gardiner T A and Stone J M 2005 *Journal of Computational Physics* **205** 509–539 (*Preprint* astro-ph/0501557)
[44] Olivares H and *et al* 2017 In preparation
[45] MacNeice P, Olson K M, Mobarry C, de Fainchtein R and Packer C 2000 *Computer Physics Communications* **126** 330–354
[46] Fryxell B, Olson K, Ricker P, Timmes F X, Zingale M, Lamb D Q, MacNeice P, Rosner R, Truran J W and Tufo H 2000 *Astrophys. J.s* **131** 273–334
[47] Zhang W and MacFadyen A 2006 *The Astrophysical Journal Supplement Series* **164** 255 URL http://iopscience.iop.org/0067-0049/164/1/255
[48] White C J, Stone J M and Gammie C F 2016 *Astrophys. J.s* **225** 22 (*Preprint* 1511.00943)
[49] Löhner R 1987 *Computer Methods in Applied Mechanics and Engineering* **61** 323–338
[50] Keppens R, Meliani Z, van Marle A J, Delmont P, Vlasis A and van der Holst B 2012 *Journal of Computational Physics* **231** 718–744
[51] Orszag S A and Tang C M 1979 *Journal of Fluid Mechanics* **90** 129–143
[52] Koren B 1993 *Numerical methods for advection–diffusion problems* Notes on numerical fluid mechanics, v. 45 (Braunschweig: Vieweg) ISBN 3528076453

IOP Conf. Series: Journal of Physics: Conf. Series **1031** (2018) 012009 doi:10.1088/1742-6596/1031/1/012009

MHD Simulations of the Evolution of the Local Interstellar Medium

Jonathan D. Slavin

Smithsonian Astrophysical Observatory, MS 83, 60 Garden Street, Cambridge, MA 02138 USA

E-mail: `jslavin@cfa.harvard.edu`

Abstract. The Sun is located inside a low density cloud known as the Local Interstellar Cloud (LIC), which is part of a group of nearby clouds known as the Complex of Local Interstellar Clouds (CLIC). All of these clouds are contained within the hot Local Bubble, which contains gas at $\sim 10^6$ K and appears to have been created by multiple supernova explosions. We present new results from our ongoing project on simulating the origins and evolution of the Local Interstellar Medium. We aim to model the origins of the CLIC, especially the way that the clouds managed to survive shock passage and reach their current warm ($T \sim 7000$ K) and low density ($n \sim 0.2$ cm^{-3}) state. We find that the magnetic field is important for maintaining the Local Bubble at a high thermal pressure and helps the clouds rebound more effectively after being shocked. Thermal conduction is necessary to make the temperature in the bubble as uniform as observed. We also find that to reach its current state, the Local Bubble requires multiple supernova explosions. The state of the CLIC and the Local Bubble provide important clues as to how supernova feedback operates in the ISM and it interaction with the different phases of the ISM.

1. Introduction

The interstellar medium (ISM) of our Galaxy contains gas at a wide range of temperatures and densities ranging from cold, molecular clouds ($T \sim 10$ K) to supernova heated very hot gas ($T \gtrsim 10^6$ K). Our local ISM (LISM) is dominated by the hot gas of the Local Bubble, a very low density region that extends 50 to 200 pc in all directions, however it also includes a collection of warm ($T \sim 7000$ K), low density clouds, known as the Complex of Local Interstellar Clouds (CLIC) in the region within 15 pc of the Sun. Understanding the nature and evolution of the LISM is important because it is so well observed and exemplifies many important processes that operate in galaxies. We present here an update on our progress in creating numerical hydrodynamical models of the evolution of the LISM [see 1, for our discussion of an earlier stage of this project].

1.1. Orientation to the Local Interstellar Medium

The extent of the Local Bubble has been determined by both absorption lines measurements, most notably of Na I and Ca II, and dust extinction. Figure 1 from Welsh et al. [2] shows a cut in the plane of the shape of the cavity as derived from the lack of Na I absorption. The bubble is much larger out of the plane as would be expected for a supernova created bubble expanding out of the Galactic disk. Bolstering the view that the Local Bubble is filled with hot gas is the observation of diffuse soft X-ray emission (energies ~ 0.25 keV and higher) seen in all directions of the sky. Since X-rays at these energies are readily absorbed by neutral H, the observation of a substantial flux in the Galactic plane is evidence for their local origin.

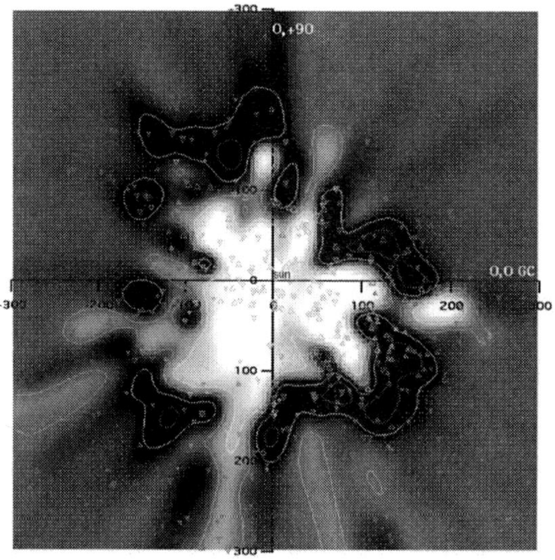

Figure 1. The Local Bubble as inferred from Na I absorption line measurements projected onto the Galactic plane [2, used by permission].

The CLIC, like the Local Bubble, was also discovered via absorption line measurements but in this case it was not the lack of absorption as it was for the bubble, but its presence. Using lines of Mg II, O I, C II and many other ions that can be detected in low column densities of neutral or partially ionized gas, the clouds have been detected toward stars that are within 15 pc of the Sun. By observing on many lines of sight and by grouping the absorption components according to their consistency with particular space vectors (and other characteristics), Redfield & Linsky [3] were able to identify the 15 separate "clouds" that make up the CLIC, some of which are shown in Figure 2 which is a projection onto the Galactic plane. These clouds have similar, though not identical, qualities such as being warm, low density, partially ionized and having moderate levels of elemental depletions.

1.2. The Properties of the CLIC

The CLIC clouds, in addition to having similar physical characteristics, also have similar space velocities relative to the local standard of rest. Figure 3 shows a projection of the space vectors of the CLIC on the Galactic plane along with a comparison of the radial velocities of absorption components vs. those predicted based on the best-fitting velocity vector for the CLIC as a whole. It's clear that while the clouds are moving as a group, there is a substantial dispersion in their velocities.

The spatial distribution of the clouds gives us some hints as to possible interactions between them. Figure 4 (left panel) shows the four clouds that are inferred to be closest to the Sun projected onto the sky. While determining their outlines is inexact given the limited number of lines of sight toward nearby stars, their shapes as determined without reference to each other are suggestive of cloud-cloud interactions. Also shown (right panel) are clouds within the CLIC that have relatively large velocities relative to each other. These results suggest possible cloud-cloud interactions within the CLIC. The fact that the group of clouds have these apparent interactions

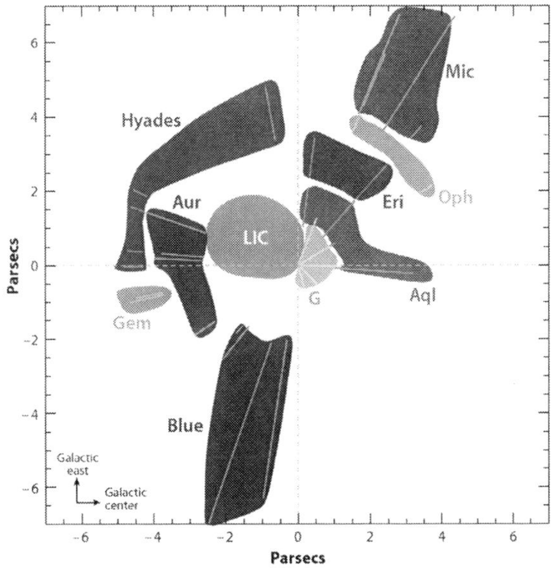

Figure 2. Schematic of those clouds in the CLIC that are close to the Galactic plane with their projected sizes and inferred shapes [from 4]. The Sun is in the center. Note that the scale is far smaller than in Figure 1. It is not known exactly how close to the edge of the LIC the Sun lies, though it must be fairly close since absorption with the LIC velocity is not seen in some parts of the sky.

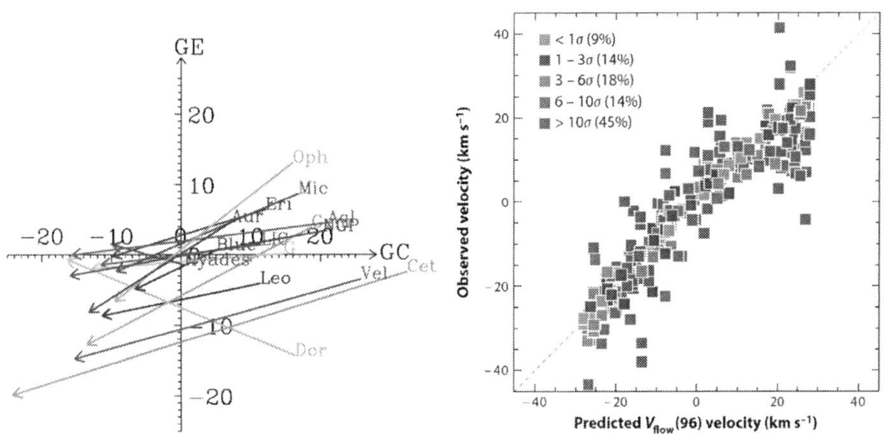

Figure 3. (*Left*) Velocity vectors for each of the clouds of the CLIC projected onto the Galactic plane [from 3]. It can be seen that the clouds are moving in generally the same direction, though with substantially deviations. (*Right*) Predicted vs. observed radial velocity for individual lines of sight for CLIC clouds. The predicted velocity is based on the best fitting velocity vector for the bulk motion of the CLIC. The points are color coded by the magnitude of their deviation from the bulk motion vector.

and relative velocities and yet each cloud seems to be coherent in velocity within itself presents a challenge for simulating the evolution of the CLIC within the Local Bubble.

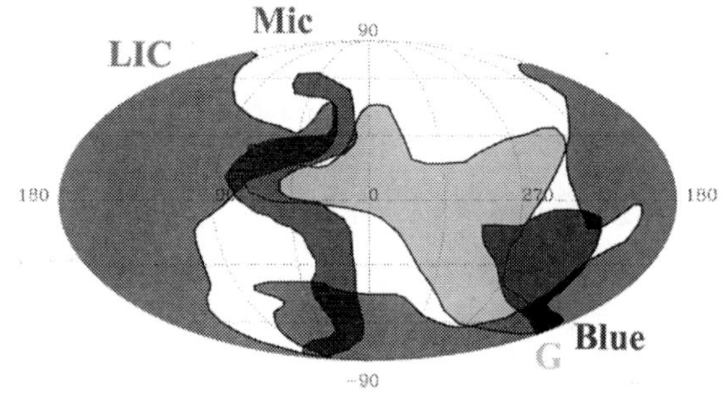

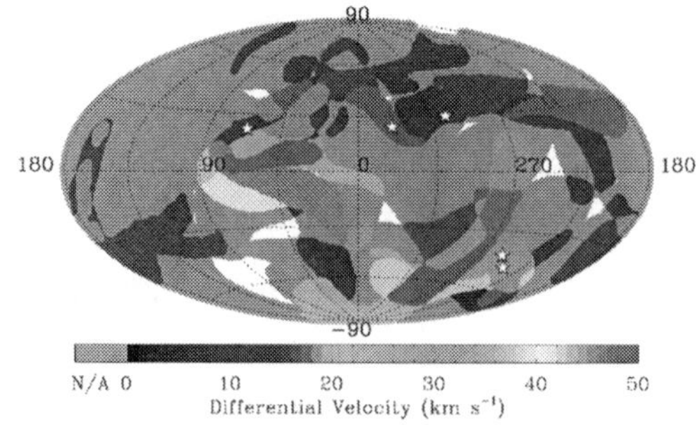

Figure 4. (*Left*) The four closest clouds in the CLIC on the sky [from 3]. Their morphology is suggestive of cloud-cloud interactions. (*Right*) CLIC on the sky highlighting regions in which multiple clouds are seen along the line of sight. The velocity differences suggest either ongoing or past interactions which could be supersonic.

1.3. The Local Interstellar Cloud

Included among the CLIC is the cloud that immediately surrounds the Solar System, known as the Local Interstellar Cloud (LIC). The LIC is the best observed interstellar cloud having been identified in absorption lines on over 80 lines of sight toward nearby stars [3, 5]. In Figure 5 we plot the observed radial velocities for lines of sight in which the LIC velocity vector has been identified vs. the angular distance from the downstream direction for the best fit LIC velocity vector. The remarkable level of consistency of the data with the simple solid body motion of the cloud relative to the Solar System indicates that there is little turbulence within the LIC. Turbulence on smaller scales is constrained by the broadening of the absorption lines and can be tested by comparing the widths of high mass ions with low mass ions, e.g. D I vs. Fe II. These determinations generally find turbulent velocities of only $1 - 3$ km s^{-1} for clouds in the CLIC.

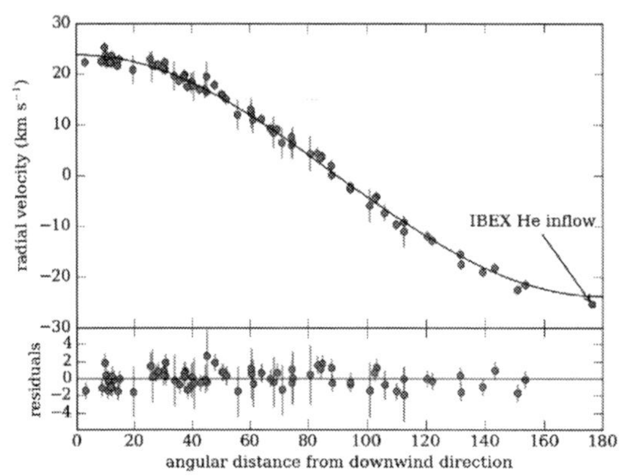

Figure 5. Observed velocities of absorption line components identified with the LIC vs. angular distance from the downstream vector. A simple solid body motion model is plotted as the solid line. The excellent fit to the model indicates that there is little turbulence in the LIC.

In addition to absorption lines, neutral atoms from the LIC that flow into the Solar System have been directly detected *in situ* by spacecraft such as Ulysses [6] and IBEX [7]. The absorption line data and *in situ* data present a consistent picture of the physical parameters of the LIC:

IOP Conf. Series: Journal of Physics: Conf. Series **1031** (2018) 012009 doi:10.1088/1742-6596/1031/1/012009

$T \sim 6300 - 7500$ K, $X(\mathrm{H}^+) \sim 20 - 30\%$, $X(\mathrm{He}^+) \sim 40\%$, $n \approx 0.25$ cm^{-3}. Heliosphere models along with data on the size of the solar wind termination shock from the Voyager spacecrafts and IBEX and Ulysses data combine to put some constraints on the strength of the magnetic field in the LIC [8, 9, 10] as well with current determinations putting it at roughly $2 - 5$ μG, fairly typical for the ISM not in molecular clouds. We note that an alternative model for the CLIC as a single cloud that includes weak shocks and flows has been put forward by Gry & Jenkins [11]. That model has in turn been disputed by Redfield & Linsky [12]. For our purposes, it is not particularly important if the clouds are contiguous or separate. The physical properties of the clouds are not in dispute.

1.4. The Local Bubble

As discussed above, the Local Bubble is inferred both by the lack of neutral, absorbing gas out to roughly 50 pc or more in all directions and by the detection of soft X-ray emission. Questions have been raised [13], however, about whether that emission is truly generated by hot gas as has been long believed. This uncertainty arises from the fact that soft X-ray emission is also generated within the Solar System through the interaction of the solar wind with neutral H and He flowing in from the ISM. This solar wind charge exchange (SWCX) emission, which results from cascades after charge exchange between the inflowing neutrals and solar wind ions, has been detected and modeled. Recently Galeazzi et al. [14] have carried observations that have shown that at least 60% of the observed soft X-ray diffuse background at energies near 1/4 keV comes from the hot gas of the Local Bubble. Liu et al. [15] have gone farther and produced all-sky maps of the background that have been cleaned of SWCX emission. From these they have derived maps of the temperature and emission measure of the hot gas and have produced estimates of the shape of the Local Bubble in 3D. These latter make very simple assumptions regarding emissivity in the hot gas, namely that it has a uniform temperature and emissivity. Nonetheless they put clear constraints on our models for the evolution of the Local Bubble, in particular that the temperature is $\sim 10^6$ K with only small variations around that value.

1.5. Modeling Goals and Challenges

Given the evidence for the origin of the Local Bubble as resulting from supernovae expanding in the ISM, the existence of the CLIC is surprising. This is because the density inferred for the clouds, $n \sim 0.1 - 0.3$ cm^{-3} is typical of what is called the intercloud medium in the ISM with so-called diffuse clouds having densities more like 10 cm^{-3} or higher. For a supernova remnant (SNR) expanding into a smooth medium the ambient ISM is either shock heated to high temperatures, $T \gtrsim 10^{5.5}$, becoming part of the hot, X-ray emitting interior, or if the shock has slowed enough that the remnant is radiative, the swept up gas becomes part of the dense, cold shell. For a cloud to be left behind as the CLIC appears to have been, it must be substantially denser than the surrounding gas. Given that the density of the current clouds is quite low, we therefore have proposed that the CLIC started as colder, denser clouds surrounded by low density intercloud gas and only after having been shocked and existing within the hot remnant for many years did the CLIC reach its current warm, low density state.

We also aim to understand the nature of the Local Bubble with our simulations. To sum up then the challenges for understanding the LISM that we hope to address include:

- How did the CLIC come to be inside the Local Bubble?
- How was the Local Bubble created? One supernova? Multiple SNe?
- How can we explain the low level of turbulence in the CLIC and the multiple cloud structure at the same time?
- What are the dominant physical processes governing the hot gas-cloud interaction? Thermal conduction, turbulent mixing, the magnetic field?

- Is the emissivity of the Local Bubble truly uniform? Is its temperature?

2. Hydro/MHD Modeling of the LISM

We have carried out our modeling of the LISM using the numerical magneto-hydrodynamical adaptive mesh refinement code FLASH [16, http://flash.uchicago.edu/site/flashcode/] (version 4.3). Up to this point the modeling has only been done in 2D assuming cylindrical symmetry (i.e. $R - Z$). For initial conditions we have a collection of cold (100 K), dense (25 cm^{-3}) clouds embedded in a warm (7500 K), lower density (0.358 cm^{-3}) intercloud medium. Supernova explosions are set off at the origin and the clouds are located close to the explosions, initially $10 - 20$ pc away. The explosions are initiated by injecting an explosions worth of thermal energy, $E_0 = 10^{51}$ ergs, in a spherical region of radius 2.25 pc. We have done runs with only one explosion and runs with two explosions with the second explosion occurring 5×10^5 yr after the first. We have included the following physical processes in our runs:

- thermal conduction including heat flux saturation using unsplit implicit diffusion,
- radiative cooling and photoionization heating using tables of coefficients,
- the magnetic field using the unsplit staggered mesh scheme which uses the constrained transport method to ensure that $\nabla \cdot \mathbf{B} = 0$.

The radiative cooling rate for simulations discussed here was taken from the solar metallicity table of Sutherland & Dopita [17]. The heating rate table was generated from data from Wolfire et al. [18] for solar metallicity and a shielding column of 10^{19} cm^{-2}. For all runs so far we have assumed a uniform B field with $B_z = 5\,\mu$G and $B_r = 0$. As of the presentation at ASTRONUM 2017 we had not yet used MHD and thermal conduction together (though since then we have successfully carried out such runs). For the runs presented here, we used AMR with a base grid of $64x64$ and up to five levels of adaptive mesh refinement resulting in a finest resolution of 0.098 pc. This is the highest spatial resolution used in a multi-dimensional simulation of the LISM to date. Previous work on the Local Bubble by de Avillez & Breitschwerdt [19], which was done in 3D, was focused on the hot gas evolution and not on the CLIC and used a resolution that was lower by a factor of 10. The high resolution needed to explore the evolution of the CLIC, along with that of the Local Bubble, makes 3D simulations very demanding. We have some guidance from our recent work on SNR evolution in a cloudy ISM [20]. In that work we carried out simulations in both 2D and 3D including thermal conduction, but not cooling. We found some evidence for an increased evaporation rate in the 3D runs, probably because of the higher surface area to volume ratio of the spherical 3D clouds as compared to the toroids of the 2D simulations, though the effects were not dramatic. We intend to carry out 3D calculations of LISM evolution in the future.

The primary goals of our modeling are, for the CLIC, to match: 1) the present day cloud temperatures, i.e. $T \sim 5000 - 10,000$ K so the initially cold gas is converted to warm gas, 2) the densities of the clouds, $n \sim 0.1 - 0.3$ cm^{-3}, 3) dynamics of the clouds, identifiable as coherent velocity structures while having some cloud-cloud differences in velocity, 4) low levels of turbulence as observed. For the Local Bubble we want: 1) temperature in the range of observations, $T \sim 10^6$ K, 2) relatively uniform intensity as observed from inside the bubble, 3) absolute level of soft X-ray intensity close to the observed level, 4) size of the low density cavity of dimensions roughly $40 - 100$ pc. This set of goals is quite constraining for both our initial conditions and the physical processes that we include.

2.1. Thermal Evolution of the CLIC

A critical factor in the thermal evolution of the CLIC is the heating and cooling rates assumed in the medium. The cooling is caused by collisional excitation (followed by a radiative cascade) and radiative recombination while heating of the clouds is primarily by photoionization. In

IOP Conf. Series: Journal of Physics: Conf. Series **1031** (2018) 012009 doi:10.1088/1742-6596/1031/1/012009

Figure 6 we illustrate the phase curve for the clouds based on our assumed heating and cooling rates. At a low pressure, like that in the ambient medium before being hit by a shock, two different temperature-density phases can coexist as indicated by the points that are intersected by the horizontal line. At high pressures, cooling exceeds heating and gas tends to move down and toward the right toward the equilibrium curve. At low pressure, the heating exceeds cooling and gas moves up and toward the left. Thermal conduction can increase the temperature of the gas at roughly a constant pressure.

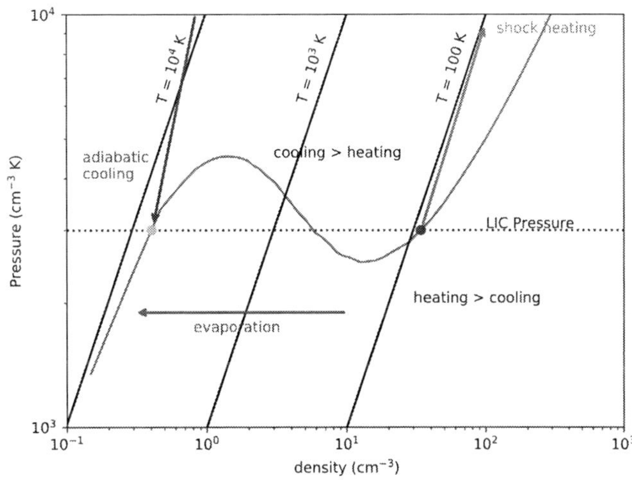

Figure 6. Phase diagram of the ISM gas given our assumed heating and cooling rates. The solid curve is the locus of pressure-density values for which the heating and cooling rates match.

After the blast wave hits the collection of cold clouds, the fast shock in the intercloud medium passes around them while slower shocks propagate through the clouds. The clouds are heated to $\sim 1-6\times10^5$ K and pressurized. They cool quickly via both radiative cooling and expansion as they rebound. At the same time the pressure in the hot gas surrounding the clouds is decreasing as the bubble expands. The expansion of the clouds to lower densities leads to slower cooling.

The effects of including the magnetic field and those of including thermal conduction are quite different as can be seen from Figure 7. These plots compare the gas inside the shock (and not including the cold shell) for two runs, one with thermal conduction and no magnetic field and the other without thermal conduction but including the magnetic field. For both runs, there were two explosions, with the second one occurring 5×10^5 yr after the first one. The plots show the state at $t = 9.5\times10^6$ yr after the first explosion. With just conduction and no magnetic field, the hot gas tends to have a relatively uniform temperature. The pressure, however can vary substantially, most likely because of the regions around the edges of the clouds where an evaporative outflow is being driven and also substantial cooling is occurring. With no conduction and a magnetic field, the hot gas has a more uniform pressure but varies considerably in temperature. In this case the cloud gas has higher pressure than in the case with thermal conduction. These results are further illustrated in Figure 8 which shows the temperature distribution for the same runs as in Figure 7. For a run with no magnetic field and no thermal conduction, the results are in between those two cases, with the lower thermal pressure as for the HD case and a substantial spread in both temperature, as for the MHD case, and pressure as for the HD case with conduction.

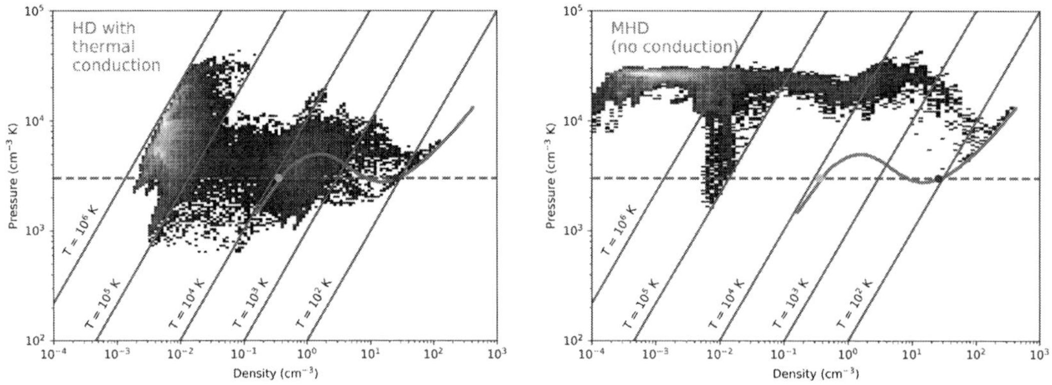

Figure 7. Phase diagrams for runs with thermal conduction, but no magnetic field (*Left*) and with MHD but no thermal conduction (*Right*). These are 2D histograms that are volume weighted and include only the region inside of the dense shell. For both runs the time is 9.5×10^5 yr after the first explosion and a second explosion occurred at $t = 5 \times 10^5$ yr. It's clear that thermal conduction leads to quite a uniform gas temperature for the hot gas, while the pressure varies considerably. For the MHD run, on the other hand, the pressure is quite uniform while the temperature of the hot gas varies over a large range. The pressure is larger for the case with no conduction because the confining pressure in the surrounding medium, including both thermal and magnetic pressure, is much higher.

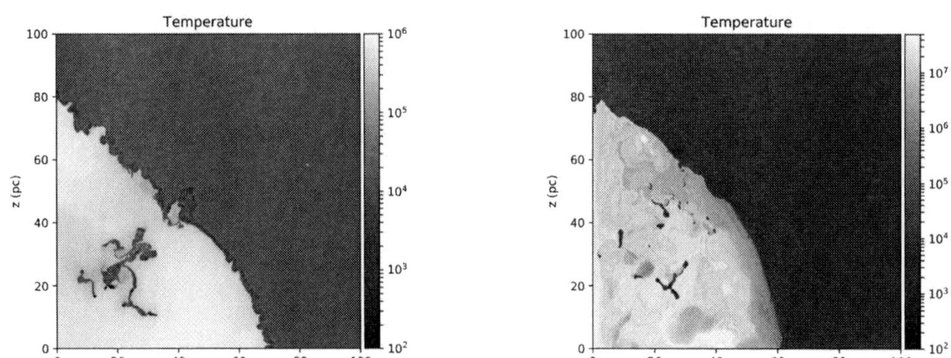

Figure 8. Temperature distributions at $t = 9.5 \times 10^5$ yr for run with thermal conduction but no magnetic field (*left*) and with magnetic field but no thermal conduction (*right*). Note that the scales are not the same for the two runs. The MHD run has substantially higher temperatures and much more temperature variation.

2.2. Preliminary Comparison With Observations

While we are still exploring parameter space with regards to the initial conditions and investigating the effects of different physical inputs, it is worthwhile to examine where we stand in comparison with the observations. Figure 9 shows a the state of the main cloud in the hydrodynamical run that includes thermal conduction.

As was illustrated in the left plot in Figure 7, the density is mostly moderate $n \sim 0.1 - 3$

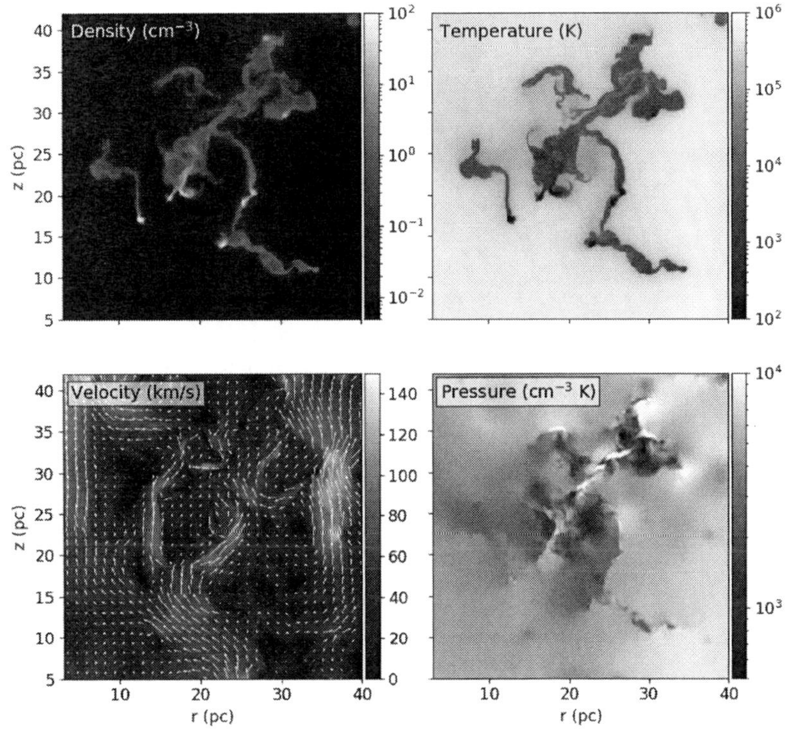

Figure 9. Properties of the main cloud complex in a hydrodynamical (i.e. $\mathbf{B} = 0$) run that includes thermal conduction at $t = 10^6$ yr. Most of the cloud gas is at moderate density, $n \sim 0.1 - 3$ cm^{-3}, and is warm, $T \sim 3 \times 10^3 - 10^4$ K, though cold dense knots remain. The hot gas surrounding the complex shows the presence of intermediate temperature gas created by thermal evaporation. The pressure pane shows that the cloud has not reached dynamical equilibrium yet.

cm^{-3}, slightly higher on average than the observed CLIC, while the temperature is mostly warm, $T \sim 3 \times 10^3 - 10^4$ K, similar to the CLIC. The lower right pane shows that the pressure has not equilibrated yet in the region of the cloud complex. This is also illustrated by the complex flow pattern shown in the lower right pane. Most of the cloud complex is moving at low velocity, but it remains to be seen if the velocities of different regions of the cloud could have the sort of velocity coherence observed for different clouds within the CLIC.

In comparing the X-ray emissivity of the simulated bubble with the observed soft X-ray background, it is clear that the MHD run with no thermal conduction is too hot to match the observations. In particular the Wisconsin B band/C band rates constrain the temperature (assuming it is uniform) to be close to 10^6 K. The run with thermal conduction appears much more promising. In Figure 10 we show the emissivity in the ROSAT R12 band. We see that even in this case where the temperature is fairly uniform, the emissivity varies by orders of magnitude. Of course when integrated across tens of parsecs we expect that such variations will be smoothed out. Nonetheless it is worth noting that the assumption of uniform emissivity that

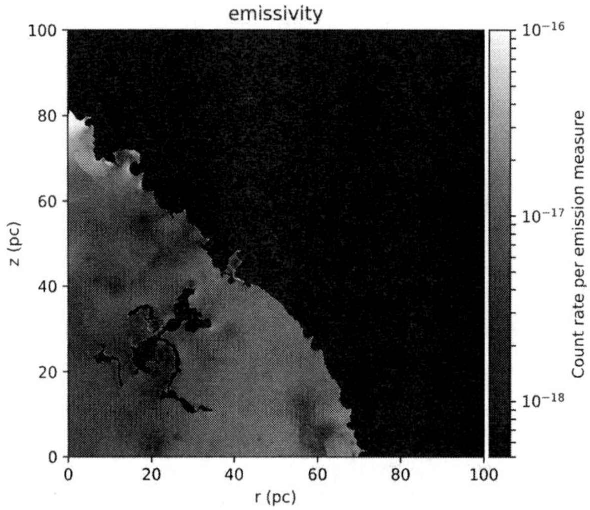

Figure 10. Emissivity in ROSAT's R12 band (in counts cm^{-1} s^{-1}) for the run with thermal conduction and no magnetic field. It is clear that, despite the fact that the temperature is fairly uniform, the emissivity varies considerably.

is generally applied to interpretations of the soft X-ray data is unlikely to be even approximately true.

3. Summary

We have carried out a series of numerical simulations aimed at improving our understanding of the local ISM. While this is a work in progress we have found some interesting results:

- The cloud complex re-fragments after the second explosion – otherwise it is too concentrated,
- The cold cloud gas is mostly converted to warm gas as desired, but cold knots persist,
- MHD runs without thermal conduction have hot gas with a wide range of temperatures, while hydro runs without a magnetic field have a relatively uniform temperature but tend to cool quickly,
- Even after 10^6 yr the bubble and clouds are not dynamically equilibrated.

We will be continuing to search for the balance that leads to the right amount of cooling with fairly uniform temperature to match the soft X-ray background results. At the same time we will be looking at how much time is needed after an explosion is needed for the clouds to become as quiescent as they appear to be. As we continue to explore more of parameter space, we intend to subject our results to more thorough testing regarding whether the hot gas distribution is consistent with the soft X-ray background and if the cloud complex has characteristics consistent with observations of the CLIC.

Even without those more stringent tests we can draw certain conclusions from our results: 1) Thermal conduction appears necessary to get the uniformity of temperature implied by the observations of the soft X-ray background, 2) The magnetic field is needed to confine the bubble and maintain the relatively high thermal pressure in the hot gas while allowing a lower thermal pressure in the clouds, 3) The tight cloud grouping combined with relative cloud-cloud motions and low turbulence in the CLIC is not easily achieved. The wealth of observations available on the CLIC and the Local Bubble provide tight constraints that can serve as a gauge of our understanding of the physics at work and our numerical methods for simulating the LISM.

Acknowledgments
This work was supported by NASA Astrophysics Theory Grant No. NNX15AI60G. The author thanks the organizers of ASTRONUM 2017 for the invitation to present this work. The author also thanks his collaborators, Priscilla Frisch and Seth Redfield for helpful conversations and the anonymous referee for comments that improved the manuscript.

References
[1] Slavin J D 2017 *Journal of Physics Conference Series* vol 837 p 012011
[2] Welsh B Y, Lallement R, Vergely J L and Raimond S 2010 *A&A* **510** A54
[3] Redfield S and Linsky J L 2008 *ApJ* **673** 283
[4] Frisch P C, Redfield S and Slavin J D 2011 *ARA&A* **49** 237
[5] Malamut C, Redfield S, Linsky J L, Wood B E and Ayres T R 2014 *ApJ* **787** 75
[6] Wood B E, Müller H R and Witte M 2015 *ApJ* **801** 62
[7] Bzowski M, Kubiak M A, Möbius E, Bochsler P, Leonard T, Heirtzler D, Kucharek H, Sokół J M, Hłond M, Crew G B, Schwadron N A, Fuselier S A and McComas D J 2012 *ApJS* **198** 12
[8] Opher M, Bibi F A, Toth G, Richardson J D, Izmodenov V V and Gombosi T I 2009 *Nature* **462** 1036–1038
[9] Izmodenov V V and Alexashov D B 2015 *ApJS* **220** 32
[10] Zirnstein E J, Heerikhuisen J, Funsten H O, Livadiotis G, McComas D J and Pogorelov N V 2016 *ApJL* **818** L18
[11] Gry C and Jenkins E B 2014 *A&A* **567** A58
[12] Redfield S and Linsky J L 2015 *ApJ* **812** 125
[13] Welsh B Y and Shelton R L 2009 *Ap&SS* **323** 1
[14] Galeazzi M, Chiao M, Collier M R, Cravens T, Koutroumpa D, Kuntz K D, Lallement R, Lepri S T, McCammon D, Morgan K, Porter F S, Robertson I P, Snowden S L, Thomas N E, Uprety Y, Ursino E and Walsh B M 2014 *Nature* **512** 171
[15] Liu W, Chiao M, Collier M R, Cravens T, Galeazzi M, Koutroumpa D, Kuntz K D, Lallement R, Lepri S T, McCammon D, Morgan K, Porter F S, Snowden S L, Thomas N E, Uprety Y, Ursino E and Walsh B M 2017 *ApJ* **834** 33
[16] Fryxell B, Olson K, Ricker P, Timmes F X, Zingale M, Lamb D Q, MacNeice P, Rosner R, Truran J W and Tufo H 2010 FLASH: Adaptive Mesh Hydrodynamics Code for Modeling Astrophysical Thermonuclear Flashes Astrophysics Source Code Library (*Preprint* 1010.082)
[17] Sutherland R S and Dopita M A 1993 *ApJS* **88** 253–327
[18] Wolfire M G, McKee C F, Hollenbach D and Tielens A G G M 2003 *ApJ* **587** 278–311 (*Preprint* astro-ph/0207098)
[19] de Avillez M A and Breitschwerdt D 2012 *A&A* **539** L1
[20] Slavin J D, Smith R K, Foster A, Winter H D, Raymond J C, Slane P O and Yamaguchi H 2017 *ApJ* **846** 77 (*Preprint* 1708.02646)

IOP Conf. Series: Journal of Physics: Conf. Series **1031** (2018) 012010 doi:10.1088/1742-6596/1031/1/012010

Contribution of supernovae-driven galactic fountain on fuel rate

K. Sung and Kyujin Kwak

Physics Department, Ulsan National Institute of Science and Technology, Ulsan 44919, Republic of Korea

E-mail: sung@unist.ac.kr

Abstract. The interstellar medium in our galaxy is a complex system which involves multiphase gas in a wide range of temperatures and densities. Accompanied by physical processes such as cooling, gravity and magnetic fields it is known that the supernovae blast waves can reshuffle the ISM by generating a galactic fountain. In our study, we mainly focus on estimating the fuel rate of the fountain which is defined as the supply rate of material from the gas falling back into the galactic disk. We take into account the effect of cooling by combining the use of non-equilibrium ionization (NEI) and collisional ionization equilibrium (CIE). Dealing with cooling in this manner, we aim to reduce computation time by using a prebuilt CIE cooling table at applicable physical conditions instead of constantly considering time dependent ion fractions for NEI cooling.

1. Introduction

First introduced by Shapiro & Field (1976), the idea that there can exist a gas flow in a galactic scale has been revisited and developed ever since. Such gas flow can be created from succesive supernovae blasts in the galactic disk by heating up the disk gas. The ejected gas cools down and condenses into atomic hydrogen clouds which in turn falls down to the disk at velocities $\sim 100\ \mathrm{km \cdot s^{-1}}$ to form a galactic fountain. Notable galactic fountain models include, analytic fountain models by Bregman (1980) and Kahn (1981), two-dimensional models by Houck & Bregman (1990) and Rosen et al. (1993), and three-dimensional models by de Avillez (2000), Joung & Mac Low (2006) and Kwak et al. (2009).

While the previous galactic fountain models mainly focus on the physical and dynamical structure of the ISM, in this study, we aim to estimate the fuel rate on the galactic disk by considering both collisional ionization equilibrium (CIE) and non-equilibrium ionization (NEI) cooling. By using a prebuilt CIE cooling table at applicable high temperatures, we avoid tracking time dependent ion fractions and therefore reduce computation time. The fuel rate describes the supply efficiency of materials from the infalling gas clouds. Such concept is carried along from our previous work regarding the infall of high velocity clouds (HVCs) on the galactic disk where we have shown that the fuel rate strongly depends on the physical configuration of the system even when the total infall rate of the materials is fixed.

In the following Section 2, we describe the basic setup for our simulations and introduce how we estimate the fuel rate. In Section 3, we only discuss the expected outcome and limitations of the simulation model in a qualitative manner as we are currently in the process of building a complete simulation model that includes physical processes such as gravity, magnetic field and cooling in a three-dimension Cartesian geometry. At the conclusion, we summarize our study in Section 4.

2. Simulation methods

2.1. Simulation setup

We use FLASH 4.4 which is a modular, adaptive-mesh, parallel multiphysics simulation code capable of handling general compressible flow problems. Message-Passing Interface (MPI) library is used for parallelization and the Adaptive Mesh Refinement (AMR) is managed by the PARAMASH library (Fryxell et al. 2000)

The computation domain is represented in a 3-D Cartesian geometry that is 10 kpc long from -5 to 5 kpc in the z-direction with a squared shape horizontal cross section of $(1\ \mathrm{kpc})^2$ and the midplane of the galactic disk positioned at z = 0. The boundary condition is periodic for the horizontal directions and outflowing for the vertical direction where the total box contains $(256)^2 \times 2560$ cube cells at the maximum refine level of 6 determined by density. This converts to a grid resolution of 3.91 pc.

As the gravitational field is plane-parallel to the center plane of the galactic disk, the disk shows a density gradient due to gravity where the average HI volume number density of the galactic disk is $1.00 \times 10^{-1}\ \mathrm{cm}^{-3}$ (Launhardt et al. 2002). The temperature of the disk is determined by simply following the ideal gas law to maintain the pressure balance with the ISM at a temperature of 10^6 K and HI volume number density of $1.00 \times 10^{-4}\ \mathrm{cm}^{-3}$.

Thermal energy of 10^{51} ergs is added for each supernova explosion, in a sphere that represents the initial supernova with a radius depending on the density of the local galactic disk. Such events will be randomly distributed inside the galactic disk at a constant rate which will further be tested as an input parameter. Collisional ionization equilibrium (CIE) cooling table is extensively used for temperatures $T > 10^{6.2}$ K. Otherwise, non-equilibrium ionization (NEI) cooling is considered.

In order to track the disk and ISM materials during the dynamical evolution of the system, scalar fields are added to the simulation by indexing the disk and ISM particles based on the initial position of a particle. Cloud materials will also be tracked using a similar method, however being indexed in a different manner based on a criteria that includes velocity, density and other surrounding physical conditions.

2.2. Fuel rate

In galactic chemical evolution models, the infall rate of material is traditionally given in the units of $\mathrm{M}_\odot \cdot \mathrm{cm}^{-2} \cdot \mathrm{yr}^{-1}$ or sometimes $\mathrm{M}_\odot \cdot \mathrm{yr}^{-1}$ in case we fix the area of gas inflow. The simplest mathematical expression for the infall rate $\dot{M}_{cloud}$ of an inflowing gas cloud is

$$\dot{M}_{cloud} = M_{cloud} \cdot \frac{v}{D} \tag{1}$$

where M_{cloud} is the mass of the infalling gas cloud in solar mass, v is the radial velocity and D is the distance from the galactic disk (Wakker et al. 2007; Thom et al. 2008). In our series of simulations, we take D as the vertical distance between the midplane of the galactic disk and the center of the gas cloud. For a typical high velocity cloud infall scenario ($v = 70\ \mathrm{km \cdot s^{-1}}$, $D = 5$ kpc, and $M = 1.16 \times 10^6 \mathrm{M}_\odot$), the infall rate is $1.66 \times 10^{-2} \mathrm{M}_\odot \cdot \mathrm{yr}^{-1}$

We define two quantities e_1 and e_2 in order to address the fuel rate which can differ from the infall rate due to the interaction of the cloud with the disk. e_1 is defined as the time variation of the disk mass fraction,

$$e_1(t) = \frac{M_{disk}(t) - M_{disk}(0)}{M_{disk}(0)} = \frac{dM_{disk}(t)}{M_{disk}(0)} \tag{2}$$

where $M_{disk}(t)$ is the mass of the disk at time t and $M_{disk}(0)$ is the initial mass of the disk. e_2 is the gas cloud infusion fraction on to the disk,

$$e_2(t) = \frac{M_{cloud,disk}(t)}{M_{cloud}} \tag{3}$$

where M_{cloud} is the total mass of the gas cloud and $M_{cloud,disk}(t)$ is the mass of gas cloud materials accreted onto the disk at time t. Defined in such way, $e_1(t)$ shows the mass variation of a patch of the galactic disk during and after the interaction with a gas cloud. In the other hand, $e_2(t)$ shows how much the materials from an infalling gas cloud is accreted onto the disk.

3. Discussion

3.1. Cooling curve

Non-equilibrium ionization (NEI) cooling is considered when the cooling rate and the recombination rate are not together in sync. More specifically, when the recombination time scale is longer than the cooling time scale, cooling rate is not simply a function of temperature but we must consider the fraction of each ions at each time step of the simulation. Otherwise, when the recombination time scale is shorter than the cooling time scale it is safe to use collisional ionization equilibrium (CIE) cooling. The temperature limit for such case is T > $10^{6.2}$ K (de Avillez 2012).

Cooling effects not only allow a galactic fountain of gas clouds after succesive supernova explosions, but also play a significant role during the interaction between the gas material and the galactic disk. As the galactic disk collides with the infalling gas cloud, a reverse shock is generated due to density contrast, proceeds backwards into the gas and drives an expansion of the gas cloud into the ISM after collision. Cooling processes will suppress such rear expansion process and increase the infusion rate of materials onto the galactic disk.

3.2. Estimating fuel rate

For this particular study, the fuel rate is addressed by considering two quantities. One is to track the time evolution of disk mass fraction and the other is to track the time evolution of infalling gas cloud material infusion fraction on the disk. Both quantities give us an idea of how gas clouds would influence the structure and composition of the disk through the hydrodynamic and thermodynamic interactions. However, it is important to identify the physical properties of the gas clouds generated after disk material outburst in order to correctly estimate the infusion rate of clouds $e_2(t)$, as the simulations in this study begin with supernovae outblasts which are different from infall-only cases.

3.3. Excluded physics

Focusing on applying the NEI and CIE combined cooling curve and investigating the ionization profile during the dynamical evolution of the system, we exclude magnetic fields in the current simulation. However, magnetic fields are known to participate in the interaction between the cloud and disk. The magnetic field lines could be compressed during the collision between the gas cloud and the galactic disk which can possibly restrain the exchange of materials between the disk and cloud.

Conduction can play a role in transferring energy, but it turns out that the energy transportation is dominated by dynamical flow over thermal conduction in our simulations because the ISM is continuously disturbed by SN. Therefore, thermal conduction is not included.

4. Summary

From our previous study of HVC infall, we carry the idea that the traditional infall rate requires assistance from a new parameter called the fuel rate and apply it to the supernovae-driven galactic fountain model. The fuel rate needs to include the supply efficiency of materials from the infalling gas clouds on the galactic disk. In order to address the fuel rate more accurately, we considered two quantities which indicate the supply efficiency into the disk, one as the mass variation fraction of the galactic disk and the other as the cloud material infusion fraction onto the galactic disk. Since the fuel rate depends heavily on the physical configuration of the infalling gas, it is important to correctly indentify the physical configuration of the gas clouds generated from cooled down materials supplied by successive supernova outbursts. In the supernova-driven fountain simulations, we use both CIE and NEI cooling depending on the temperature of the gas. We apply CIE cooling at temperatures T $> 10^{6.2}$ K, while NEI cooling needs to be used at temperatures T $< 10^{6.2}$ K in order to trace the cooling process more accurately.

Acknowledgments

This work has been supported by NRF (National Research Foundation of Korea) Grant funded by the Korean Ministry of Education (NRF-2015H1A2A1031629-Global Ph.D. Fellowship Program and No. 2016R1A5A1013277, 2016R1D1A1B03936169 and 2014R1A1A1006050). The software which will be used in this work was in part developed by the DOE-supported ASC/Alliance Center for Astrophysical Thermonuclear Flashes at the University of Chicago.

References

[1] Avillez M A 2000 *MNRAS.* **315** 479
[2] Avillez M A, Breitschwerdt D, Asgekar A and Spitoni E 2012 *Proc. of IAU* **10(H16)** p 606
[3] Bregman J N 1980 *ApJ* **236** 577
[4] Fryxell B, Olson K, Ricker P, Timmes F X, Zingale M, Lamb D Q, MacNeice P, Rosner R, Truran J W and Tufo H 2000 *ApJS* **131** 1
[5] Houck J C and Bregman J N 1990 *ApJ* **352** 506
[6] Joung M K R and Mac Low M-M 2006 *ApJ* **653** 2
[7] Kahn F D 1981 *in Investigating the Universe* ed. F D Kahn and D Reidel (Dordrecht:Kluwer), 1
[8] Kwak K, Shelton R and Raley E A 2009 *ApJ* **699** 2
[9] Launhardt R, Zylka R and Mezger P G 2002 *A&A* **384** 112
[10] Rosen A, Bregman J N and Norman M L 1993 *ApJ* **413** 137
[11] Shapiro P R and Moore R T 1976 *ApJ.* **207** 460
[12] Thom C, Peek J E G, Putman M E, Heiles C, Peek K M G and Wilhelm R 2008 *ApJ* **684** 364
[13] Wakker B P et al. 2007 *ApJL* **670** 113

IOP Conf. Series: Journal of Physics: Conf. Series **1031** (2018) 012011 doi:10.1088/1742-6596/1031/1/012011

Simulating the interaction between supra-thermal particles and the magnetic field in astrophysical shocks

A. J. van Marle[1], F. Casse[1] & A. Marcowith[2]
[1]Laboratoire AstroParticle & Cosmologie (APC), Université Paris Diderot, CNRS/IN2P3, CEA/Irfu, Observatoire de Paris,
Sorbonne Paris Cité, 10, rue Alice Domon et Leonie Duquet, F-75205 Paris Cedex 13, France.
[2]Laboratoire Univers et Particules de Montpellier (LUPM) Université Montpellier, CNRS/IN2P3, CC72, place Eugène Bataillon,
34095, Montpellier Cedex 5, France.

E-mail: `vanmarle@apc.univ-paris7.fr`

Abstract. In order to model the magnetic field amplification and particle acceleration that takes place in astrophysical shocks, we need a code that can efficiently model the large-scale structure of the shock, while still taking the kinetic aspect of supra-thermal particles into account. Starting from the proven MPI-AMRVAC magnetohydrodynamics code we have created a code that combines the kinetic treatment of the Particle-in-Cell (PIC) method for supra-thermal particles with the large-scale efficiency of grid-based hydrodynamics (MHD) to model the thermal plasma, including the use of adaptive mesh refinement. Using this code we simulate astrophysical shocks, varying both the Mach-number and the angle between the magnetic field and the shock to test our code against existing results and study both the evolution of the shock and the behaviour of supra-thermal particles. We find that the combined PIC-MHD method can accurately recover the results that were previously obtained with pure PIC codes. Furthermore, the efficiency of the code allows us to explore the available parameter space to a larger degree than has been done in previous work. Our results suggest that efficient particle acceleration can take place in near-oblique shocks were the magnetic field makes a large angle with the direction of the flow.

1. Introduction

Astrophysical shocks accelerate particles to high velocities in a process known as Fermi-acceleration, which involves repeated crossings of the shock. We observe these particles as cosmic rays (CRs). This process also influence the evolution of the shock, causing turbulence and triggering instabilities in the magnetic field in both the pre-shock and post-shock areas (see [1] for a review.) Simulating this process numerically is difficult because of the different length scales involved. Astrophysical shocks are best modelled using grid-based magnetohydrodynamics (MHD), which is a computationally efficient method that allows for the modelling of large scale structures by treating the gas as a fluid that can be represented by average values. Unfortunately, MHD does not include particle physics and can therefore not be used to simulate the behaviour of particles that behave in a supra-thermal fashion. The movement of individual particles in a gas can be modelled using the Particle-in-Cell (PIC) approach, which treats the gas a collection of individual particles. However, although the PIC method can accurately model the acceleration of individual particles, it is computationally expensive and difficult to apply to large scale structures.

In order to be able to model both the shock and the particle acceleration, we combine these two methods in a a single code, using the prescription by [2]. This model is based on the assumption that an astrophysical plasma can be described as being primarily thermal, with a relatively small supra-thermal component. The thermal part of the plasma can be modelled using grid-based MHD. The supra-thermal part can be modelled using the PIC method. The effect of the thermal plasma on the particles can be expressed using the electromagnetic field generated by the thermal plasma. For the influence of the supra-thermal particles on the thermal component, [2] used a modified version of Ohm's law. We now present a computer code that uses a similar approach. The main advantage of this approach over the more traditional PIC and PIC-hybrid codes lies in its computational efficiency. Because the thermal gas can be modelled as a fluid, rather than as individual particles, the total number of particles in the simulation can be greatly reduced, which in turn allows us to focus on those particles that behave in a supra-thermal fashion.

We have used this code to repeat simulations of a high-Mach shock travelling into a weakly magnetized gas done by [2] who used the same approach. Simulations of this situation were done previously by [3, 4] using a spherical harmonic expansion of the Vlasov-Fokker-Planck equation and by [5–7] using the PIC-hybrid approach.

2. Particles in MHD cells II: physics

The underlying assumption for the Particles in MHD Cells method is that the plasma can be described as a thermal plasma with a (relatively) small correction term caused by the presence of supra-thermal particles. The thermal plasma can be described by a series of conservation equations for mass, momentum and energy and the magnetic induction equation, as is the case in classical MHD. However, the equations contain additional terms to incorporate the effect that the presence of the supra-thermal articles have on the thermal plasma.

The new conservation equations for the thermal gas can be written as follows:

$$\frac{\partial \rho}{\partial t} + \nabla \cdot (\rho \mathbf{v}) = 0, \tag{1}$$

where ρ and $\mathbf{v}$ are the mass density and velocity of the thermal plasma,

$$\frac{\partial \rho \mathbf{v}}{\partial t} + \nabla \cdot \left(\rho \mathbf{v} \otimes \mathbf{v} - \frac{\mathbf{B} \otimes \mathbf{B}}{4\pi} + P_{tot} \mathbb{1} \right) = - \mathbf{F}_{\rm part}, \tag{2}$$

with $\mathbf{B}$ the magnetic field and $P_{\rm tot} = P + B^2/8\pi$ the total pressure. The energy equations is

$$\frac{\partial e}{\partial t} + \nabla \cdot \left((e + P_{\rm tot}) \mathbf{v} + (\mathbf{E} - \mathbf{E}_0) \times \frac{\mathbf{B}}{4\pi} \right) = - \mathbf{u}_{\rm part} \cdot \mathbf{F}_{\rm part} \tag{3}$$

where e is the total energy density of the thermal plasma, $\mathbf{E}_0 = -\mathbf{v}/c \times \mathbf{B}$ is the electric field generated by the motion of the thermal plasma, and $\mathbf{E}$ is the electric field generated by the thermal plasma and the supra-thermal particles combined. $\mathbf{E}$ is obtained through a modified version of Ohm's law:

$$c\,\mathbf{E} = -\left((1-R)\,\mathbf{v} + R\,\mathbf{u}_{\rm part}\right) \times \mathbf{B}, \tag{4}$$

where R is a direct measure of the relative density of supra-thermal particles compared to the total charge density and $\mathbf{u}_{\rm part}$ is the average supra-thermal particle velocity. The new terms appearing on the right hand sides of Eqs. 2 and 3 are obtained from the opposite of the force density applied by the electromagnetic field upon the supra-thermal particles $\mathbf{F}_{\rm part}$, which is defined as

$$\mathbf{F}_{\rm part} = (1-R)\left(n_{\rm part} q \mathbf{E}_0 + \frac{\mathbf{J}_{\rm part}}{c} \times \mathbf{B} \right), \quad (5)$$

with $n_{\rm part}q$ the particle charge density and $\mathbf{J}_{\rm part}$ the current generated by the supra-thermal particles. We close this set of equations with the induction equation:

$$\frac{\partial \mathbf{B}}{\partial t} = c\nabla \times \mathbf{E}, \quad (6)$$

and an equation of state, for which we assume an ideal gas.
Defined this way, equations 1-3 and 6 can be solved together in a self-consistent way. The only outside information required, are the charge and current density generated by the supra-thermal particles.

The motion of the supra-thermal particles can be calculated through the momentum equation of a charged particle in an electromagnetic field:

$$\frac{\partial \mathbf{p}_{\alpha,j}}{\partial t} = q_j \left(\mathbf{E} + \frac{\mathbf{u}_{\alpha,j}}{c} \times \mathbf{B} \right) \quad (7)$$

with $\mathbf{p}_{\alpha,j}$, q_j and $\mathbf{u}_{\alpha,j}$ the momentum, charge, and velocity of a particle with index j and type α. (In practice we will treat all particles as protons in this paper, but the code allows for multiple particle species.) Note that there is no direct interaction between the particles because we assume that they behave in a non-collisional fashion.

3. Particles in MHD cells II: numerics

In order to combine the PIC and MHD methods into a single code we start with an existing grid-based finite volume MHD code: `MPI-AMRVAC` [8]. This is a well-tested code that uses the `OCTREE` system of adaptive mesh refinement [9] and can run in parallel on distributed memory platforms. We retain the existing MHD module but add the extra terms described in the previous section to the conservation equations. In addition we have written a new module to handle the particle motion. This module uses the existing MHD grid, including the adaptive mesh refinement and adds particles. At the beginning of each time step, the MHD quantities of the thermal gas are updated according to the conservation equations. Simultaneously, the particles are moved according to the momentum equation (Eq. 7) by means of a relativistic Boris-pusher [10], which uses the electromagnetic field as it was at the beginning of the time-step. Once the particles have been moved, their effective charge and current is interpolated onto the cell-centres of the MHD grid, where they will contribute to the integration of the conservation equations for the next time step.

In order to maintain numerical stability we limit the time-step using the CFL condition:

$$\Delta t \leq C \min\left(\frac{\Delta \mathrm{x}}{\mathrm{v}_{\rm max}} \right) \quad (8)$$

with $\Delta\, t$ the time step, C a constant (usually less than one), $\Delta\, x$ the size of a gridcell and $v_{\rm max}$ the maximum speed at which any kind of signal travels through the gas. The latter is calculated as the maximum value of the sum of the bulk motion of the thermal fluid various wave velocities, such as sound speed and Alfvén speed. This is then compared to the maximum particle velocity. In addition we limit the time-step to a predetermined fraction of the gyro-time to ensure that we capture the motion of individual particles.

A problem that occurs when combining PIC and MHD is that the local magnetic field deviates from being divergence free as a result of local charge concentrations. We have implemented a Constrained Transport algorithm based on [11] to ensure that the magnetic field remains divergence free.

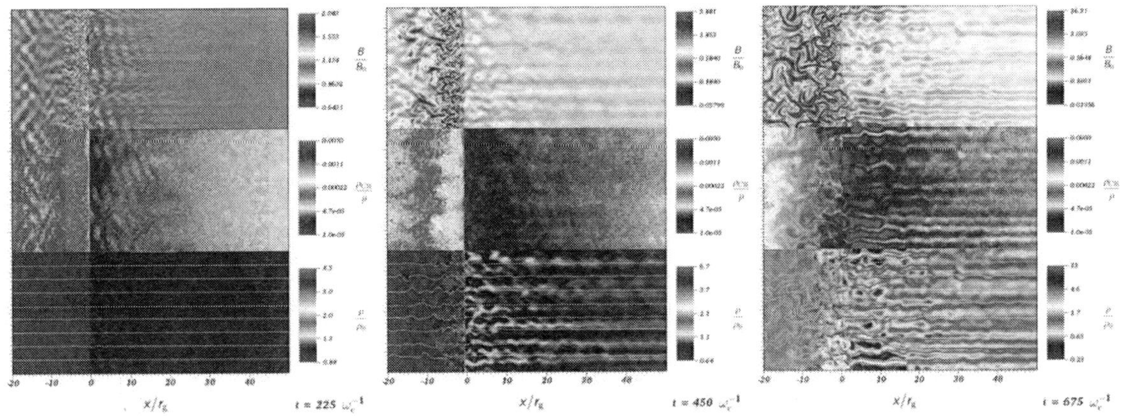

Figure 1. The magnetic field amplitude, cosmic-ray density and plasma density in the early (left), intermediate (centre), and late (right) stages of the parallel shock simulation. From top to bottom, these figures show the magnetic field strength relative to the original magnetic field, supra-thermal particle charge density relative to the thermal gas charge density, and the thermal gas mass density relative to the upstream density at the start of the simulation, combined with streamlines following the magnetic field.
Initially, ($t = 225\omega_c^{-1}$) the upstream medium shows the start of the streaming instability, while the downstream medium shows the onset of turbulence. As time passes, both the downstream turbulence and the upstream filamentation increase, until they distort the shock. Axis are normalized to the gyro radius of injected particles in the upstream medium.

4. Simulations

We test our code by performing two simulations of moving shocks in a weak magnetic field. For the first simulation de duplicate the input parameters used by [2]: The shock moves parallel to the magnetic field with a shock velocity equal to 30 times the Alfvén speed. Unlike [2] we simulate in the frame of reference of the shock in order to reduce the size of the simulation. At the start of the simulation we fill the gas with a thermal plasma conform the Rankine-Hugoniot conditions for a standing shock with an upstream velocity of $V_{\rm sh} = 3.10^{-3}c$ along the x-axis, flowing from the right x-boundary toward the left x-boundary. Furthermore the gas contains a magnetic field corresponding to an Alfvén speed of $V_{\rm A0} = 10^{-4}c$ and the upstream medium is considered to follow equipartition. For a pure MHD simulation this is a stable situation, which should only display minor deviation at the shock due to the size of individual gridcells.

Once the simulation starts we introduce individual particles directly behind the shock at such a rate that they contain 0.2 percent of the total mass that flows through the shock, conform [2], which in turn derived this percentage from [5–7]. These particles are introduced at 3 times the shock velocity and move in a random direction relative to the speed of the post-shock gas.

For the second simulation, we repeat the model described previously, but with the magnetic field at a 70 degree angle with the direction of the flow in the pre-shock medium. Note that this means that the post-shock magnetic field is stronger than the pre-shock field as a result of compression of perpendicular component of the field in the shock. This configuration has not yet been explored using the combined PIC-MHD method, but similar simulations have been performed using the PIC-hybrid method by [5–7].

4.1. Results: 1 parallel model

The results of our first simulation are shown in Fig. 1, which shows the magnetic field strength, thermal gas density, and relative supra-thermal particle density to thermal gas density at three

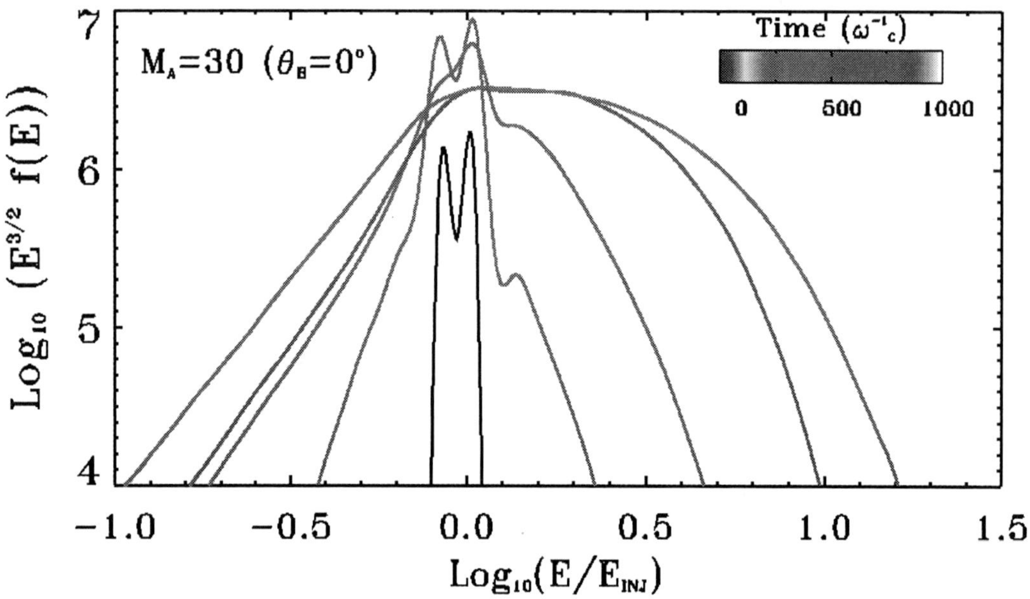

Figure 2. Time evolution of the energy spectrum of supra-thermal particles injected at energy $E_{\rm inj}/m_i c^2 = 4 \times 10^{-5}$, for the simulation shown in Fig. 1. In the late stages of the simulation a supra-thermal tail is forms, which tends to a power-law spectrum in agreement with a diffusive shock acceleration process.

different moments in the simulation: $t = 225, 450$ and $657\,\omega_c^{-1}$, with ω_c^{-1} the ion cyclotron frequency. In the upstream medium, the particles trigger the non-resonant streaming instability [12], which manifests itself as a filamentary structure parallel to the magnetic field and the flow. By contrast, in the downstream medium, the flow becomes turbulent without a preferential direction. This is the result of the subsonic nature of the downstream medium, which allows for signals to travel perpendicular to the flow in the form of sound waves. Eventually, the disturbance in the magnetic field starts to manifest itself in the density, temperature, and velocity of the thermal gas as well, causing a disturbance of the shock front, which start to fall apart into multiple shocks. At this stage we stop the simulation, because the injection mechanism for the particles, which depends on the shock conditions, becomes invalid.

As a result of the disturbance in the magnetic field, particles start to move back and forth across the shock, being accelerated in the process. This is shown in Fig. 2, which shows the particle spectra at various moments in time. As time progresses, the number of high velocity particles increases and they form the kind of spectral energy distribution that is expected from Fermi-acceleration.

4.2. Results: 2 near-perpendicular model

The results of the second simulation show a different behaviour compared to the first simulation, as shown in Fig. 3. Initially, a long wavelength pattern appears in the upstream medium. This pattern is the result of small-scale corrugation of the shock, which causes particles to move upstream, not as a smooth flow, but in the form of localized streams. As this pattern interacts with the shock, the shock corrugation increases in strength and downstream turbulence appears. Eventually, she shock itself is disturbed to the point where we have to stop the simulation.

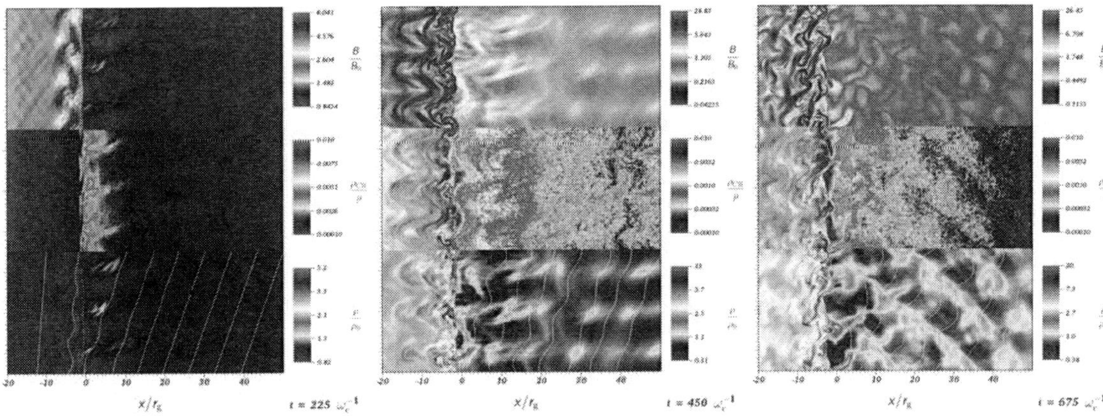

Figure 3. Similar to Fig. 1, this figure demonstrates tge evolution of the gas for the near-perpendicular shock (70 degree angle between flow and magnetic field.) From an early stage, the upstream medium shows a long-wavelength pattern caused by particle streams moving upstream from the shock. Over time this phenomenon causes both increased shock corrugation as well as downstream turbulence.

The particle acceleration for this model follows as different pattern as well as shown in Fig. 4. Initially, only a limited acceleration occurs. This is the result of particles circling the streamlines near the shock and crossing the shock repeatedly as they move [13]. Only when the upstream and downstream magnetic disturbance increases does the pattern of Fermi-acceleration emerge.

5. Discussion

Our parallel simulation recovers the results obtained by [2] as well as the PIC-hybrid simulations by [5–7], proving the validity of our code. However, our results for the near-perpendicular case differ radically from thoset obtained by [6] for a similar situation. Upon closer investigation we can identify two reasons for this: Firstly, the PIC-hybrid models have trouble generating a sufficient number of high velocity particles. Although the absolute number of particles in these simulations is much higher than in our own, which typically contains no more than 10 million particle at any given time, a PIC-hybrid code uses most of its particles to simulate the thermal plasma, leaving only a small fraction to behave in a supra-thermal fashion. By contrast, all our particles behave supra-thermally, because the thermal gas is represented by the MHD part of the code. This leaves us with a larger *effective* particle population. Without such a population, it is impossible to generate an upstream disturbance of the magnetic field. The second reason lies in the size of the simulation box. The progression of events in the near-perpendicular model relies on the formation of the long-wavelength pattern in the upstream medium. The PIC-hybrid models did not have sufficient space in the perpendicular (y) direction to resolve this wave, being smaller than the wavelength. This demonstrates the advantage of the combined PIC-MHD approach, which allows for larger simulations without becoming prohibitively computationally expensive.

6. Conclusions

We have shown the initial results obtained with our new combined PIC-MHD code. For a parallel shock we have recovered the results obtained with several models ranging from combined PIC-MHD, through PIC-hybrid, to Vlasov-Fokker-Planck. In the case of the near-perpendicular shock model, the computational efficiency of the combined PIC-MHD method has allowed us to increase the scale of the simulations, which has led to a new result, showing that particle

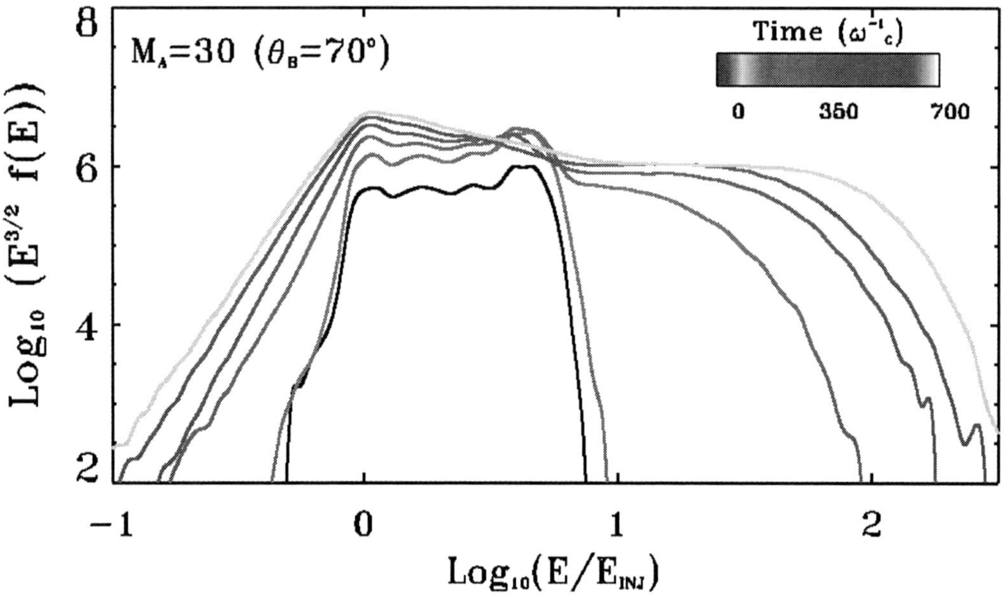

Figure 4. Similar to Fig. 2 but for the simulation shown in Fig. 3. Initially, the particles show only limited acceleration as they circle back and forth across the shock. In the later stages a plateau corresponding to Fermi-acceleration appears.

acceleration can occur even in near-oblique shocks. A full description of our code and results will soon be published and has appeared in pre-print [14].

Acknowledgements

We are grateful for fruitful discussions with: A.R. Bell, C.A. Bret, A. Bykov, M. Dieckmann, E. D'humières, M. Grech, L. Gremillet, R. Keppens, B. Lembège, M. Lemoine, G.Pelletier, I. Plotnikov, V. Tikhonchuk. This work is supported by the ANR-14-CE33-0019 MACH project. We would also like to thank D. Caprioli and B. Reville for helpful comments. This work acknowledges financial support from the UnivEarthS Labex program at Sorbonne Paris Cité (ANR-10-LABX-0023 and ANR-11-IDEX-0005-02). This work was granted access to HPC resources of CINES under the allocation A0020410126 made by GENCI (Grand Equipement National de Calcul Intensif).

References

[1] Marcowith A, Bret A, Bykov A, Dieckman M E, O'C Drury L, Lembège B, Lemoine M, Morlino G, Murphy G, Pelletier G, Plotnikov I, Reville B, Riquelme M, Sironi L and Stockem Novo A 2016 *Reports on Progress in Physics* **79** 046901 (*Preprint* `1604.00318`)
[2] Bai X N, Caprioli D, Sironi L and Spitkovsky A 2015 *ApJ* **809** 55 (*Preprint* `1412.1087`)
[3] Reville B and Bell A R 2012 *MNRAS* **419** 2433–2440 (*Preprint* `1109.5690`)
[4] Reville B and Bell A R 2013 *MNRAS* **430** 2873–2884 (*Preprint* `1301.3173`)
[5] Caprioli D and Spitkovsky A 2013 *ApJl* **765** L20 (*Preprint* `1211.6765`)
[6] Caprioli D and Spitkovsky A 2014 *ApJ* **783** 91 (*Preprint* `1310.2943`)
[7] Caprioli D and Spitkovsky A 2014 *ApJ* **794** 46 (*Preprint* `1401.7679`)
[8] van der Holst B, Keppens R and Meliani Z 2008 *Comp. Phys. Commun.* **179** 617–627
[9] Shephard M S and Georges M K 1991 *International Journal for Numerical Methods in Engineering* **32** 709–749

[10] Birdsall C K and Langdon A B 1991 *Plasma Physics via Computer Simulation*
[11] Balsara D S and Spicer D S 1999 *Journal of Computational Physics* **149** 270–292
[12] Bell A R 2004 *MNRAS* **353** 550–558
[13] Decker R B 1988 *SSR* **48** 195–262
[14] van Marle A J, Casse F and Marcowith A 2017 *ArXiv e-prints* (*Preprint* 1709.08482)

IOP Conf. Series: Journal of Physics: Conf. Series **1031** (2018) 012012 doi:10.1088/1742-6596/1031/1/012012

Exploring the Kipenhahn-Schlüter model

Carlos D. Vigh,[123] **Rafael Gonzalez,**[4] **and Diego Rial**[5]

[1]Instituto de Ciencias, Universidad Nacional de General Sarmiento, Los Polvorines, Buenos Aires, Argentina;
[2]Universidad de Buenos Aires. Facultad de Ciencias Exactas y Naturales. Departamento de Física. Buenos Aires, Argentina;
[3]CONICET-Universidad de Buenos Aires. Instituto de Física del Plasma (INFIP). Buenos Aires, Argentina;
[4]Instituto de Desarrollo Humano, Universidad Nacional de General Sarmiento, Los Polvorines, Buenos Aires, Argentina;
[5]Universidad de Buenos Aires. Facultad de Ciencias Exactas y Naturales. Departamento de Matemática. Buenos Aires, Argentina;

E-mail: `cvigh@ungs.edu.ar`

Abstract. We explored the generalised Kippenhan-Schlüter model non-isothermal stationary states for solar prominences where we made some improvements. The most important is to use recent observed values of radiative losses to build the piecewise heat balance equation. We explicity found several stationary states without and with magnetic shear solutions for the system of equations in the beta plasma range between 0 and 1,6. Finally, we computed magnetic streamlines field for the different shears. Joining all these elements we obtained a catalogue to study the stability of the prominence candidates.

1. Introduction

Quiescent solar prominences are structures extended into the solar corona, are denser and colder than coronal neighborhood. They have a small width compared to their length and height. Also, they have lifetimes from minutes to months and present relative small velocity flows. Therefore it implies that structures must be close to energy balance equilibrium [1]. The magnetic field is one of the agents of thermal isolation respect its environment [2]. Costa et al. [3] made an stability analysis using thermodynamic irreversible energy principles for different solar ambients (For examples, see [4] and [5]). The goal in that work was extend and generalize the MHD energy principle of Bernstein [6] using a general procedure considering irreversible thermodynamics processes where Bernstein's principle is a particular case. This method includes dissipation and it shows explicitly the contributions of radiation, heat mechanisms and magnetic energy (see equations 36 and 37 in [3]). Applied this method to solar prominences shows an alternative point of view to understand which mechanisms participates into its developing and stability structure. Recently, several works used the isothermal Kippenhan-Schlüter (KS) model [7] to study the prominence dynamics where were performed numerical simulations and compared with observations. (To see some examples [8], [9], [10], [11] and [12]) In particular, Costa et al [3] used this general energy principle onto the isothermal KS model to study oscillation wave modes. In this way is possible to identify which is the relative importance between different sources like radiation and thermal heating and also is possible to compare with observational data periods corresponding to typical oscillation of the cromosphere and photosphere which can be explained as internal prominence modes. A natural extension of that work is to examine the non-isothermal case, *i.e.* the Generalized Kippenhan-Schlüter model (KS-G). The KS-G model has the advantage to bring solutions into a wide range of β plasma. This variation of β allows to determine the relevance of the different contributions to the prominence stability and its internal wave modes *i.e. magnetic energy, thermal heating and radiative losses*, comparing for example, with force-free models (low β range) where is most difficult to identify it. To do this, it is necessary to recompute previous results of this simple model upgrading and revalidating the dynamical solutions of KS-G model for quiescent prominences. In that way, in a next work, will be possible to study stability and modal structure corresponding to perturbations from the equilibria using the appointed sources before. According the beta plasma definition, Milne et al. [13], [14] found that in the range 0 and 1.6 exists stationary solutions for different heating functions models. For a better description we took sixteen regions cooling of radiative values from recent works [15] and recompute profiles of temperature, density, pressure and magnetic fields. We extend the analysis to the whole range $-1 \leq x \leq 1$ checking the symmetry of the solutions. Finally, we computed the magnetic field streamlines. This is a working progress where the next stage of our project will be recompute the variational principle using the collection of solutions obtained here to generalize the aforementioned study.

1.1. Kippenhahn-Schlüter Model

Following Milne et al. [13] we worked onto the extension of Kippenhahn-Schlüter model taking into account thermal effects. For convenience we used the set of equations in adimensional form. Firstly, the equations that describes balance between hydrostatic, gas and magnetic pressure in a stationary regime are

$$\beta \frac{dp}{dx} = -2B_z \frac{dB_z}{dx} \tag{1}$$

$$\frac{dB_z}{dx} = \frac{1}{2}\beta\rho \tag{2}$$

where $\beta = 2\mu p_1/B_0^2$ is the "beta" plasma, being the characteristical pressure is $p_1 = 1.67 \times 10^{-13}\mathrm{Kg/m^3}$. μ is the magnetic permeability. B_0 is the characteristical magnetic field in the x coordinate, p is the gas pressure and B_z the magnetic field in the z direction as shown in Figure 1. Pressure p and temperature T are related by the state equation of an ideal gas

$$p = \rho T \tag{3}$$

The heat balance between mechanical heating, radiation losses and thermal conductivity is

$$\frac{d}{dx}\left(\frac{T^{5/2}}{B^2}\frac{dT}{dx}\right) = C\rho^2 T^\alpha - C_1(T)\rho \tag{4}$$

Finally, the magnetic field is:

$$B^2 = 1 + B_y^2 + B_z^2 \tag{5}$$

where B is the magnetic field in cartesian coordinates, using as normalization $B_x = B_0$. By other way x is the normalized position, being $x = 0$ at the prominence center and $x = 1$ its external boundary in the solar corona. ρ, is the gas density and T the temperature.
The constant C is defined like:

$$C = \frac{\chi\, p_1^2\, T_1^{\alpha-3.5}}{g^2\, \kappa} \tag{6}$$

IOP Conf. Series: Journal of Physics: Conf. Series **1031** (2018) 012012 doi:10.1088/1742-6596/1031/1/012012

Table 1. Values of radiative losses from CHIANTI using sixteen cooling regions [15]

Temperature (K)	χ	α
$T < 1.26 \times 10^4$	2.02×10^{-15}	8.06
$1.26 \times 10^4 \leq T < 1.58 \times 10^4$	5.60×10^{-2}	4.78
$1.58 \times 10^4 \leq T < 2.51 \times 10^4$	1.36×10^{24}	−1.26
$2.51 \times 10^4 \leq T < 3.16 \times 10^4$	1.46×10^{17}	0.32
$3.16 \times 10^4 \leq T < 7.9 \times 10^4$	3.11×10^{11}	1.58
$7.9 \times 10^4 \leq T < 1.0 \times 10^5$	4.44×10^{16}	0.53
$1.0 \times 10^5 \leq T < 1.25 \times 10^5$	2.31×10^{20}	−0.22
$1.25 \times 10^5 \leq T < 2.0 \times 10^5$	1.44×10^{17}	0.41
$2.0 \times 10^5 \leq T < 2.51 \times 10^5$	1.20×10^{19}	0.05
$2.51 \times 10^5 \leq T < 3.98 \times 10^5$	2.02×10^{27}	−1.47
$3.98 \times 10^5 \leq T < 7.94 \times 10^5$	6.38×10^{17}	0.22
$7.94 \times 10^5 \leq T < 1.0 \times 10^6$	1.40×10^{19}	0.0
$1.0 \times 10^6 \leq T < 2.0 \times 10^6$	1.26×10^{24}	−0.82
$2.0 \times 10^6 \leq T < 3.98 \times 10^6$	4.14×10^{28}	−1.54
$3.98 \times 10^6 \leq T < 1.0 \times 10^7$	7.74×10^{16}	0.23
$1.0 \times 10^7 \leq T < 3.16 \times 10^7$	2.06×10^{25}	−0.98
$T < 3.16 \times 10^7$	3.20×10^{16}	0.20

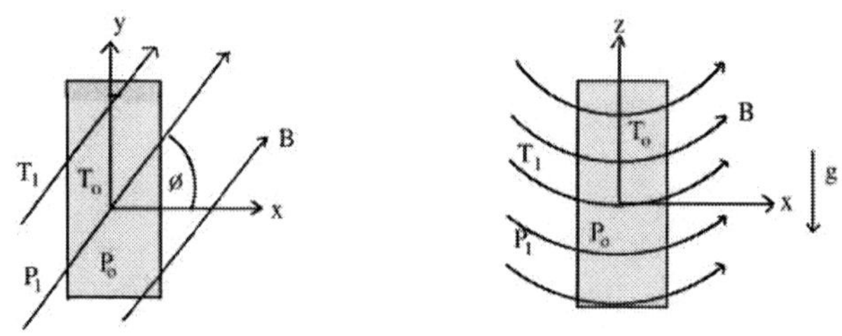

Figure 1. Schematic representation of a prominence. The x-axis is normal to prominence sheet, while the y-axis runs along the length of the prominence. The quantity ϕ is the shear angle between the prominence normal and the horizontal magnetic field. Values at the center are denoted by the subscript 0, while in the external corona are denoted by the subscript 1, see [13].

where T_1 is the characteristical temperature $T_1 = 2 \times 10^6$K, the gravity $g = 274\text{m/s}^2$ and the thermal conductivity $\kappa = 3 \times 10^{-11}$. C is a temperature piecewise function and C_1 is:

$$C_1(T) = \begin{cases} 0 & \text{si } T < 0.1; \\ C(T = p = x = 1) & \text{si } T \geq 0.1. \end{cases}$$

We defined a zero heat region near $x \sim 0$, the center prominence and in $x = 1$ we assured no-heat exchange with the external environment. The values of χ and α will be different in each region as shown in the Table 1. This is the most relevant improvement respect the work of Milne et al [13], where they used the Hildner's six regions cooling [16].

1.2. Boundary conditions

To solve the system is necesary to specify β and B_z and also four boundary conditions, the first two are according to maintain the symmetry as shown in Figure 1:

$$B_z = 0 \qquad at\ x = 0 \tag{7}$$

$$\frac{dT}{dx} = 0 \qquad at\ x = 0 \tag{8}$$

The next two conditions are stablished to assure the continuity with the coronal values

$$\rho = 1 \qquad at\ x = 1 \tag{9}$$

$$T = 1 \qquad at\ x = 1 \tag{10}$$

In this way, the shear angle is directly:

$$\Phi = arctg(B_y) \tag{11}$$

2. Semi-analytical solutions

The system equations 1 to 4 constitutes a two-point boundary-value problem, and it has not got analytical solution. Then, it is necessary to solve it numerically. Also, the character of piecewise functions $C_1(T)$ and $\alpha(T)$ inserts into equation 4 strong gradients between cooling regions being a source of numerical errors. Another source of troubles is that the model we are using is not valid beyond $x = 1$ where vertical variations become important [13]. This system could be solved via Runge-Kutta method building an elementary code or using numerical solvers tools for ODE equations [17]. The boundary conditions are fullfiled by shooting, it means, fixing ρ_0 and T_0 at $x = 0$ and varying β and B_y until satisfy the boundary conditions at $x = 1$. This strategy is easy and fast to obtain the collection of solutions that we are searching for a theoretical comprehension explained in the Introduction. Increase a 3D model requires a more sofisticated tool. In this way it is possible to obtain *temperature, magnetic field, pressure and density.* Also, compare and check it with semi-analytical expressions it is important (see equations 11 and 12 in [13]). Therefore, we took the temperature profile from the numerical solution and we use it as input for the following analytical expressions.

$$\frac{dB_z}{dx} - \frac{1}{2}\frac{\beta p}{T} = 0 \tag{12}$$

$$\frac{d}{dx}\left(T\frac{dB_z}{dx}\right) + \frac{d}{dx}\left(\frac{1}{2}B_z^2\right) = 0 \tag{13}$$

Then, the solutions are:

$$B_z(x) = \sqrt{\beta p_0}\ tanh\left[\frac{1}{2}\ell(x)\sqrt{\beta p_0}\right] \tag{14}$$

$$p(x) = p_0\ sech^2\left[\frac{1}{2}\ell(x)\sqrt{\beta p_0}\right] \tag{15}$$

where

$$\ell(x) = \int_0^x \frac{dy}{T(y)} \tag{16}$$

If $T = 1$ we recover the Kipenhahn-Schlüter model (1957).

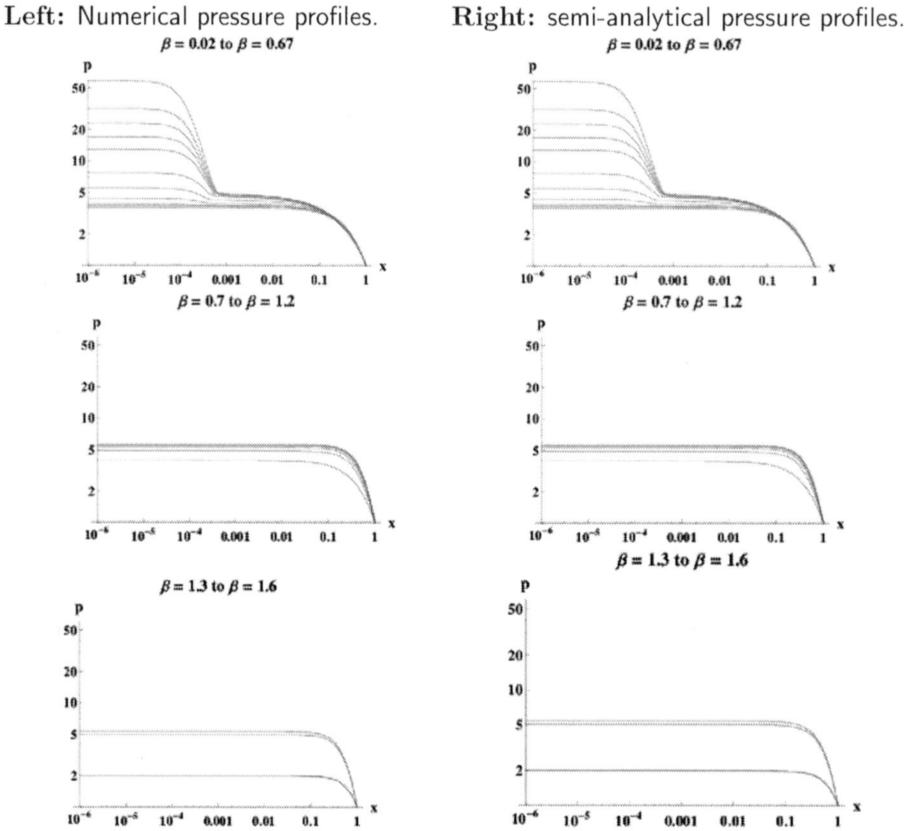

Figure 2. Top: β is in the range $[0.02 - 0.67]$, inferior curves, lower β, upper curves, higher values. Experiments are setted as β= 0.02; 0.04; 0.06; 0.08; 0.1; 0.2; 0.3; 0.4; 0.5; 0.6 y 0.67 from bottom to top curves. **Medium:** β= 0.7; 0.8; 0.9; 1.0; 1.1 y 1.2. The qualitative change is for $\beta = 0.7$, values greater that 0.8, tends to isothermic behaviour. **Bottom:** β: 1.3; 1.4; 1.5 and 1.6.

3. Checking consistency

We checked numerical and semi-analytical solutions comparing between different β, here we present the pressure evolution. Also for convenience, the profiles are in the range $0 \leq x \leq 1$, but the solutions are completely symmetric in the range $-1 \leq x \leq 0$.

3.1. Numerical setting

In Table 2 we showed the input values to set the experiments and the outputs in the boundaries. If we divided the values of T_0 in cold ($T \leq 0.1$) and hot solutions ($T \geq 0.1$), we noted that for higher shears the stationary states corresponds to hot solutions.

3.2. Pressure Evolution

We divided the solutions in three ranges, at the top of Figure 2 solutions for $0.02 < \beta < 0.67$, at medium of Figure 2 for $0.7 < \beta < 1.2$, and at the bottom for $1.3 < \beta < 1.6$. We noted a qualitative change of behaviour around $\beta = 0.67$ for higher values the profiles tends to isothermal solutions.

IOP Conf. Series: Journal of Physics: Conf. Series **1031** (2018) 012012 doi:10.1088/1742-6596/1031/1/012012

Table 2. Input and outputs values of the numerical experiments.

By	β	T_0	ρ_0	$\rho(1)$	$T(1)$	$Bz(1)$
0	0.02	0.000081	58.4	1.00684	1.00611	1.07138
0	0.6	0.00136	3.595	1.00125	1.00195	1.2475
0	1.6	0.999999	2.000	0.98292	1.00326	1.27567
0.5	0.025	0.000089	42.12	1.00784	0.99481	1.01387
0.5	0.6	0.001045	3.593	1.00093	1.00997	1.24709
0.5	1.6	0.999	2.12	1.00245	1.00394	1.33719
5.8	0.025	0.000175	1.2	1.00555	1.02614	0.0697222
5.8	0.6	0.47	1.51	1.00443	1.00085	0.550763
5.8	1.6	0.935	2.3	1.00233	1.05758	1.44093
10	0.025	0.54	1.03	1.018	1.005	0.017
10	0.6	0.73	1.31	1.007	1.011	0.426
10	1.6	0.95	1.9	0.985	1.118	1.21

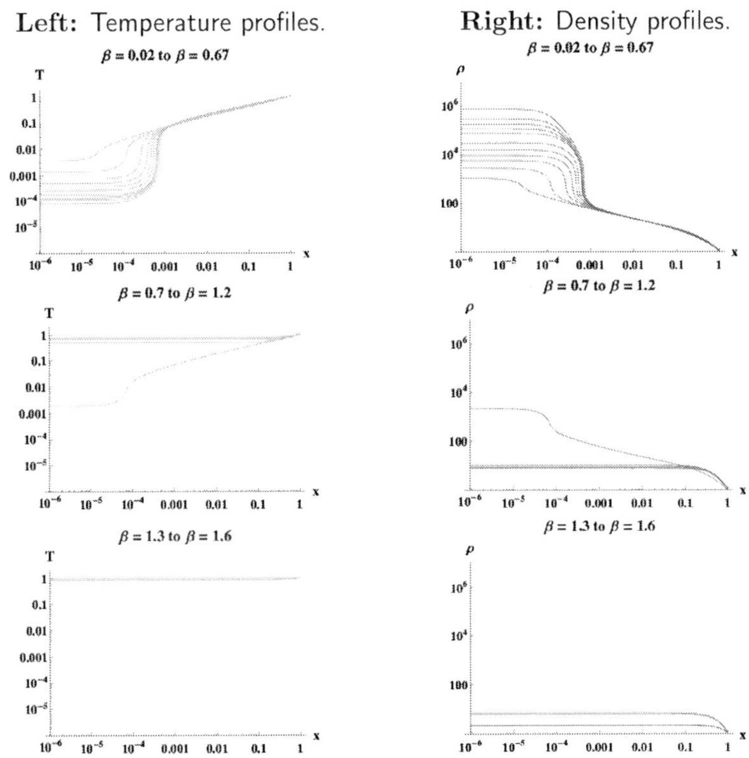

Figure 3. Top: left, temperature and right, density. β is in the range $[0.02 - 0.67]$, inferior curves, lower β, upper curves, higher values. Experiments are setted as β= 0.02; 0.04; 0.06; 0.08; 0.1; 0.2; 0.3; 0.4; 0.5; 0.6 y 0.67 from bottom to top curves. **Medium:** β= 0.7; 0.8; 0.9; 1.0; 1.1 y 1.2. The qualitative change is for $\beta = 0.7$, values greater that 0.8, tends to isothermic behaviour. **Bottom:** β: 1.3; 1.4; 1.5 and 1.6.

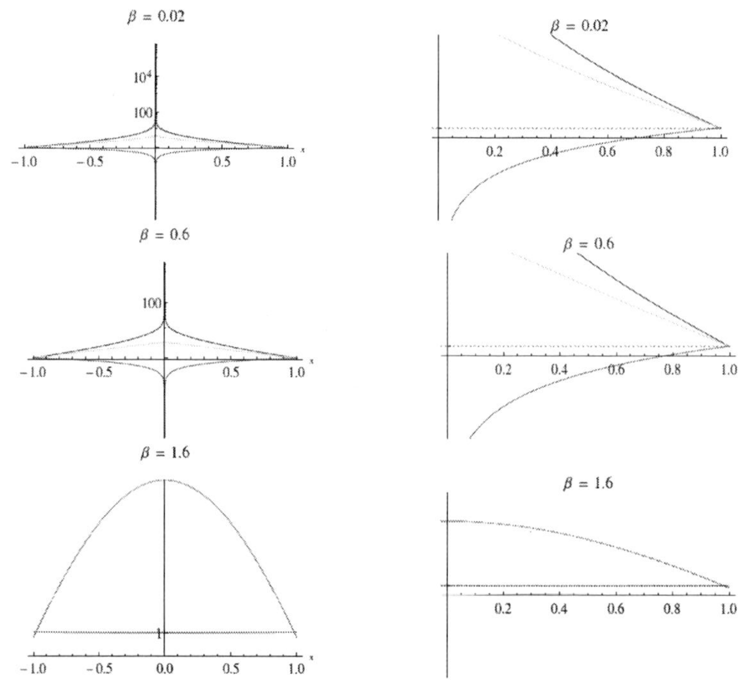

Figure 4. Left: Density (blue), pressure (green) and temperature (red) profiles.**Right:** Same magnitudes near $x = 1$.

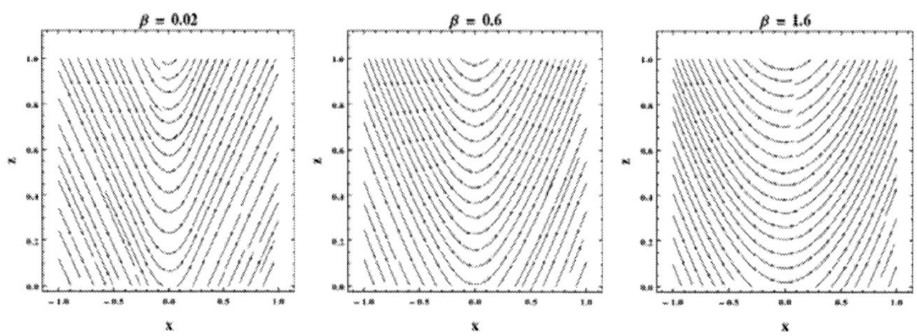

Figure 5. Magnetic field streamlines with $B_y = 0, \Phi = 0$

3.3. Profiles of Temperature and Density

In Figure 2 we checked that the semi-analytical and numerical solutions have a good degree of coincidence. In the Figure 3 we showed the corresponding profiles of temperature and density for different values of β. Here is possible to see explicitly that the solutions are consistent with the fact that prominences are denser and closer in its interior.

3.4. Checking the state equation of an ideal gas

Last quality test is to check boundary conditions and the state equation for an ideal gas. Here we selected for $\beta = 0.02, 0.6$ and 1.6; we showed for the whole range for values of x in the Figure 4 (left) and near the boundaries (right) where the boundary conditions are fulfilled numerically.

IOP Conf. Series: Journal of Physics: Conf. Series **1031** (2018) 012012 doi:10.1088/1742-6596/1031/1/012012

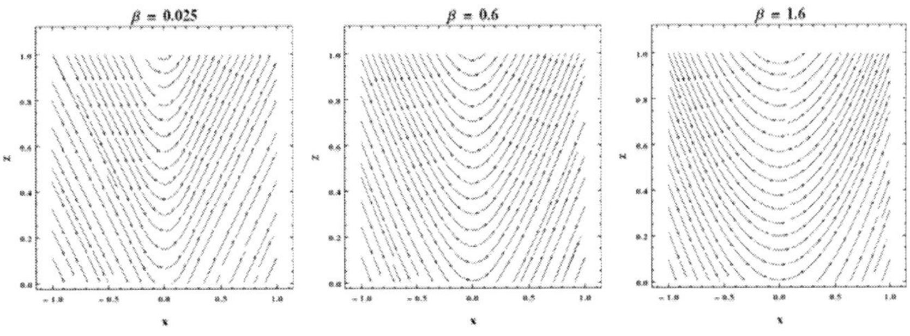

Figure 6. Magnetic field streamlines with $B_y = 0.5, \Phi = 28^o$

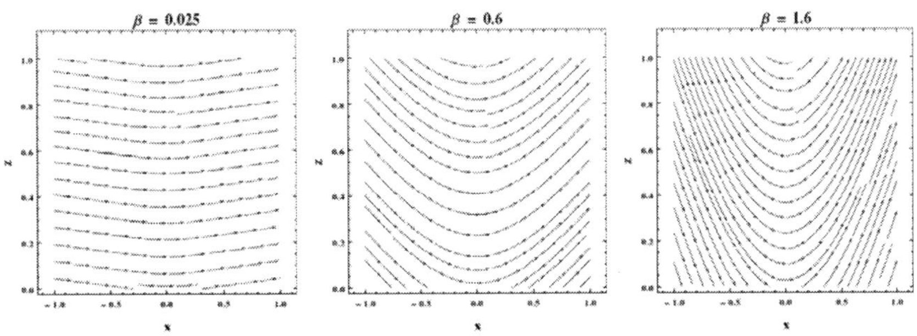

Figure 7. Magnetic field streamlines with $B_y = 5.8, \Phi = 80^o$

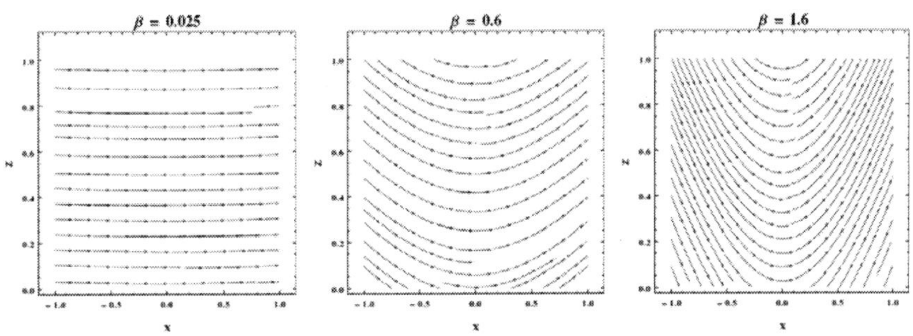

Figure 8. Magnetic field streamlines with $B_y = 10, \Phi = 84^o$

4. Magnetic field streamlines for non-isothermal cases

We selected for three values of β different magnetic shears to see how to affect the magnetic field streamlines. $\Phi = \{0^o, 28^o, 80^o, 84^o\}$ which corresponds respectively to $B_y = \{0, 0.5, 5.8, 10\}$ The Figure 5 is the reference case without shear, the behaviour of streamlines increasing β is qualitative similar. However, in the Figures 6, 7 and 8, we can see when the shears increase, the streamlines tends to alligned into direction $\hat{y}$. The case $\beta = 0.025$ is the most sensitive to change with shear. For $\beta = 0.6$, we have change with shear too, but is lower respect to $\beta = 0.025$. Finally, for $\beta = 1.6$ the effect is weak comparing the another two cases.

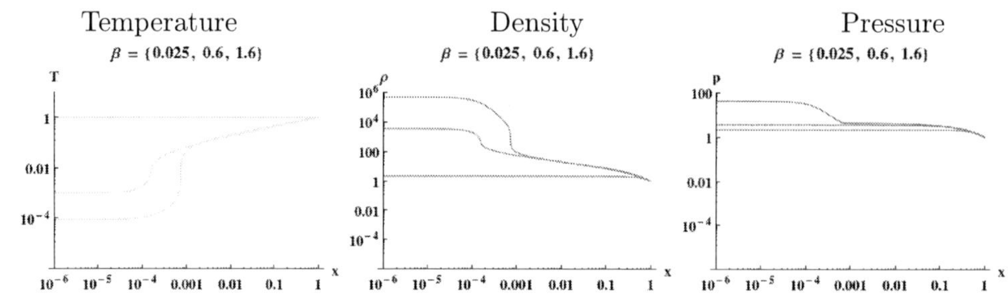

Figure 9. Case $B_y = 0.5, \Phi = 28^o$. For the temperature lower curve is for $\beta = 0.025$, intermediate curve for $\beta = 0.6$ and upper one for $\beta = 1.6$. For density and pressure the curves are upper $\beta = 0.025$, intermediate $\beta = 0.6$ and lower $\beta = 1.6$.

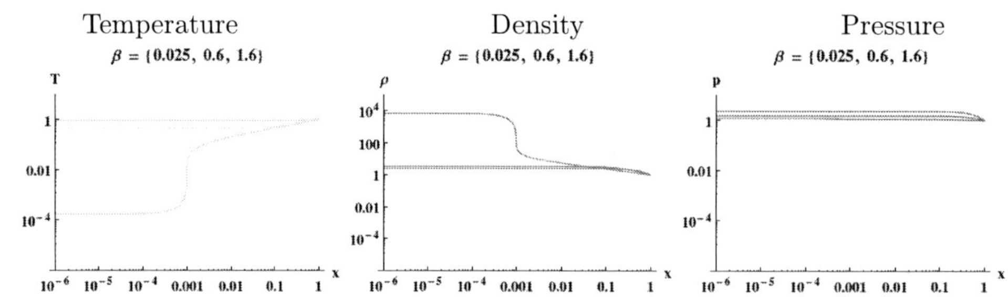

Figure 10. Case $B_y = 5.8, \Phi = 80^o$. For the temperature lower curve is for $\beta = 0.025$, intermediate curve for $\beta = 0.6$ and upper one for $\beta = 1.6$. For density and pressure the curves are upper $\beta = 0.025$, intermediate $\beta = 0.6$ and lower $\beta = 1.6$.

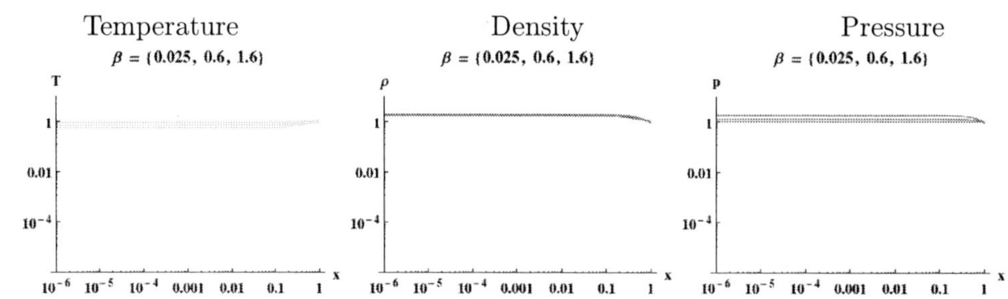

Figure 11. Case $B_y = 10, \Phi = 84^o$. For the temperature lower curve is for $\beta = 0.025$, intermediate curve for $\beta = 0.6$ and upper one for $\beta = 1.6$. For density and pressure the curves are upper $\beta = 0.025$, intermediate $\beta = 0.6$ and lower $\beta = 1.6$.

4.1. Temperature, density and pressure profiles for different shears

In this paragraph we saw how the shear affects temperature, density and pressure profiles for the same values of β. The case $B_y = 0$ is shown in subsection 3.4. From Figures 9, 10 and 11 is clear that for $\beta = 0.025$ is the most sensitive for the change of shear. More specifically when increase the shear the system has a tendency to be isothermic.

5. Conclusions and next steps

In this work we explored for Kippenhan-Schlüter model extended to non-isothermal stationary states. We upgraded the piecewise heat equation according the values of Table 1. For these reason we used recent observed values to improve the heat equation subdividing in sixteen cooling regions in place to six cooling regions due to Hildner's values. Before to make the numerical experiments for several cases we checked consistency between analytical and semi-analytical solutions. After that, we began to analyse the behaviour of different profiles for several values of β, in this way we explicitly showed temperature, density and pressure profiles. We found several stationary states without and with magnetic shear solutions for the system of equations in the β range between 0 and 1.6 which corresponds to solar prominences. Beyond this value of β we can not find stationary solutions which satisfies boundary conditions in the solar corona.

After that, we selected for $\beta = 0.025, 0.6$ and 1.6 cases to study the behaviour of magnetic field streamlines for different magnetic shears. In all cases the increasing of shear modifies the streamlines tends to allign in the y direction. The case $\beta = 0.025$ is the most sensitive with the change of shear.We obtain non-isothermal solutions for values of β lower than 0.7, meanwhile for higher values ($\beta > 0.7$) solutions corresponds to Kippenhan-Schlüter isothermic model and it is independent of the shear.

By another way, if B_y component of magnetic field is higher than B_x the system will have a tendency to be isothermic, independently of the values of β.

This work is a first step to perform a catalogue to analyse the stability of this kind of structures using variational principles. Joining all these elements we have a catalogue to study in a next stage of our project the stability of the prominence candidates in this minimalistic unidimensional context.

Acknowledgments

CV is grateful for partial financial support of Instituto de Ciencias, UNGS, Los Polvorines.

References

[1] Anzer U and Heinzel P 1999 *Astron.Astrophys.* **349** 974–984
[2] Galindo-Trejo J 2006 *RMxAA* **42** 89–98
[3] Costa A, González R and Sicardi-Scifino A 2004 *Astron.Astrophys.* **427** 353–361
[4] Costa A and González R 2006 *Astron.Astrophys.* **458** 953–963
[5] Costa A and González R 2008 *Astron.Astrophys.* **489** 755–762
[6] Bernstein I, Frieman E, Kruskal M and Kulsrud R 1958 *Proc.Roy.Soc A* **244** 17
[7] Kippenhahn R and Schlüter A 1957 *Zs.Ap.* **43** 36
[8] Hillier A, Isobe H, Shibata K and Berger T 2011 MHD simulations of quiescent prominence upflows in the Kippenhahn-Schlüter prominence model *Astronomical Society of India Conference Series* (*Astronomical Society of India Conference Series* vol 2) ed Chouduri A and Banerjee D pp 331–336
[9] Petrie G and Low B The internal structures and dynamics of solar quiescent prominences *Astronomical Society of the Pacific Conference Series* Astronomical Society of the Pacific Conference Series
[10] Low B and Petrie G 2005 *ApJ* **626** 551–562
[11] Hillier A, Shibata K and Isobe H 2010 *Publ.Astron.Soc.Japan* **62** 1231–1237
[12] Hillier A, Isobe H, Berger T and Shibata K 2012 *ApJ* **746** 120
[13] Milne A, Priest E and Roberts B 1979 *ApJ* **232** 304
[14] Priest E 1982 *Solar Magnetohydrodynamics* 4th ed (*Geophysics and Astrophysics Monographs* no 21) (Dordrecht, Holland: Kluwer Academic Publishers)
[15] Soler R, Ballester J and Parenti S 2012 *Astron.Astrophys.* **540** id.A7,6pp.
[16] Hildner E 1974 *Solar Phys.* **35** 123
[17] Wolfram 2014 *Mathematica* version 10.0 ed (Champaign, Illinois: Wolfram Research, Inc)

IOP Conf. Series: Journal of Physics: Conf. Series **1031** (2018) 012013 doi:10.1088/1742-6596/1031/1/012013

Numerical simulation of DA white dwarf surface convection

[1]F. Zaussinger, [2]F. Kupka, [3]M. Montgomery, [1]Ch. Egbers

[1] BTU Cottbus-Senftenberg, Departement for Aerodynamics and Fluid Mechanics, Cottbus, Germany

[2] Institute for Astrophysics, University of Göttingen, Friedrich-Hund-Platz 1, 37077 Göttingen, Germany

[3] Department of Astronomy and McDonald Observatory, University of Texas at Austin, Austin, TX, 78712, USA

E-mail: florian.zaussinger@b-tu.de

Abstract. White dwarfs are compact objects with masses comparable to our Sun, but a radius similar to our Earth. They are the final evolutionary stage for about 95% of all stars in the Galaxy, i.e., for all stars that have a final mass less than the Chandrasekhar mass (about 1.4 times the solar mass), the upper mass limit for which hydrostatic equilibrium can be maintained by the degeneracy pressure of electrons at very high densities. The outermost shell of most white dwarfs contains a convective layer. Even if the latter is very thin ($\lesssim$ 10 km), it is important for mixing properties, observed radiation, and pulsational stability of the whole object. During a long phase white dwarfs have effective temperatures $T_{\rm eff}$ of about 10000 K $\sim$ 14000 K, since the time scale to reach such temperatures by cooling is already $\approx 10^9$ years. Here, we focus on DA (hydrogen-rich) white dwarfs with $T_{\rm eff} \approx 12000$ K. This is at the transition from shallow to deep convection zones. Due to very high gravitational acceleration ($\sim 10^6$ g at the surface) the material is overturned about five times per second over the distance of a few kilometers. Numerical simulations of such objects have to be done for a compressible flow and feature highly turbulent granules at the surface, which are qualitatively comparable to the convection cells observed at the surface of the Sun. For this study we compare three white dwarf surface simulations with realistic microphysical properties and full 3D radiative transport. The simulations differ in effective temperature, namely, $T_{\rm eff}$ = 11800 K, 12100 K, and 12400 K. A statistical analysis of the convective processes as function of $T_{\rm eff}$ is presented.

1. Introduction

About 20 years ago a first effort in performing numerical hydrodynamical simulations of the surface layers of DA type white dwarfs was made, [1]. It had to be limited to just two spatial dimensions and rather low numerical resolution, with typically less than 100 grid points per direction. After having constructed a new code (CO5BOLD, for a review see [2]), the same group performed numerical hydrodynamical simulations for these objects in three spatial dimensions, [11]. With a resolution of 150 grid points per direction these simulations aimed at constructing a grid of models as a function of effective temperature $T_{\rm eff}$ and surface gravity to investigate problems of observational spectroscopy, [11, 12]. More recently, these simulations have also been used to investigate the physics of overshooting underneath the convection zone and calibrate simple convection models and estimates of convection induced mixing, [10]. The present numerical simulations differ from previous work, among others, in the following aspects. First of all, ANTARES [9, 13, 4] has a much lower intrinsic numerical viscosity (fifth order WENO) than the numerical schemes used in actual runs with CO5BOLD (typically second order with shock and subgrid scale viscosity). Secondly, we suggest simulations to be performed which cover a much deeper and wider domain with typically around 400 grid points per direction. Together these two features should allow a more accurate quantification of convective mixing and convective overshooting and more reliable statistical analyses. Thirdly, the grid refinement capability of ANTARES optionally allows a fully turbulent flow simulation for limited areas of the simulation domain. Previous work had to remain limited to simulations of a laminar flow at the grid scale with turbulence represented only through subgrid scale modelling, [6]. Finally, since ANTARES is parallelized according to the MPI paradigm (and some parts in a hybrid parallelization based on both MPI and OpenMP), much larger simulations on supercomputers based on more varied hardware are possible compared to a code parallelized through OpenMP only. Thus, ANTARES is ideally suited for the specific challenges posed by the numerical simulations on super computer clusters required to be done to gain further insight into the physics of overshooting and mixing in DA white dwarf stars.

1.1. Settling of heavy elements and overshooting

Due to their high surface gravities diffusion of elements in the upper part of white dwarf envelopes occurs on short time scales. DA white dwarfs have a residual amount of hydrogen which after a phase of gravitational settling "floats" on top of fluid consisting of heavier elements (helium around a core chiefly made of carbon and oxygen). Hence, their atmospheres consist of nearly pure hydrogen. Indeed, for a non-convective atmosphere calculations show that heavy elements like Ca or Mg should settle out of the photosphere in one or two weeks. This is important because actually Ca and Mg are observed in the photospheres of up to 25% of all DA WDs [3, 5]. It has been suggested that these elements are continuously accreted. To calculate accretion rates the settling rates have to be known in turn. This is straightforward to do if the star is not convective, but if there is a surface convection zone, the relevant settling rate is the rate at the base of the convection zone. Since the important physical process in this context is mixing due to convection, the overshooting region located underneath the convectively unstable region is considered part of the convection zone for this calculation.

Obviously, how fast overshooting velocities decay with depth is important for this study, since even velocities much smaller than those inside the linearly convectively unstable region may be able to provide enough mixing so that settling cannot occur. This can have a large effect on the settling rate of heavy elements in a WD surface convection zone, and this in turn has a large effect on the inferred accretion rate of metals such as Mg or Ca. To measure the decay of overshooting, the presented simulations are analyzed statistically. Since the thermal stratification in the overshooting region can be estimated only inaccurately when constructing a starting model the relaxation time is a significant fraction of its associated thermal time scale. The analysis of our pre-calculated 2D simulations showed stable statistical results after a relaxation period of $\approx 100\,\mathrm{scrt}$, where a sound crossing time (scrt) is defined as the traveling time of a sound wave passing the vertical extent of the simulation box. Two dimensional pre-runs as described in [8] have not been performed for the two more recent simulations, since the relaxation period in 2D has to be limited to avoid the mean stratification to drift away from its 3D counterpart. Instead, 3D relaxation from the beginning and, as a faster alternative, 3D relaxation of a model simulated at lower resolution followed by an increase in resolution after initial relaxation have been investigated as alternatives.

IOP Conf. Series: Journal of Physics: Conf. Series **1031** (2018) 012013 doi:10.1088/1742-6596/1031/1/012013

2. Governing Equations

The ANTARES (A Numerical Tool for Astrophysical RESearch) code is designed to numerically solve the fluid dynamical equations. Among several options it can do so on an equidistant, three dimensional rectangular grid following the popular box-in-a-star ansatz (see Fig. 2). It can also perform compressible hydrodynamic simulations with radiative transfer in all three spatial dimensions. The source code is written in Fortran90 and parallelized in a hybrid way with MPI and OpenMP, [13]. To avoid high computational costs for global high resolution simulations, a local grid refinement is implemented, too. The order of spatial and temporal discretization can be chosen arbitrarily. However, a 5th order weighted essentially non-oscillatory scheme (WENO5) is favored for the mentioned type of simulations, [9]. This scheme calculates the boundary fluxes of the characteristic variables in their eigenspace and hence guarantees stable up-winding, even at high Mach numbers as they occur in the photosphere of a DA white dwarf. Turbulence is modeled with a Smagorinsky subgrid scale model with a constant of 0.2 and a turbulent Prandtl number of 0.72. The parallel output is realized with HDF5 which is based on MPI. Best parallel performance is expected when each core has to calculate about 200,000 grid points, which typically results in columns of 25×25 points horizontally and, e.g., 320 points vertically. The MPI-decomposition for simulations with radiative transfer is available only horizontally, since the solution process of the radiative transport equation (short characteristics method) cannot be parallelized vertically without massive overhead. The compressible formulation of the problem brings only local boundary communication between the columns, which affects the MPI overhead positively. Hence, no global communication is needed, except for some routines at the beginning of the calculation. The equations governing the dynamics of granulation and convection in the white dwarf stars are as follows:

$$\frac{\partial \rho}{\partial \mathrm{t}} + \nabla \cdot (\rho \vec{u}) = 0, \tag{1}$$

$$\frac{\partial \rho \vec{u}}{\partial \mathrm{t}} + \nabla \cdot (\rho \vec{u}\vec{u} + \mathrm{p}\mathbf{I}) = \nabla \cdot \tau + \rho \mathbf{g}, \tag{2}$$

$$\frac{\partial \rho \mathrm{E}}{\partial \mathrm{t}} + \nabla \cdot (\vec{u}(\mathrm{E} + \mathrm{p})) = \nabla \cdot (\tilde{\mathrm{u}} \cdot \tau) + \rho(\tilde{\mathrm{u}} \cdot \tilde{\mathrm{g}}) + \mathrm{Q}_{\mathrm{rad}}, \tag{3}$$

where, besides standard notation, the total (kinetic + internal) energy is E and the radiative heating and cooling rate is $\mathrm{Q}_{\mathrm{rad}}$. It can be calculated from the radiative energy flux F_ν, [9], according to where an optical depth of 1 is reached. Thus,

$$\mathrm{Q}_{\mathrm{rad}} = -\int_\nu (\nabla \cdot \mathrm{F}_\nu) \mathrm{d}\nu. \tag{4}$$

We are using slip boundary conditions at the bottom and Dirichlet boundaries for the density at the bottom, too. The energy at the radiative lower boundary is fed into the domain by a heat flux $\mathrm{F}_{\mathrm{rad,bottom}} = \sigma \mathrm{T}^4_{\mathrm{eff}}$. The top boundary conditions are open in the sense that fluid can leave and enter the domain but sound- and shock waves are absorbed. This avoids reflecting waves which would heat the fluid locally. The computational time is counted in sound crossing times (scrt), which measures the traveling time of a sound wave, passing the vertical distance. Typical extensions of such a simulation box are $8 - 15\,\mathrm{km}$ horizontally at a height of $6 - 8\,\mathrm{km}$. One scrt is about 0.2 seconds. A meaningful simulation — for statistical post-processing — needs at least 60 scrt in 3D or 12 seconds of physical time, respectively. The numerical time step is limited by the compressible CFL number which is set to 0.25. All three initial 1D stellar model envelopes for the different values of $\mathrm{T}_{\mathrm{eff}}$ are constructed from 1D stellar structure models (see [8] for details). The microphysical properties (equation of state) and input properties for the radiative transfer (opacities) are available as tabulated values (see [9, 8]).

The calculation of the radiative flux needs about 90% of the computational cost. To balance the performance of the entire solver towards the much faster hydro-part, the MPI-boundaries of the radiative flux are communicated from the last stage of the actual time step. This lowers the order of the numerical error, but is stable. Though, this error is small compared to the error obtained by the finite angular resolution. This procedure reduces the MPI overhead significantly and lets the solver scale up to 2048 cores with a scaling efficiency of about 90%. Communicating the latest radiative MPI boundaries stops efficient scaling at 512 cores, depending on the cluster architecture and the MPI domain size. However, the boundary error is only visible at the initialization of the simulations. Here, the MPI communication of F_{rad} may lead to a regular pattern depicting the MPI decomposition. In order to overcome this an adapted initial perturbation routine has been implemented. Now, the perturbation function has the same amount of modes as MPI cores in each direction and acts only in the upper convective zone at the maximum of the super-adiabatic peak. The initial perturbations a simple trigonometric function which is perturbed by a random noise. The noise itself lets the base function fluctuate at most 1%.

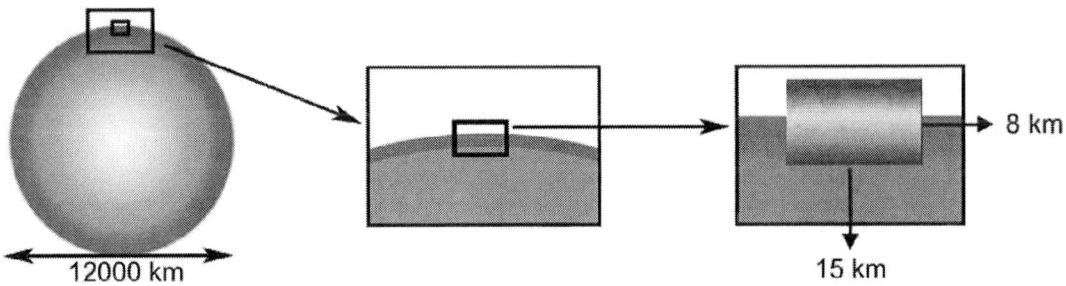

Figure 1. We use the box-in-a-star ansatz available in the ANTARES code suite for the surface convection zone simulation of a hydrogen rich white dwarf (DA). The typical extent of a three dimensional simulation box is 14–15 km. The depth ranges between 7–9 km.

3. Results

The case of $T_{eff} = 11800\,K$ has been investigated in detail in [8]. We here first focus on two more recent simulation runs.

3.1. Case study with $T_{eff} = 12100\,K$

The simulation case with $T_{eff} = 12100\,K$ is calculated in a box with a volume of $8.49 \times 14.2 \times 14.2\,km^3$. Here, the first value is the vertical extension and the following values are the horizontal extension. The maximum temperature at the bottom is $1.2 \cdot 10^5 K$. The grid resolution is set to $211 \times 461 \times 461$ points, respectively. The case was calculated for 63 scrt which corresponds to 16 seconds. Especially, the initial condition has been tested carefully. We found that deep vertical perturbations (> 4 km) lead to long lasting thermal fluctuations which yield to peaks in the radiative flux. It would take significant fractions of the thermal time scale of the simulation box to flatten these peaks (see [7, 8]).

3.2. Case study with $T_{eff} = 12400\,K$

The second case has a resolution of $211 \times 432 \times 432$ points, and a physical domain size of $8.45 \times 14.1 \times 14.1\,km^3$. Except for the bottom temperature and the larger total flux as specified by T_{eff}, and hence the resulting hydrostatic equilibrium stratification, the properties of the

IOP Conf. Series: Journal of Physics: Conf. Series **1031** (2018) 012013 doi:10.1088/1742-6596/1031/1/012013

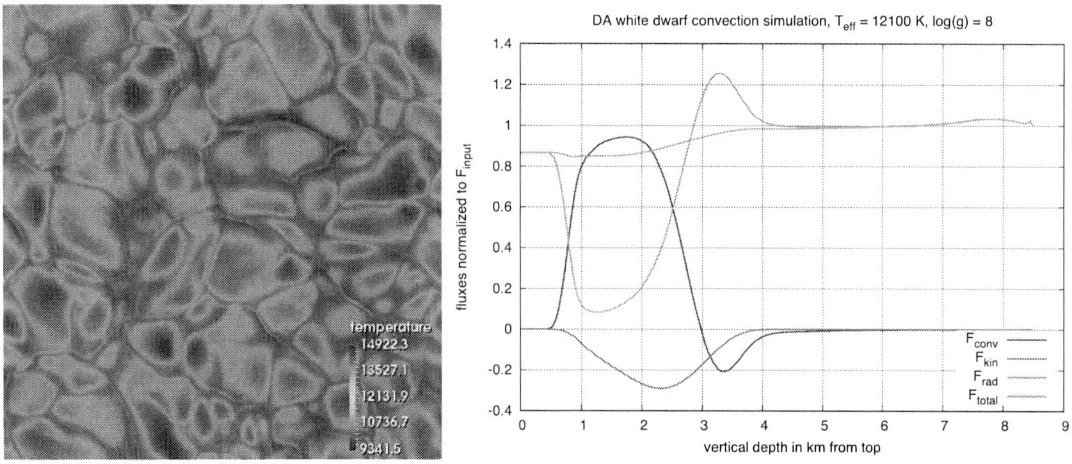

Figure 2. (left) Temperature field for a horizontal slice measured from the top at 520m where the horizontally averaged temperature is $T_{eff} = 12100\,K$. (right) Convective, kinetic, radiative and total flux normalized to the input flux $F_{input} = \sigma T_{eff}^4$ at the bottom.

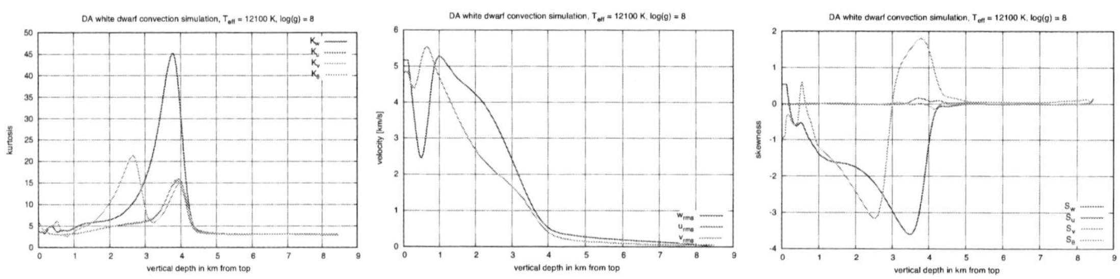

Figure 3. (left) kurtosis, (middle) velocity (rms), (right) skewness for $T_{eff} = 12100\,K$

simulation setup are identical to the case $T_{eff} = 12100\,K$. However, because the deepest layer, in which F_{rad} notably differs from the total flux, is much closer to the surface for a T_{eff} of 12400 K (compare Fig. 4 with Fig. 2), its associated Kelvin-Helmholtz time scale is also much smaller and indeed, just as explained in [7, 8], thermal relaxation is much faster for this model. Indeed, its total agrees with the input flux to within $\lesssim 2\%$ (see Fig. 4).

4. Conclusion

Comparing root mean square (rms) vertical and horizontal velocities from Fig. 2–5 for the cases of $T_{eff} = 12100\,K$ and 12400 K, respectively, with those of Fig. 3 and Fig. 6 to 9 from [8], it becomes clear that the two cooler models are rather similar in terms of convective efficiency, maximum velocity, profile of horizontal and vertical velocity as a function of depth, despite the model with $T_{eff} = 12100\,K$ is not yet sufficiently well relaxed. The hottest model with $T_{eff} = 12400\,K$ in turn is excellently relaxed. However, the region underneath what is called the wave-dominated region in [8], where $F_{conv} \to 0$ and $F_{rad} \to F_{total} \approx F_{input}$, is distinguished by much higher horizontal velocities than vertical ones. This is the opposite from what is found for the 12100 K and the 11800 K models. We suggest it is related to its lower convective efficiency (and hence lower maximum of F_{conv}, as the simulations are all deep with no indications for interaction with the bottom boundary over several pressure scale heights as a function of depth). If we compare the skewness of the different components of the velocity field (vertical

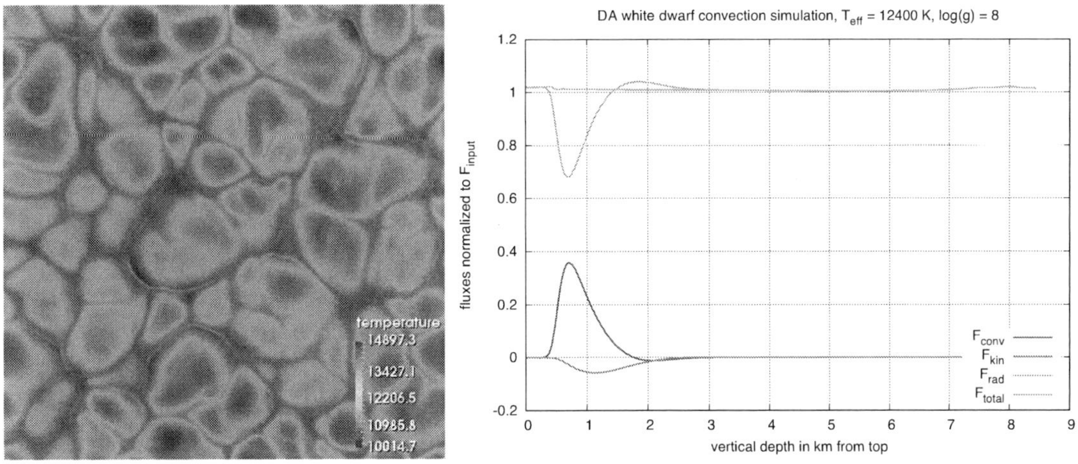

Figure 4. (left) Temperature field for a horizontal slice measured from the top at 520m where the horizontally averaged temperature is $T_{eff} = 12400\,K$. (right) Convective, kinetic, radiative and total flux normalized to the input flux $F_{input} = \sigma T_{eff}^4$ at the bottom.

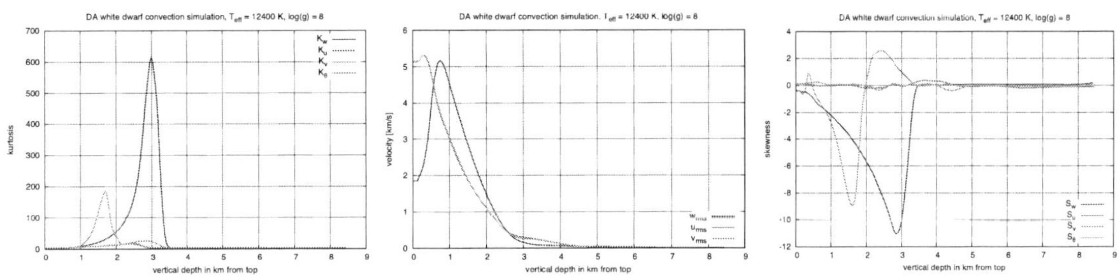

Figure 5. (left) kurtosis, (middle) velocity (rms), (right) skewness for $T_{eff} = 12400\,K$

one, w, as well as horizontal ones, u and v) and the temperature field (θ) from Fig. 3 and 5 with Fig. 11 from [8], we again find a close agreement between the two cooler models (with $T_{eff} = 11800\,K$ showing the larger absolute values). The hottest model features more extreme absolute values and a drop in S_w without a point of inflection. The same can be concluded for the kurtosis of these fields (Fig. 3 and 5 in this paper and Fig. 12 and 13 from [8]), but for one important difference: for the hottest model the horizontal velocity fields have their maxima much further inside. This has to be related to its smaller convective fluxes and also to the smaller drop in horizontal velocities. Thus, the basic structure of a convectively unstable region on top of a countergradient region, followed by a plume-dominated region where plumes are hotter than their surroundings and the convective flux is negative, and finally a wave-dominated region with essentially zero non-radiative heat flux is the same for all three models. However, the low efficiency (small Peclet number) of the hottest model apparently leads to a different behavior of the horizontal velocities in the wave-dominated region and most likely to differences in the kind of waves being formed there and the mixing efficiency. An extensive comparison with further simulation cases is currently in preparation.

Acknowledgments. FK is grateful for support by the Austrian Science Fund FWF projects P25229 and P29172. MHM gratefully acknowledges support from the United States Department of Energy under grant DE-SC0010623 and the National Science Foundation under grant AST-1312983 and AST-1707419. The simulations were performed mainly at the Northern German

Network for High-Performance Computing (project number bbi00008). Preparation simulations have been performed on the Vienna Scientific Cluster on VSC-2 (project p70708) and the cluster nodes of the Faculty of Mathematics, Univ. of Vienna, on VSC-3.

References

[1] B. Freytag, H.-G. Ludwig, and M. Steffen. Hydrodynamical models of stellar convection. The role of overshoot in DA white dwarfs, A-type stars, and the Sun. *Astron. Astrophys.*, 313:497–516, 1996.

[2] B. Freytag, M. Steffen, H.-G. Ludwig, S. Wedemeyer-Böhm, W. Schaffenberger, and O. Steiner. Simulations of stellar convection with CO5BOLD. *Journal of Computational Physics*, 231:919–959, February 2012.

[3] A. Gianninas, P. Dufour, M. Kilic, W. R. Brown, P. Bergeron, and J. J. Hermes. Precise atmospheric parameters for the shortest-period binary white dwarfs: Gravitational waves, metals, and pulsations. *Astrophys. Jour.*, 794:35, 2014.

[4] H. Grimm-Strele, F. Kupka, B. Löw-Baselli, E. Mundprecht, F. Zaussinger, and P. Schiansky. Realistic simulations of stellar surface convection with ANTARES: I. Boundary conditions and model relaxation. *New Astron.*, 34:278–293, 2015.

[5] D. Koester, B. T. Gänsicke, and J. Farihi. The frequency of planetary debris around young white dwarfs. *Astron. Astrophys.*, 566:A34, 20 pp., 2014.

[6] F. Kupka. Turbulent convection and numerical simulations in solar and stellar astrophysics. In W. Hillebrandt and F. Kupka, editors, *Interdisciplinary Aspects of Turbulence*, volume 756 of *Lecture Notes in Physics*, pages 49–105. Springer, Berlin, 2009.

[7] F. Kupka and H. J. Muthsam. Modelling of stellar convection. *Living Rev. Comput. Astrophys.*, 3:1:159 pp., 2017.

[8] F. Kupka, F. Zaussinger, and M. H. Montgomery. Mixing and overshooting in surface convection zones of da white dwarfs: first results from ANTARES. *Mon. Not. Roy. Astron. Soc.*, 474:4660–4671, 2018.

[9] H.J. Muthsam, F. Kupka, B. Löw-Baselli, C. Obertscheider, M. Langer, and P. Lenz. ANTARES – A Numerical Tool for Astrophysical RESearch with applications to solar granulation. *New Astron.*, 15:460–475, 2010.

[10] P.-E. Tremblay, H.-G. Ludwig, B. Freytag, G. Fontaine, M. Steffen, and P. Brassard. Calibration of the mixing-length theory for convective white dwarf envelopes. *Astrophys. Jour.*, 799:142 (14 pp.), 2015.

[11] P.-E. Tremblay, H.-G. Ludwig, M. Steffen, P. Bergeron, and B. Freytag. Solution to the problem of the surface gravity distribution of cool da white dwarfs from improved 3d model atmospheres. *Astron. Astrophys.*, 531:L19 (5 pp.), 2011.

[12] P.-E. Tremblay, H.-G. Ludwig, M. Steffen, and B. Freytag. Pure-hydrogen 3D model atmospheres of cool white dwarfs. *Astron. Astrophys.*, 552:A13 (13 pp.), 2013.

[13] Zaussinger, F. and Spruit, H. C. Semiconvection: numerical simulations. *Astron. Astrophys.*, 554:A119, 2013.

IOP Conf. Series: Journal of Physics: Conf. Series **1031** (2018) 012014 doi:10.1088/1742-6596/1031/1/012014

On the problem of two-tail heliosphere/astrospheres

Vladislav Izmodenov[1,2,3,4], **Dmitry Alexashov** [2,3]
[1] Lomonosov Moscow State University
[2] Space Research Institute (IKI) Russian Academy of Sciences
[3] Institute for Problems in Mechanics Russian Academy of Sciences
[4] National Research University Higher School of Economics

E-mail: izmod@iki.rssi.ru

Abstract. We consider the effect of the solar/stellar wind collimation towards the solar/stellar wind rotation axis in the heliosheath between the termination shock and the heliopause/astropause. The collimation is due to the magnetic force produced by the toroidal component of the solar/stellar magnetic field. The collimation leads to formation of a two-jet structure and change of topology of the heliopause. Tube-like shape of the heliopause/astropause is formed instead of the commonly accepted sheet-like shape.

Three different situations are explored in this paper: (1) the Sun/star is at the rest with respect to local interstellar medium (LISM), (2) the Sun/star moves with respect to fully ionized LISM, (3) the Sun/star moves with respect to partially ionized LISM. 3D **non-dissipative** MHD model results have shown that the tube-like structure is formed in the first two cases. The thickness of the heliosheath depends on the model parameters strongly. The case of the partially ionized LISM is the most realistic one for the heliosphere. In this case the collimation towards the solar poles is also observed in the results of 3D kinetic-MHD modeling. However, the tube-like structure of the heliopause is not seen in the numerical results. We argue that this is due to charge exchange between proton and H atom components that results in displacement of the stagnation points in further downwind of the Sun. Note also that dissipative effects (e.g. reconnection at the current sheet in the non-stationary heliosphere) or instabilities could destroy the effects of the solar wind collimation.

1. Introduction

Starting from the pioneering papers [1, 2] it was conventionally accepted that the heliopause (HP), i.e tangential discontinuity separating the solar wind (SW) plasma from the plasma of the local interstellar medium (LISM), is topologically equivalent to a plane (see Figure 1a). The sheet-like topology remains to the distances ~3000 AU into the tail, where the heliopause - as a tangential discontinuity - disappears. This happens due to charge exchange with interstellar atoms of hydrogen, which interact with protons on both sides of the heliopause. Momentum and energy exchange between neutral and ionized components occurs due to the interaction. As a result, the solar wind and interstellar medium parameters on the both sides of the heliopause approach each other. As it has been shown in [3, 4], the jump in the plasma parameters across the heliopause disappears at ~3000 AU counted along the axis determined by the interstellar wind velocity vector.

Very recently, this 'classical' picture underwent a major revision. In 2015 [5] and [6] have shown that the heliopause may in fact have a tube-like shape (Fig. 1b). Such a shape has been obtained in [5] in the frame of their numerical 3D multi-fluid MHD code for the case when the

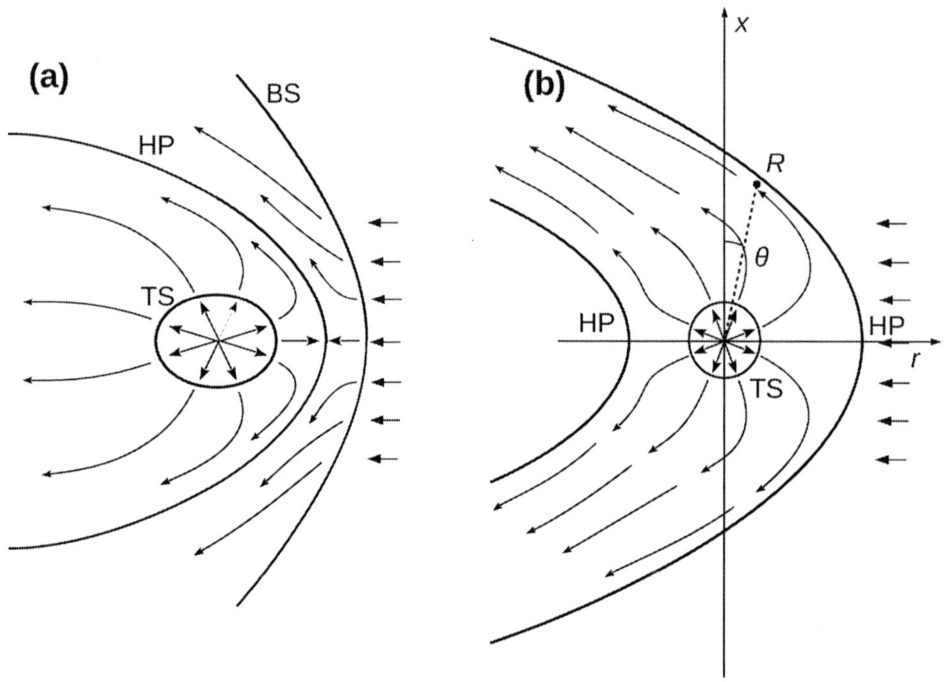

Figure 1. Schematic picture of the heliospheric interface with a sheet-like topology (a) of the heliopause (HP) and a tube-like topology (b).

interstellar gas flows with respect to the star. Later this result has been discussed by [7] and [8] and will be discussed later in the paper.

A simpler case, when the solar wind flows into the homogeneous interstellar gas at rest, has been considered by [6] and later by [9, 10]. To understand the effects of the stellar (heliospheric) magnetic field qualitatively, let us start with the 'classical' stationary model of supersonic point source flow into the surrounding gas with non-zero pressure at rest (e.g. [1]). There is a shock transition in this solution (i.e. the termination shock) at $R_{TS} \sim \sqrt{\frac{\dot{M}V_0}{4\pi p_\infty}}$, where R_{TS} is the heliocentric distance to the termination shock, $\dot{M}$ is the stellar mass loss rate, V_0 is the terminal velocity of the supersonic stellar wind, p_∞ is the interstellar gas pressure. In the supersonic SW (for $R < R_{TS}$) the solution is $V \sim V_0$, $\rho \sim 1/R^2$ and $p \sim 1/R^{2\gamma}$, where R is the distance to the Sun/star. In the subsonic region ($R > R_{TS}$) the gas may be considered incompressible and the solution is $V \sim 1/R^2$, $\rho \sim \rho_\infty$ and $p \sim p_\infty$.

This solution can be used to calculate the frozen-in magnetic field in the kinematic approximation. Solving $\nabla \times [\mathbf{V} \times \mathbf{B}] = 0$ and assuming that magnetic field is parallel to the velocity vector at the Sun:

$$R < R_{TS} : B_R \sim 1/R^2, \quad B_\phi \sim (1/R)\sin\theta, B_\theta = 0 \tag{1}$$

The solution above for $R < R_{TS}$ has been obtained in [11].

$$R > R_{TS} : \quad B_R \sim 1/R^2, \ B_\phi \sim R\sin\theta, B_\theta = 0. \tag{2}$$

Here θ is the polar angle counted from the stellar rotational axis (x-axis), ϕ is the azimuthal angle.

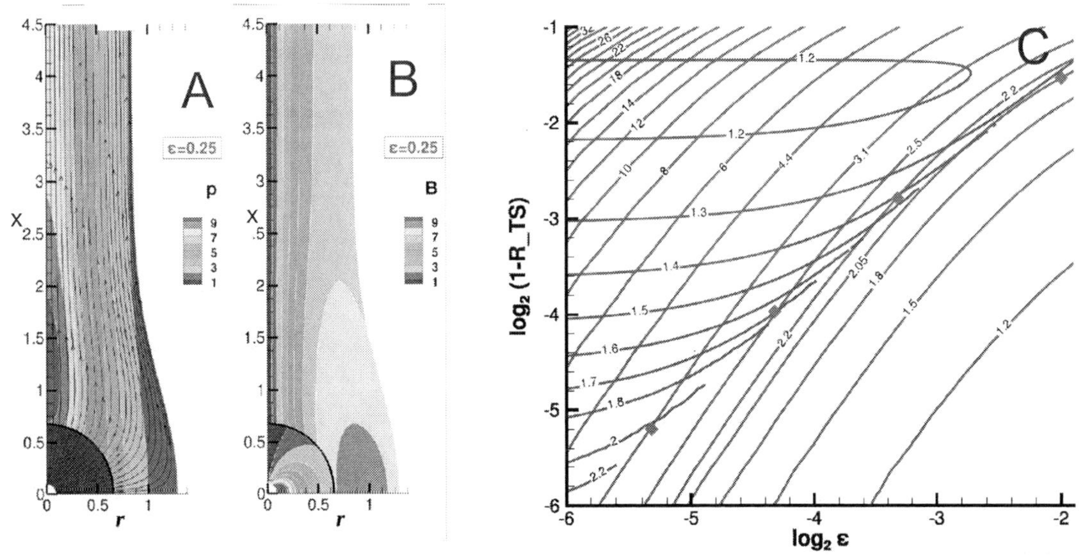

Figure 2. Panels A and B present numerical results obtained in [9]: (A) the plasma streamlines and density isolines, (B) magnetic field contour plot. Panel C present contour plots of $R_{HP,0}/R(TS)$ (red curves) and r_{jet}/R_{TS} (blue plots) that can be obtained from the first integrals and ODE solution along the jet. Brown diamonds are the results of numerical solution.

In the subsonic wind the magnetic field grows proportionally to $r = R\sin\theta$, that is, the distance to the axis of stellar rotation. Alfvénic Mach number $A = \sqrt{(4\pi\rho V^2)/B^2} \sim 1/(R^3 sin\theta)$ in the subsonic wind, so it decreases with distance rapidly. In the supersonic wind A remains constant. For the Sun, for example, the constant is about 15. At the strong shock A decreases by a factor of $((\gamma-1)/(\gamma+1))^{3/2}$ that is equal to 1/8 for $\gamma = 5/3$. Downstream of the TS $A \approx 15/8 \approx 1.9$ and then it decreases in the equatorial plane as $1/R^3$. The Alfvénic Mach number becomes on the order of unity at the distances of $\sim 1.23 R_{TS}$. Therefore at these distances and further one can expect a strong influence of the magnetic field on the plasma flow.

Magnetic force $\mathbf{F}_{mag} = ([\nabla \times \mathbf{B}] \times \mathbf{B})/(4\pi)$ has the main component in r-direction (in the cylindrical (x, r, ϕ) coordinate system where x-axis is the axis of stellar rotation). As a result, the stellar wind deflects from the original radial direction and flows towards x-axis. Hence the two-jets structure of the flow is formed.

2. Two-jet structure of the stellar wind flow in the interstellar medium at rest

In [9, 10] we explored in detail the two-jet scenario for a simplified astrosphere in which 1) the star is at rest with respect to the local interstellar medium (LISM), 2) radial magnetic field is neglected as compared with the azimuthal component, 3) the stellar wind outflow is assumed to be hypersonic (both the Mach number and the Alfvénic Mach number are much greater than unity at the inflow boundary).

It has been shown that the problem in its dimensionless form depends only on one parameter - the Alfvénic Mach number at the inflow boundary, $A = \sqrt{((4\pi\rho V^2))/B^2}$. Sometimes, it is more convenient to use $1/A$ that is called as Alfvénic number and denoted as ε in [9, 10]. Numerical parametric studies have been performed for the parameter ε varying in the range from 0.01 to 0.5. The numerical results clearly show (Figure 2A,B) the formation of the tube-like heliopause and the two jets along the stellar rotation axis. According to the numerical results obtained in [9] the distances from the axis of stellar rotation to the heliopause at the equatorial plane (x = 0) and in the jets (x $\rightarrow \infty$) are both proportional to $\varepsilon^{-1/3}$ as long A is small enough ($\varepsilon < 0.1$). Some analytical argumentation for such a dependence is given in [10].

Note that numerical results for $\varepsilon < 0.01$ and for $\varepsilon > 0.5$, were not presented in the papers because of complex vortex structure beyond the termination shock formed for these cases. Such solutions require more detailed studies.

Besides the numerical solution, analytical consideration of the three first integrals of the MHD equations allow to establish [9] analytical relations between three parameters – (1) the distance to the termination shock, $R_{TS,0}$, (2) the distance to the heliopause, $R_{HP,0}$ (both are in the stellar equatorial plane i.e. the plane perpendicular to the axis of the stellar rotation), and (3) the parameter ε. Therefore, for a given value of $R_{TS,0}$ one can obtain $R_{HP,0}$ as a function of ε.

The distribution of the plasma parameters in the jet as well as the size of the jet have been obtained in [9] as a solution of an ordinary differential equation obtained under assumptions of the hypersonic stellar wind outflow and the spherically symmetric termination shock.

These analytical considerations allow to estimate the magnitude of the stellar magnetic field from the geometric picture of a two-jet astrosphere. In particular, the knowledge (obtained, for example, from observations) of two ratios – R_{HP}/R_{TS} in the equatorial plane and r_{jet}/R_{TS} – allows (see Figure 2C) to determine the dimensionless distance to the termination shock $\hat{R}_{TS}$ and ε. Then, knowing ε and the actual dimensional distance to the TS one can derive the magnitude of the stellar magnetic field at any given distance from the star.

3. Two-jet structure of the stellar wind flow in the moving magnetized LISM (no interstellar H atoms)

In this section we demonstrate the first and *preliminary* results of the numerical 3D ideal MHD model of the SW/LISM interaction. In this model we assume that the interstellar medium is moving with respect to the Sun. The calculations were performed by employing the kinetic-MHD numerical model [8] and neglecting the effect of interstellar H atom on the plasma component. Therefore, the "kinetic" part of the model was abandoned for the present calculations.

Since in this paper we focus on qualitative picture, the results of the calculations will be presented in dimensionless form. In dimensionless formulation, the problem depends on the following 5 parameters: gas-dynamical and Alfvénic Mach numbers, $M = v/\sqrt{\gamma p/\rho}$ and $A = \sqrt{(4\pi\rho V^2)/B^2}$, in the supersonic solar wind at the inner boundary (M_s, A_s) and in the undisturbed interstellar wind (M_∞, A_∞), and the angle α that determines the angle between the direction of the interstellar velocity vector and the direction of the interstellar magnetic field. Below we present the results for $\alpha = 0$ and $\alpha = \pi/2$.

The results of our calculations are presented in Figures 3 and 4 in the following coordinate system: axis Z is directed towards the interstellar flow direction, axis X coincides with the axis of stellar rotation, and axis Y is chosen to complete the right-handed coordinate system. Here we assume that the velocity vector of the pristine LISM is perpendicular to the axis of solar rotation.

Figure 3 presents the results of calculations with the following set of parameters: $M_s = 10$, $A_s = 4.5$, $M_\infty = 1.5$, $A_\infty = 1.18$, $\alpha = 0$. Panels A and C show the plasma streamlines and density in the ZX- and ZY- planes, respectively. Panels B and D present the magnetic field lines and magnitude in the same planes. The discontinuities - the heliopause (HP), the termination shock (TS) and the bow shock (BS) - are shown as thick white curves. All parameters as well as heliocentric distances are shown in dimensionless units.

There is no doubt that the tube-like structure is present in this case. Note, however, that 1) the tube is strongly deflected towards the tail, 2) the downwind stagnation point is much (at least twice) further from the Sun than the upwind point. The stagnation points/lines are the points/lines at the heliopause at which the plasma velocity is zero.

The thickness of the heliosheath depend strongly on the orientation of the interstellar magnetic field. Panels A and B in Figure 3 show results of the calculations for the model with $\alpha = \pi/2$, i.e. when the interstellar magnetic field vector is perpendicular to the LISM velocity vector. In these calculations we assume that the magnetic field is in the XZ-plane. It is seen from the figure that the stagnation point in downwind is much further in this case as compared with the case of parallel magnetic field presented in Figure 3. Note, also, that the jumps of density and magnetic field at the heliopause are quite small in this case. So, despite the two-tail structure formally exists, the obtained picture is quite similar to the classical picture of the heliosphere.

Figure 4 demonstrates how the structure of the SW/LISM interaction depends on the other dimensionless parameters. Panels C and D demonstrate the results for $A = 0.55$ which corresponds to stronger interstellar magnetic field as compared with results presented in Figure 4A,B. It is seen that the stronger magnetic field makes the heliospheric tube more collimated. The heliopause jets become even more collimated (in dimensionless units) in the tail for smaller gas-dynamic Mach number as shown in Figure 4E,F.

Therefore, we conclude at the end of this section that our calculations in the frame of 3D ideal MHD model demonstrate, without any doubts, that the heliopause has tube-like structure. However, the degree of its collimation (or, in other words, the thickness of the heliosheath between the termination shock and the heliopause) strongly depend on model parameters. Detailed analysis of the dependence will be done elsewhere.

4. Are the jet features present in the model of Alexashov and Izmodenov (2015)?

The effect of solar wind collimation towards the solar rotation axis and formation of the two-jet structure is actually seen in the results of the 3D kinetic-MHD model developed in [8]. This model of the solar wind interaction with the two-component (plasma and H atoms) local interstellar medium takes into account both the interstellar and heliospheric magnetic fields. So the effects of the heliospheric magnetic field should be observed.

In order to explore these effects, the solar flux (ρV) is shown in Figure 5 in projection on a closed surface located in the inner heliosheath. The surface has been constructed by adding a fixed distance to the TS distance in every direction. Panel A demonstrates the solar wind mass flux in the case when heliospheric magnetic field is not taken into account. It is seen that the mass flux reaches its maximum for $\theta \sim 100^o$-110° and $\phi \sim 180^o$. This nearly crosswind direction corresponds to the direction in which all of the solar mass flux from the upwind hemisphere passes into the tail. The maxima at $\phi \sim 180^o$ correspond to that part of the (BV) plane where the distance between the TS and HP is minimal. Conversely, panel B demonstrates that there are two mass-flux maxima (i.e. jets) in the north and south directions. These jets are a clear result of the magnetic tension.

Therefore, in this respect, the effect of magnetic field collimation is clearly observed. Nevertheless, the conclusion on the tube-like shape of the heliopause proposed in [5] and confirmed by our MHD results presented in previous section is not supported in [8]. The most likely reason of the discrepancy is connected with the effect of charge exchange of solar protons and interstellar H atoms. Indeed, having large mean free path the interstellar H atoms penetrate

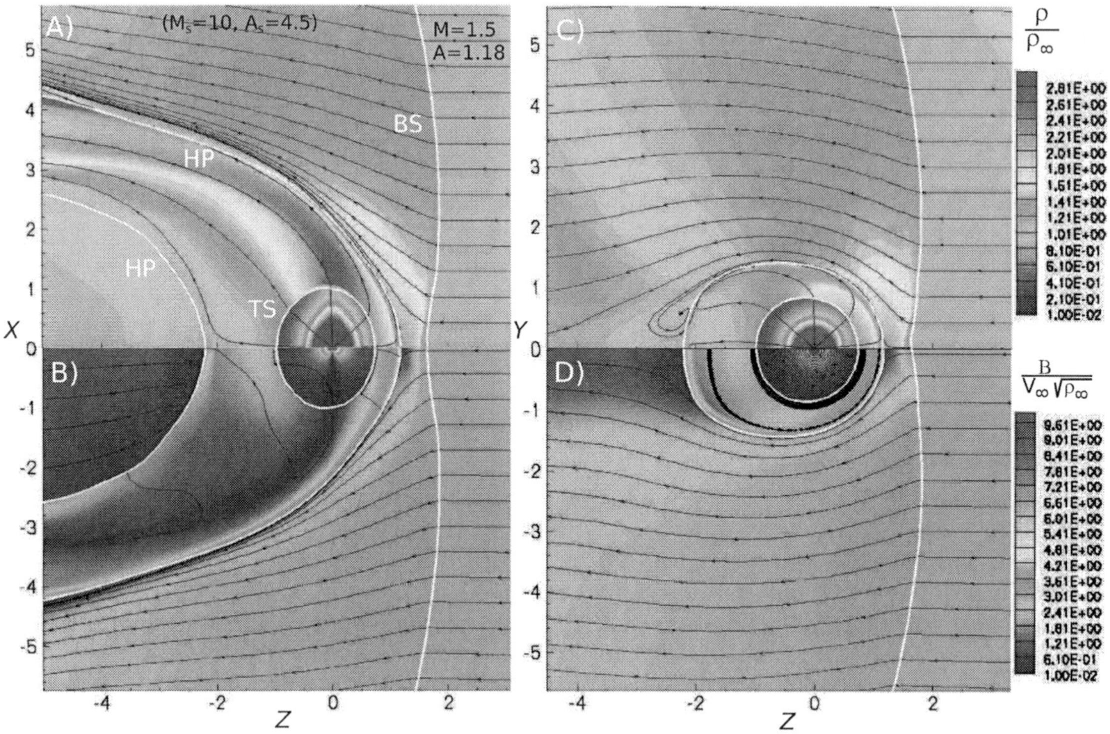

Figure 3. Panels A and C show the plasma streamlines and density in the ZX- and ZY- planes, respectively. Panels B and D present magnetic field lines and magnitude in the same planes. The discontinuities - the heliopause (HP), the termination shock (TS) and the bow shock (BS) - are shown as white thick curves. All parameters are given in dimensionless units.

into the vicinity of the downwind/upwind stagnation points. There the atoms exchange charge with the protons. As a result of the charge exchange, momentum is transferred from the neutral component into the plasma component. The momentum transfer results in diminished heliocentric distance to the heliopause in the upwind side of the heliosphere. The effect is very well known for the traditional sheet-like heliosphere (see, e.g., Figure 2 in [12]). In the tube-like heliosphere, the same momentum-transfer effect should result in increase of the heliocentric distance to the heliopause in the downwind direction. The heliopause, in principle, could move further and further downwind, and so it can be treated as a sheet-like structure in the considered computational domain.

In principle, the effect of charge exchange is taken into account in the multi-fluid approach of [5]. However, as it has been shown in [13], the multi-fluid and kinetic approaches may have qualitatively different results. It is also shown in [7] that a multi-fluid treatment of charge exchange may create a region in the tail where charge exchange is very much suppressed as compared with the kinetic-MHD model results. This may explain different shape of the heliopause obtained in [5] and in [8].

5. Summary and conclusions

In this paper we present results of the 3D ideal-MHD model of SW/LISM interaction which demonstrate the effect of solar wind collimation in the region of inner heliosheath between the termination shock and the heliopause due to toroidal component of the heliospheric magnetic

IOP Conf. Series: Journal of Physics: Conf. Series **1031** (2018) 012014 doi:10.1088/1742-6596/1031/1/012014

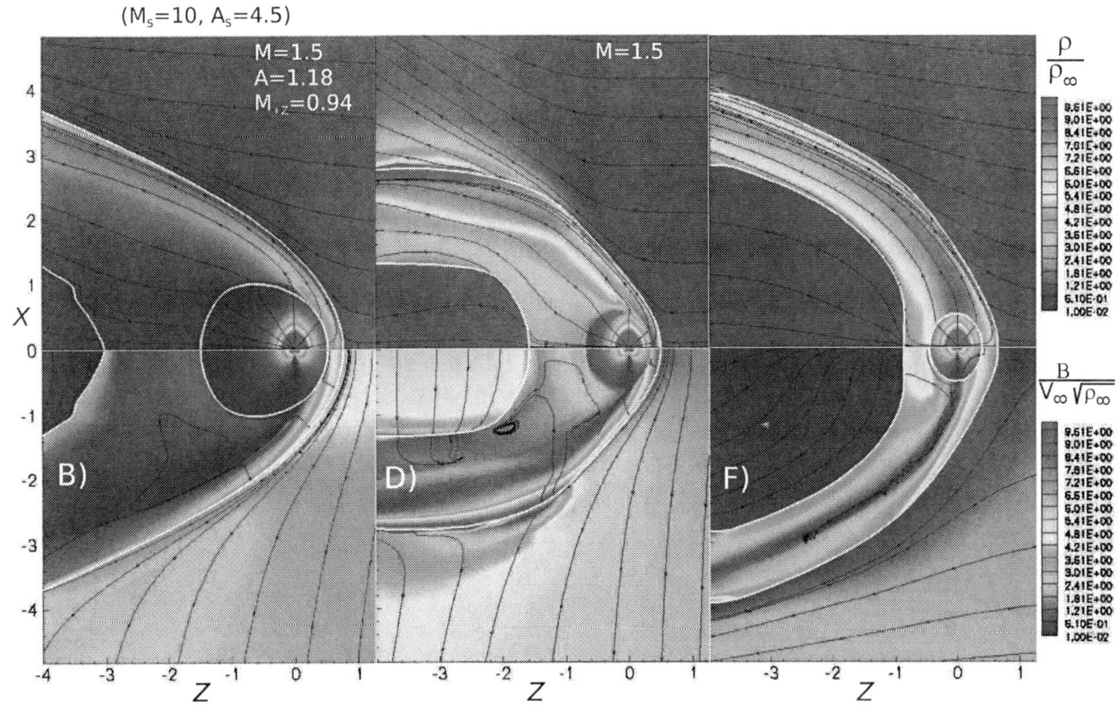

Figure 4. Panels A, C, E show the plasma streamlines and density in the ZX-plane. Panels B, D, F show the magnetic field lines and magnitude. Panels A and B show results for $M_\infty = 1.5, A = 1.18$, and $\alpha = \pi/2$. Panels C and D show results for $M_\infty = 1.5, A = 0.55$ and $\alpha = \pi/2$. Panels E and F show results for $M_\infty = 0.5, A = 1.77$ and $\alpha = 0$. The discontinuities are shown as white thick curves. All parameters are given in dimensionless units.

field. The collimation leads to the formation of the so-called tube-like shape of the heliopause (Figure 1b) contrary to the commonly accepted sheet-like shape.

The tube-like or two-jet structure of the heliosphere has been suggested in [5]. At present we do not have any doubt that such a structure exists when the solar wind interacts with fully ionized local interstellar medium. This conclusion has been made both (1) for the model with the Sun at the rest with respect to the interstellar medium and (2) for the model when the Sun moves in the LISM. In the latter case, the thickness of the heliosheath is 2-3 times larger downwind as compared to upwind. The thickness depends on the model parameters.

In the case of partially ionized local interstellar medium when interstellar H atoms interact with protons by charge exchange, the kinetic-MHD model of [8] does show the tube-like structure of the heliopause while the effect of solar wind collimation towards solar poles is clearly observed in the results. We explain the absence of the tube-like heliopause by momentum transfer between proton and H atom components at the downwind stagnation point. The momentum added to the plasma component in charge exchange moves the heliopause further away from the Sun into the downwind direction. Finally, although the tube-like structure could potentially exist in this case, the thickness of the heliosheath becomes so large in downwind of the Sun that locally there is no difference between the sheet-like or tube-like heliosphere.

Note that the explanation given above is qualitative and based on the results obtained in the cases examined. More detailed quantitative proof requires very computationally expensive parametric studies in the frame of 3D kinetic-MHD model.

It is also very important to note that all of the results presented in this paper as well as in

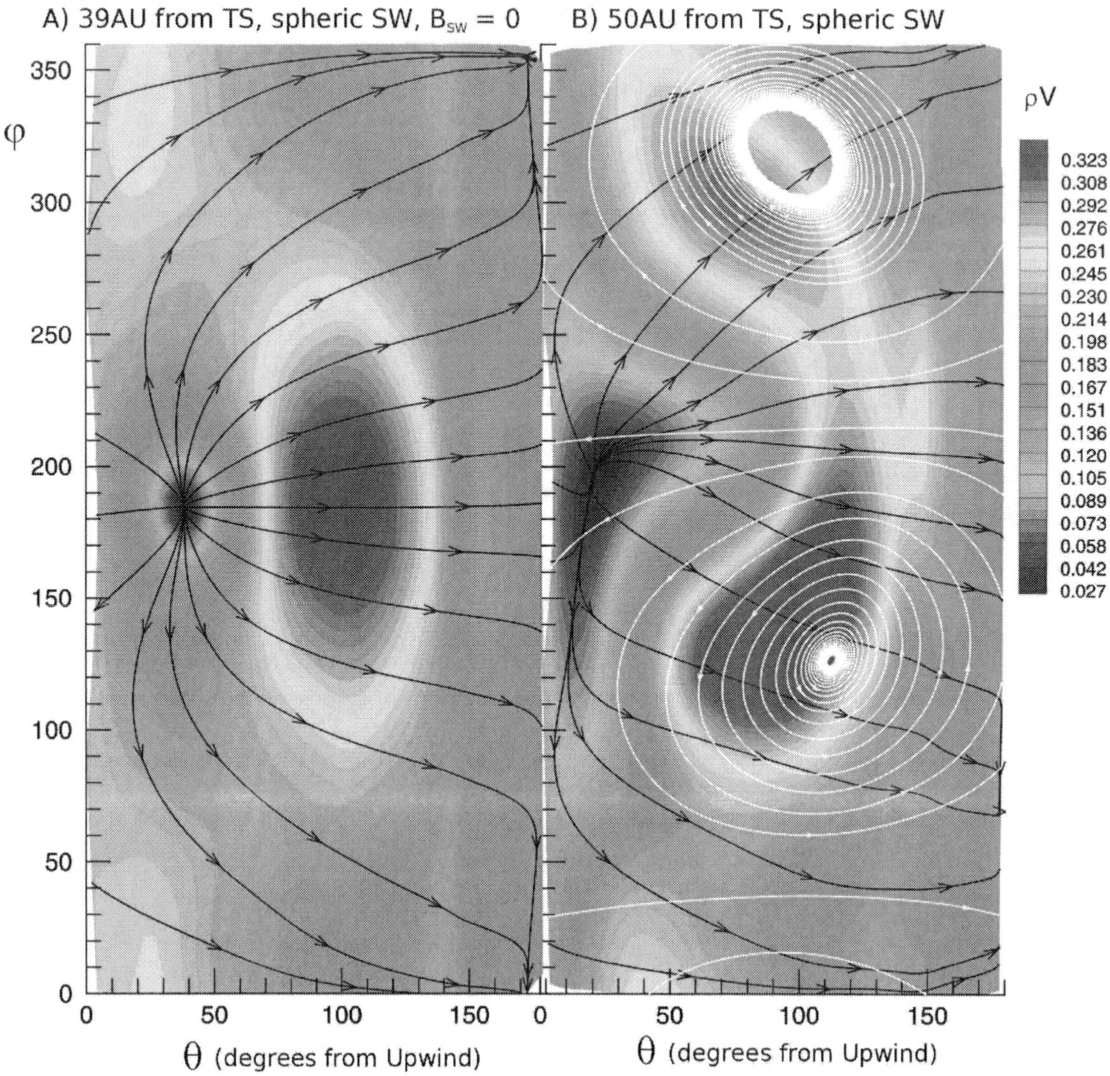

Figure 5. Solar wind mass flux (ρV) projected on a closed surface located in the inner heliosheath at equal distances from the heliopause. Results of the model without heliospheric magnetic field are presented in panel A. Results of the model where the heliospheric magnetic field has been taken into acccount are presented in panel A. The angles are θ and ϕ. Black curves are the projections of the streamlines into the surface. White curves are the projections of the heliospheric magnetic field lines. [8]

[5] have been obtained in the frame of the so-called unipolar approach for the solar magnetic field. This approach is based on the fact that the terms responsible for the influence of the magnetic field in the ideal MHD equations do not depend on the orientation of the magnetic field. Also, in the frame of the ideal MHD equations, we are not interested in the structure of the heliospheric current sheet assuming that it is a discontinuity where the magnetic field changes its orientation. Taking this into account the approach assumes unipolarity of the heliospheric magnetic field in MHD calculations that allows to avoid difficulties connected with requirements of numerical resolution of the current sheet structure. An MHD model with a current sheet would

either require very high resolution (impossible for modern computers) or have large numerical dissipation which could lead in unphysical results.

In addition, the reconnection of the heliospheric and interstellar magnetic field at the heliopause is not possible in the frame of the ideal MHD approach. We (in our model here and in [8]) consider the heliopause as a tangential discontinuity where $B_n = 0$ on both sides of the heliopause. This allows us to avoid numerical reconnection at the heliopause.

Overall, the solution of the considered problem does not depend on the orientation of the inner and outer magnetic fields along their magnetic field lines. Therefore, to perform numerical calculations, we can assume unipolar heliospheric magnetic field at the inner boundary: $B_R = -B_E\left(\frac{R_E}{R}\right)^2$. The actual polarity of the heliospheric magnetic field can be calculated after the self-consistent solution has been obtained. This unipolar magnetic field approach helps us to avoid problems with unphysical numerical reconnection and related unphysical dissipation.

The ideal MHD is not necessarily a valid approach in the heliosheath. Most likely, physical reconnection appears at the heliopause. In addition, the SW/LISM interaction region is not a stationary structure. Magnetic field change its polarity during the solar cycle. It has been argued in [7, 15] that the reconnection and solar cycle effects strongly reduce the solar wind collimation by the heliospheric magnetic field. It is also interesting to note the following results shown in [14, 15]: 1) the bent-tube structure of the heliopause disappears even within unipolar field, if solar-cycle effects are taken into account, 2) the development of kink instability obtained in numerical simulations qualitatively agrees with prediction made more than 40 years ago in [16] . So the kink instability could be another reason that reduces the solar wind collimation. Therefore, most likely that the heliopause has a "classical" sheet-like topology rather than a tube-like.

Overall, we conclude that the effect of magnetic field collimation is certainly important and should be considered in the models of global structures of heliosphere and astrospheres. Nevertheless, in case of the heliosphere, there are effects that may potentially destroy the jet-like structure of the heliosphere in the tail. The main effect explored by us is charge exchange. Other effects such as reconnection at the heliopause and time-dependence have been explored in [7, 15]. Another important aspect of the considered problem that needs to be examined is the stability of the obtained two-jet solution.

6. Acknowledgments

Global heliospheric/astrospheric modeling part of this work has been done in the frame of Russian Science Foundation grant 14-12-01096.

References

[1] Parker E.N., ApJ., 134, 20 (1961).
[2] Baranov V.B., Krasnobaev K.V., Kulikovskii A.G., Doklady Akademiia Nauk SSSR, Ser. Mat. Fiz., 194, 41 (1970).
[3] Izmodenov V.V., Alexashov D.B., Astronomy Letters 9, 58-63 (2003)
[4] Alexashov D.B., Izmodenov V.V., AIP Conf. Proc. 679, 218-221 (2003).
[5] Opher M., Drake J.F., Zieger B., Gombosi T.I., 2015, ApJ., 800, L28
[6] Drake J.F., Swisdak M., Opher M., ApJ., 808, L44 (2015).
[7] Pogorelov N.V. et al., ApJL 812, L6 (2015).
[8] Izmodenov V.V., Alexashov D.B., Ap. J. Suppl, 220, 32 (2015).
[9] Golikov E.A., Izmodenov V.V., Alexashov D.B., Belov N.A., MNRAS 464, 1065-1076 (2017).
[10] Golikov E.A., Izmodenov V.V., Alexashov D.B., Belov N.A., J. Phys. Conf. Ser. 815, 1, article id. 012035 (2017).
[11] Parker E.N., ApJ., 128, 664 (1958).
[12] Izmodenov V. V., Ap&SS, 274, 55 (2000).
[13] Alexashov D. B., Izmodenov V. V., A&A, 439, 1171 (2005).
[14] Pogorelov N.V. et al., J of Physics Conf. Ser. 719, 012013 (2016).

[15] Pogorelov N.V. et al., Space Sci. Rev. 212, 193 (2017).
[16] Yu G., Ap.J. 194, 187 (1974)

A Versatile Numerical Method for the Multi-Fluid Plasma Model in Partially- and Fully-Ionized Plasmas

A. Alvarez-Laguna[1,2], N. Ozak[1], A. Lani[2], N. N. Mansour[3], H. Deconinck[2], and S. Poedts[1]

[1] Centre for mathematical Plasma Astrophysics, KU Leuven, Celestijnenlaan 200B, 3001 Leuven, Belgium
[2] von Karman Institute for Fluid Dynamics, Waterloosesteenweg 72, 1640 Sint Genesius Rode, Belgium
[3] NASA Advanced Supercomputing Division, NASA Ames Research Center, Moffett Blvd, Mountain View, CA 94035, USA

E-mail: alaguna@vki.ac.be

Abstract. We present an innovative numerical method that solves for the multi-fluid plasma equations, including the transport, frictional, and chemical reactions terms, coupled to full Maxwell's equations. The numerical method features a scheme for the electromagnetic field with a proper scaling for the numerical dissipation, a scheme that solves flows at all speeds regimes (from subsonic to supersonic), and implicit time integration to tackle the stiffness of the system. Verification of the numerical scheme is also presented in a wide variety of plasma conditions.

1. Introduction

The dynamic interaction between magnetic fields and plasmas underlies a great variety of fascinating phenomena that occur in the universe. Most of the numerical simulations of magnetized plasmas have focused on solving the Magnetohydrodynamic (MHD) equations (e.g., [1]). However, astrophysical and laboratory plasmas are rich in non-ideal processes that are not contained in the ideal MHD approximation. These effects include the collisional resistivity, the anisotropy caused by the gyration of ions and electrons around the magnetic fields lines, the decoupling between the motion of the plasma species, the presence of neutral atoms or molecules, turbulent transport, etc [2].

The most complete and universal description of plasma is contained in the kinetic equation that, together with Maxwell's equations, describes the evolution of the distribution function in the phase space for each plasma species. Based on the kinetic equation, Particle-in-Cell (PIC) (e.g.,[3, 4, 5]) are mostly used for fully-ionized low-collisional plasmas such as the ones in space weather applications [6] and nuclear fusion plasmas (e.g., [7]). Alternatively, the direct solution to the Vlasov equation ([8] and references therein) in a six-dimensional grid is proposed. However, the numerical solution of these approaches is computationally very expensive and unaffordable to simulate full-domain high-density plasma phenomena.

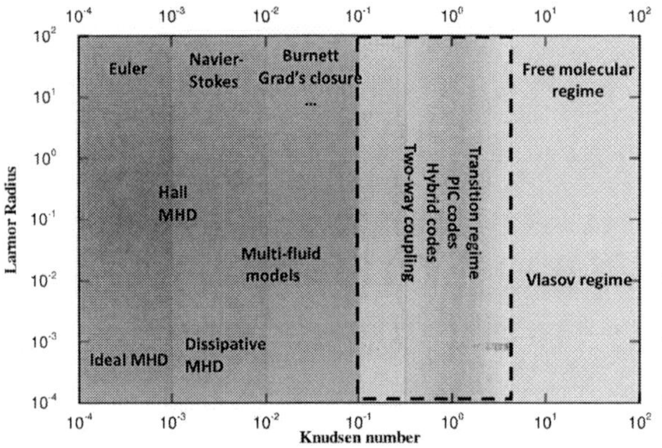

Figure 1: Region of applicability of different plasma models based on the collisionality (Knudsen number calculate as the ratio between the characteristic mean free path and the characteristic macroscopic length scale) and the magnetization (Larmor radius normalized to the characteristic macroscopic length scale). (Adapted from Liu et al. [14]).

The MHD approach is able to simulate full-scale dynamics although it only deals with large-scale low-frequency phenomena and therefore is not able to represent some of the important phenomena. Different modifications of the MHD equations are proposed in order to tackle the non-ideal processes. The Hall-MHD approach extends the induction equation of the MHD equations in order to tackle finite ion gyroradius effects [1]. For partially-ionized plasma, the effect of the neutrals is tackled with an ambipolar diffusion (Perdersen dissipation) term in the induction equation [9].

Other methods that combine the kinetic and the fluid approach are proposed to capture the non-ideal dynamics. Hybrid codes use a hydrodynamic model for the electrons while the ions are represented with a PIC model (e.g., [10]). Alternatively, the two-way coupling of MHD simulations with PIC is recently proposed. In this approach, a limited part of the domain evolves with particle methods whereas the remaining is represented with a MHD approach [11, 12, 13]. In Fig. 1, we summarize the different plasma models as function of the collisionality and the magnetization of the plasma.

In this article, we propose a multi-fluid solver as an alternative to kinetic and MHD models. The multi-fluid plasma description is obtained by taking moments are of the Boltzmann equation for each species. Therefore, mass, momentum, and energy conservation laws are solved for each species within the plasma. The different species are hence treated as separate fluids that can interact among each other through collisions and reactions, while the charged species interact with the electromagnetic field.

Recently, different numerical methods that solve for the multi-fluid equations coupled to the Maxwell's equations have been proposed. The ANTHEM code [15, 16] first carried out simulations of the fully-ionized ideal two-fluid (electron+ion) model. More recently, a high-order discontinuous Galerkin method have been proposed for the ideal two-fluid model by Loverich & Shumlak [17], Loverich et al. [18], Srinivasan & Shumlak [19], and Sousa & Shumlak [20]. Baboolal [21], Baboolal & Bharuthram [22], and Kumar & Mishra [23] propose finite difference methods for the ideal two-fluid (electron+ion) model. Alternatively, finite volume discretizations are developed by Shumlak & Loverich [24], and Hakim et al. [25]. For partially-ionized plasmas, Meier [26] proposes a two-fluid model that considers neutrals and ionized particles as separate fluids, interacting by means of chemical reactions, collisional momentum and energy exchanges. Numerical solutions in chromospheric conditions are obtained by Leake et al. [27, 28], Murphy & Lukin [29], Maneva et al. [30], and Alvarez-Laguna et al. [31].

In the present paper, we present a novel finite volume method that is based on upwind schemes for discretizing the multi-fluid equations coupled to the full Maxwell's equations. Compared to previous multi-fluid finite volume schemes, the present scheme rescales the numerical dissipation of the numerical flux of the Maxwell's equations. By doing this, the numerical flux takes into account the effect of the source terms, responsible for the coupling between the fluid and the electromagnetic dynamics. Similarly, the numerical scheme uses an AUSM$^+$-up (Advection Upstream Splitting Method) for the fluid equations, which provides an accurate solution in all velocity regimes from subsonic to supersonic. Second-order accuracy is obtained by total variation diminishing (TVD) reconstruction with a Venkatakrishnan limiter for near-smooth flows [32]. Finally, the last main difference to previous schemes is the fully-implicit time discretization that provides numerical stability for large CFL numbers. These novelties result in a robust scheme that is able to reproduce different regimes, from ideal MHD to multi-fluid equations both in the collisional and collisionless regime.

2. The multi-fluid plasma equations

The general multi-fluid plasma equations consider mass, momentum and total energy balance equations for the plasma species α, as follows,

$$\frac{\partial \rho_\alpha}{\partial t} + \vec{\nabla} \cdot (\rho_\alpha \vec{u}_\alpha) = \dot{\rho}_\alpha, \tag{1}$$

$$\frac{\partial \rho_\alpha \vec{u}_\alpha}{\partial t} + \vec{\nabla} \cdot (\rho_\alpha \vec{u}_\alpha \vec{u}_\alpha + p_\alpha \bar{\bar{I}}) = \vec{\nabla} \cdot \bar{\bar{\pi}}_\alpha + \vec{F}_\alpha + \sum_{\beta \neq \alpha} \vec{R}^{elastic}_{\alpha\beta} + \vec{R}^{react}_\alpha, \tag{2}$$

$$\frac{\partial}{\partial t}\left[\mathcal{U}_\alpha + \rho_\alpha \frac{u_\alpha^2}{2}\right] + \vec{\nabla} \cdot \left[\left(\mathcal{U}_\alpha + \rho_\alpha \frac{u_\alpha^2}{2}\right)\vec{u}_\alpha\right] = \vec{\nabla} \cdot (\vec{u}_\alpha \cdot \bar{\bar{\pi}}_\alpha - \vec{q}_\alpha - p_\alpha \vec{u}_\alpha) + \rho_\alpha \vec{F}_\alpha \cdot \vec{u}_\alpha + \sum_{\beta \neq \alpha} \dot{Q}^{elastic}_{\alpha\beta} + \dot{Q}^{react}_\alpha. \tag{3}$$

Here, ρ_α is the mass density, $\vec{u}_\alpha$ is the macroscopic velocity, p_α is the scalar pressure, $\bar{\bar{\pi}}_\alpha$ is the stress tensor, $\vec{q}_\alpha$ is the heat flux, $\mathcal{U}_\alpha$ is the internal energy, $\vec{F}_\alpha$ is the external force, $\vec{R}^{elastic}_{\alpha\beta}$ and $\dot{Q}^{elastic}_{\alpha\beta}$ are the exchanges of momentum and energy due to elastic binary collisions, and $\dot{\rho_\alpha}$ and $\dot{Q}^{react}_\alpha$ are the mass and energy production due to chemical reactions.

The external force in presence of electromagnetic fields reads $\vec{F}_\alpha = m_\alpha e_\alpha \left(\vec{E}_\alpha + \vec{u}_\alpha \times \vec{B}\right)$. The electric and magnetic fields $\vec{E}$ and $\vec{B}$ are determined by Maxwell's equations as

$$\frac{\partial \vec{B}}{\partial t} + \vec{\nabla} \times \vec{E} = 0, \tag{4}$$

$$\frac{\partial \vec{E}}{\partial t} - c^2 \vec{\nabla} \times \vec{B} = -\frac{\vec{j}}{\epsilon_0}, \tag{5}$$

$$\vec{\nabla} \cdot \vec{B} = 0, \tag{6}$$

$$\vec{\nabla} \cdot \vec{E} = \frac{\rho_c}{\epsilon_0}, \tag{7}$$

where c is the speed of light, ϵ_0 is the permittivity of free space, the total electric current density $\vec{j} = \sum_\alpha n_\alpha e_\alpha \vec{v}_\alpha$, and the total charge density $\rho_c = \sum_\alpha e_\alpha n_\alpha$.

3. Multi-dimensional finite volume method on unstructured meshes

We summarize the finite volume method described in Alvarez-Laguna et al. [33, 34]. We propose a discretization that is based on the finite volume method for the purely hyperbolic equations Maxwell's equations [35] coupled to the general multi-fluid equations (1)-(3).

We apply the spatial discretization:

$$\frac{d\mathbf{U}_i}{dt}|\Omega_i| + \sum_{j\in\mathscr{D}_i} \mathbf{H}_{ij}|\partial\Omega_{ij}| = \sum_{j\in\mathscr{D}_i} \mathbf{G}_{ij}|\partial\Omega_{ij}| + \mathbf{S}_i|\Omega_i|, \tag{8}$$

where the cell average $\mathbf{U}_i$ is the array of unknowns, $\mathbf{H}_{ij}$ and $\mathbf{G}_{ij}$ are respectively the numerical inviscid and diffusive fluxes at the interface between cells i and j, and $\mathbf{S}_i$ is the cell-average source term (external forces, reaction and collisional terms). $|\Omega_{ij}|$ is the cell volume and $|\partial\Omega_i|$ is the face area. In the following sections, we will present the numerical schemes that are used in the present work to approximate the inviscid numerical flux. These are a modified-CIR (Courant, Isaacson, and Rees) for Maxwell's equations and a generalization of the AUSM$^+$-up (Advection Upstream Splitting Method) for multi-fluid plasma equations.

3.1. Modified-CIR scheme for Maxwell's equations

Numerical dissipation is necessary to ensure stable solutions without spurious numerical oscillations [36]. However, an excessive numerical dissipation can hinder the accuracy of the method.

The CIR scheme [37] is an upwind discretization that reads:

$$\mathbf{H}_{ij} = \mathbf{A}_n^+\mathbf{U}_L + \mathbf{A}_n^-\mathbf{U}_R = \frac{\mathbf{H}(\mathbf{U}_L) + \mathbf{H}(\mathbf{U}_R)}{2} - \frac{1}{2}|\mathbf{A}_n|(\mathbf{U}_R - \mathbf{U}_L), \tag{9}$$

where $\mathbf{U}_{L/R} = (B_x, B_y, B_z, E_x, E_y, E_z, \Psi, \Phi)_{L/R}$ are the variables on the left and right of the face, $\mathbf{A}_n^\pm = \mathbf{L}\mathbf{\Lambda}_\pm\mathbf{L}^{-1}$ and $\mathbf{\Lambda}_\pm$ is the diagonal matrix that contains either the positive or negative eigenvalues of $\mathbf{\Lambda}$ respectively. In this case, the eigenvalues are $\lambda = \pm c, \pm c, \pm\gamma c, \pm\chi c$. The matrix $|\mathbf{A}_n| = \mathbf{L}|\mathbf{\Lambda}|\mathbf{L}^{-1}$ is the numerical dissipation, where $|\mathbf{\Lambda}| = \frac{\mathbf{\Lambda}_+ - \mathbf{\Lambda}_-}{2}$. Expression of the matrices $\mathbf{A}_n^\pm$ can be found in Munz et al. [35].

In the multi-fluid plasma model, Maxwell's equations are coupled to the multi-fluid equations through the source terms. The CIR scheme is based on the eigenvalue structure of the numerical flux and does not take into account the scales of the fluid-electromagnetic coupled problem.

The numerical dissipation of the original CIR scheme reads:

$$|\mathbf{A}_n| = \left[\begin{array}{c|c|c} |\mathbf{A}_n|^B & 0 & 0 \\ \hline 0 & |\mathbf{A}_n|^E & 0 \\ \hline 0 & 0 & |\mathbf{A}_n|^{HDC} \end{array}\right], \tag{10}$$

where

$$|\mathbf{A}_n|^B = c\mathbf{D}^B, \quad |\mathbf{A}_n|^E = c\mathbf{D}^E, \quad \text{and} \quad |\mathbf{A}_n|^{HDC} = c\mathbf{D}^{HDC}. \tag{11}$$

The matrices $\mathbf{D}^B$, $\mathbf{D}^E$, and $\mathbf{D}^{HDC}$, defined in Alvarez-Laguna et al. [34], are function of the unit normal vector to the face $\vec{n}$ and the divergence cleaning constants γ and χ.

The equivalent equation including the numerical dissipation, as shown by Alvarez-Laguna et al. [33], reads

$$\frac{d}{dt}\int_{\Omega_i}\begin{pmatrix}\vec{B}\\ \vec{E}\\ \hline \Psi\\ \Phi\end{pmatrix} d\Omega + \oint_{\partial\Omega_i}\begin{pmatrix}\bar{\bar{I}}\times\vec{E}+\gamma^2\Psi\bar{\bar{I}}\\ -c^2\bar{\bar{I}}\times\vec{B}+(\chi c)^2\Psi\bar{\bar{I}}\\ \hline c^2\vec{B}^T\\ \vec{E}^T\end{pmatrix}\cdot\vec{n}\, d\Sigma$$
$$+\oint_{\partial\Omega_i}\begin{pmatrix}c\frac{\Delta x}{2}\left(\vec{\nabla}\vec{B}_{\parallel}+\gamma\vec{\nabla}\vec{B}_{\perp}\right)\\ c\frac{\Delta x}{2}\left(\vec{\nabla}\vec{E}_{\parallel}+\chi\vec{\nabla}\vec{E}_{\perp}\right)\\ \hline \gamma c\frac{\Delta x}{2}\vec{\nabla}\Psi\\ \chi c\frac{\Delta x}{2}\vec{\nabla}\Phi\end{pmatrix}\cdot\vec{n}\, d\Sigma = \int_{\Omega_i}\begin{pmatrix}0\\ -\frac{\vec{j}}{\epsilon_0}\\ \hline 0\\ \frac{\rho_c}{\epsilon_0}\end{pmatrix} d\Omega. \quad (12)$$

As noted by MacCormack [38, 39] and Alvarez-Laguna et al. [33, 34], this numerical dissipation is not balanced with the flux terms when the equations are coupled to the dynamics of a conducting fluid. The reason for that is that the CIR scheme is built upon the Riemann solution to a purely electromagnetic problem. However, the exact solution for the Riemann problem that couples the multi-fluid equations to Maxwell's equations is not known. Owing to that reason, we propose to rescale the numerical dissipation in order to account for the scales of the fluid-electromagnetics coupled problem.

The propagation of waves depends on the multi-fluid model in consideration. Two different cases are studied in this article: 1) the two-fluid (ion+neutral) model for partially ionized plasmas that considers electrons and ions coupled and assumes charge neutrality [27], and 2) the ideal two-fluid (electron+ion) model [24].

We note that when charge neutrality is assumed and the electric field follows resistive Ohm's law, the behaviour of the electromagnetic waves is similar to the MHD model. In Alvarez-Laguna et al. [33], we propose a similar scaling to MacCormack's [38]. This results in the following numerical dissipation matrices:

$$|\mathbf{A}_n|^B = \mathbf{D}^B, \quad |\mathbf{A}_n|^E = c^2\mathbf{D}^E, \quad \text{and} \quad |\mathbf{A}_n|^{HDC} = \begin{bmatrix}1 & 0\\ 0 & c^2\end{bmatrix}\mathbf{D}^{HDC}. \quad (13)$$

Alternatively, the ideal two-fluid plasma model contains high frequency dispersive waves, such as the plasma waves. As noted by Goedbloed & Poedts [40], the symmetry between the electric field and the magnetic field is lost in the two-fluid model. While the magnetic field waves are in the MHD regime (i.e., Alfvén and magnetosonic waves), the electric field fluctuates in much higher frequency, i.e., the plasma frequency. Owing to that reason, Alvarez-Laguna et al. [34] propose the following rescaling of the numerical dissipation:

$$|\mathbf{A}_n|^B = \hat{r}_D^* v_A\mathbf{D}^B, \quad |\mathbf{A}_n|^E = c\mathbf{D}^E, \quad \text{and} \quad |\mathbf{A}_n|^{HDC} = c\mathbf{D}^{HDC}. \quad (14)$$

where $\hat{r}_D^* = \max(r_D/l_0, \Delta x/l_0)$, r_D is a characteristic value of the Debye length in the background plasma, and l_0 is a characteristic macroscopic length.

Note that the rescaling of the numerical dissipation that is given in Eqs. (13) and (14) maintains the linearity of Maxwell's equations. This choice reduces the number of algebraic operations of the algorithm and improves the numerical convergence of the linear system in the implicit solver. Additionally, the use of implicit time discretization that allows for long time scales helps to damp the high-frequency waves (e.g., pure electromagnetic waves) that does not need to be resolved. More complex methods, such as computing the roots of the dispersion relation of the coupled fluid-Maxwell system in each interface, as proposed by Amano [41], is a more sophisticated alternative to our choice.

3.2. AUSM$^+$-up for all speeds extended to multi-fluid equations

The AUSM method splits the numerical inviscid flux into advective and pressure flux terms, as follows:

$$\mathbf{H}_s(\mathbf{U}) = \vec{\mathbf{F}}_s^{(c)} \cdot \vec{n} = \begin{pmatrix} \rho_s \vec{u}_s \cdot \vec{n} \\ \rho_s \vec{u}_s \vec{u}_s \cdot \vec{n} + p\vec{n} \\ \rho_s H_s \vec{u}_s \cdot \vec{n} \end{pmatrix} =$$

$$\vec{u}_s \cdot \vec{n} \begin{pmatrix} \rho_s \\ \rho_s \vec{u}_s \\ \rho_s H_s \end{pmatrix} + p_s \begin{pmatrix} 0 \\ \vec{n} \\ 0 \end{pmatrix} = M_s a_s \begin{pmatrix} \rho_s \\ \rho_s \vec{u}_s \\ \rho_s H_s \end{pmatrix} + p_s \begin{pmatrix} 0 \\ \vec{n} \\ 0 \end{pmatrix}, \quad (15)$$

where $M_s = \vec{u}_s \cdot \vec{n}/a_s$. The AUSM scheme discretize the numerical flux for species s at the interface $|\partial\Omega_{ij}|$ as:

$$\mathbf{H}_{ij,s} = \mathbf{H}_{ij,s}^{(A)} + \mathbf{H}_{ij,s}^{(P)}, \quad (16)$$

where $\mathbf{H}_{ij,s}^{(A)}$ and $\mathbf{H}_{ij,s}^{(P)}$ are the upwinded advective and pressure fluxes, respectively. The former reads:

$$\mathbf{H}_{ij,s}^{(A)} = M_{1/2,s} a_{1/2,s} \begin{cases} \begin{pmatrix} \rho_s \\ \rho_s \vec{u}_s \\ \rho_s H_s \end{pmatrix}_L, & \text{if } M_{1/2,s} > 0 \\ \begin{pmatrix} \rho_s \\ \rho_s \vec{u}_s \\ \rho_s H_s \end{pmatrix}_R, & \text{otherwise.} \end{cases} \quad \text{and} \quad \mathbf{H}_{ij,s}^{(P)} = p_{1/2,s} \begin{pmatrix} 0 \\ \vec{n}_{ij} \\ 0 \end{pmatrix}. \quad (17)$$

The expressions for $M_{1/2,s}$, $a_{1/2,s}$, and $p_{1/2,s}$ are given in Alvarez-Laguna et al. [33]. The advantage of AUSM is its ability to achieve shock resolution, monotonicity, positivity, and entropy-satisfying properties. Additionally, it is designed to improve accuracy for low Mach number regime. This property is important in the multi-fluid model as the electron speed of sound is much greater than this of ions.

3.3. Total variation diminishing (TVD) reconstruction and implicit time stepping

Second-order accuracy is obtained with a weighted linear least square reconstruction [42] with Venkatakrishnan's limiter for near-smooth flow regions [32]. We observe that the choice of the limiter has a strong impact in the ideal two-fluid (ion+electron) model. Owing to the presence of dispersive waves, the use of a limiter with a correction for near-smooth regions reduces the numerical dissipation.

The time discretization uses a fully-implicit three-point backward Euler scheme, as explained in Alvarez-Laguna et al. [33].

3.4. COOLFluiD framework

The numerical method has been implemented into COOLFluiD [43], a modern world-class high performance computing (HPC) platform that allows a proper efficiency on massively-parallel architectures up to several tens of thousands of cores (see Fig. 2). The multi-fluid/Maxwell code presented in this article has been tested on a number of supercomputer facilities. These include the ThinKing cluster of the Flemish Supercomputer Centre (VSC), NASA's Pleiades supercomputer, the XC40 Hazel Hen system at HLRS, the Blue Gene/Q JUQUEEN at JSC,

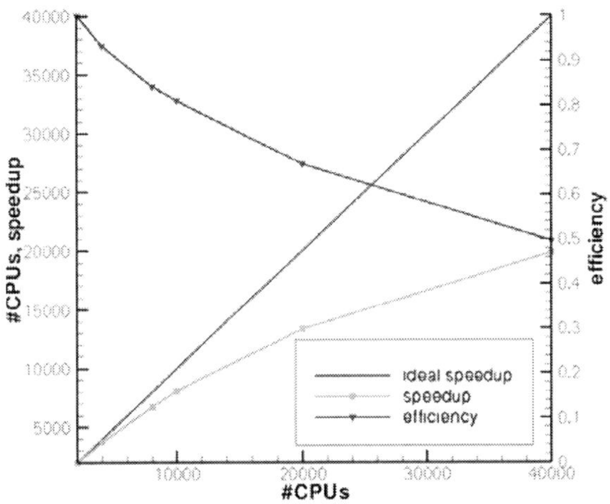

Figure 2: Strong scaling on NASA Pleiades for 0.5 billion-cells 3D grid. In this case, as the size of the mesh is kept constant during the test, the scaling of the algorithm is limited by communication.

and VKI's local clusters. Additionally, in a work that is not included in this article, the algorithm has been ported to graphics processing units (GPUs).

The multi-fluid/Maxwell numerical method and validation is explained in detail in Alvarez-Laguna et al. [33, 34]. Additionally, it is used to study the propagation of waves in the chromosphere in Maneva et al. [30] and the impact of radiation on chromospheric magnetic reconnection [31].

4. Numerical results

4.1. Accuracy analysis

As seen in Section 3.1, an excessive numerical dissipation can damage the accuracy of the numerical method. We perform an accuracy analysis in order to demonstrate that we achieve second-order accuracy in space with the TVD scheme explained above.

The scheme in the charge-neutrality limit is benchmarked in the simulation of a MHD iso-density vortex that is advected at 45° to the horizontal direction [44, 45, 46]. In Fig. 3a, the L_2 norm of the error is shown for the electric field (solid black), the magnetic field (dashed black), and the velocity field (red) in different meshes. We observe that second-order is achieved with the proposed scheme when we compare the previous curves with the slope corresponding to the second-order accuracy that is shown in blue.

Alternatively, we assess the accuracy of the scheme for the ideal two-fluid (electron+ion) model where charge neutrality is not assumed. We study a two-fluid circularly polarized wave, as described in Alvarez-Laguna et al. [34]. The results of this study are shown in Fig. 3b. In this figure, the L_2 norm of the error is shown for the electric field (solid black), the magnetic field (dashed black), the velocity field of electrons (red), and ions (blue) in different meshes that are compared to the second-order slope. Second-order accuracy is also achieved with the proposed rescaling of the electromagnetic field numerical dissipation without the charge-neutrality assumption. In Alvarez-Laguna et al. [34] it is shown that second-order accuracy is not achieved when the numerical dissipation is not rescaled.

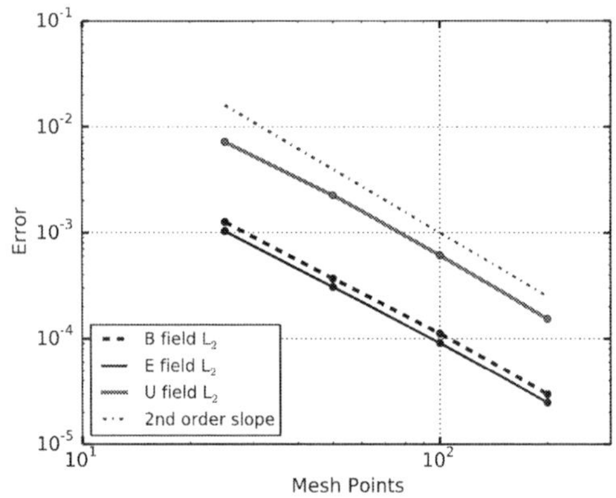

(a) MHD iso-density vortex

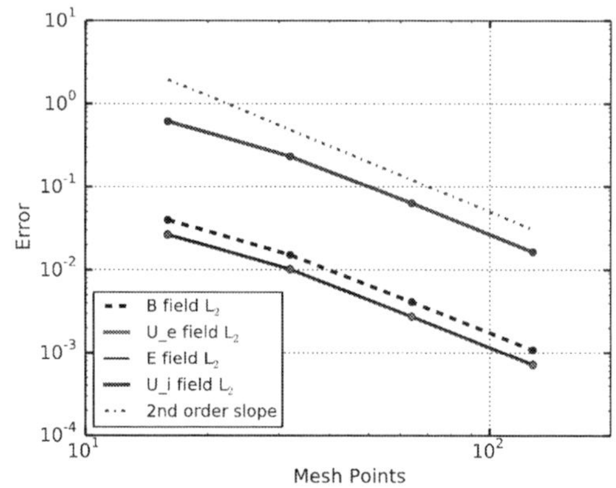

(b) Two-fluid circularly polarized wave

Figure 3: Accuracy analysis: a) in the charge-neutrality limit, and b) in the two-fluid (electron+ion) model.

4.2. MHD shocks resolution

We test the divergence cleaning method and the ability to capture shocks in two ideal-MHD cases, namely, the Orszag-Tang vortex [47] and the MHD rotor [48]. In Fig. 4, we show the evolution of the thermal pressure in the Orszag-Tang vortex. The comparison with a constrained-transport MHD code shows that the shocks are resolved by the AUSM method. Similarly, in Fig. 5, the result for the magnetic pressure of the MHD rotor shows that the proposed method is able to properly capture the MHD shocks while fulfilling the divergence constraint.

4.3. Stiff source terms associated to elastic collisions and chemical reactions

The proposed method, as it uses fully-implicit time integration, is able to handle very stiff source terms. The chemical reactions and elastic collisions terms introduce a wide range of time scales which may result in very restricted time steps for explicit discretizations. In order to test a model that considers stiff source terms, we solve for the reactive and collisional partially-ionized two-fluid model that is proposed by Leake et al. [27]. This model includes chemical reactions, collisional momentum and energy exchange, transport fluxes, and anisotropic heat conduction.

In Alvarez-Laguna et al. [31], we study magnetic reconnection under chromospheric conditions for different ionization levels. In Fig. 6, we show the magnetic field lines and electric current for two cases with initial ionization level of $\Psi_0 = 0.5\%$ (top) and $\Psi_0 = 50\%$ (bottom). Collisional and chemical reaction terms are computed with the local plasma parameters. The reconnection in the chromosphere strongly depends on the role played by ambipolar diffusion that is produced by the presence of neutrals. As seen in Fig. 6, the weakly ionized case ($\Psi_0 = 0.5\%$) presents a very sharp current sheet as compared to the more ionized case ($\Psi_0 = 50\%$). The collisions between neutrals and ions is responsible for the ambipolar diffusion which narrows the current sheet and enhances reconnection. In Fig. 7, we show the reconnection rate for a series of simulation for different ionization levels between $\Psi_0 = 0.5 - 50\%$. The reconnection rate is larger for low ionization cases.

In order to correctly capture the separation between the motions of ions and neutrals, as well as the chemical non-equilibrium in chromospheric magnetic reconnection, multi-fluid models are needed.

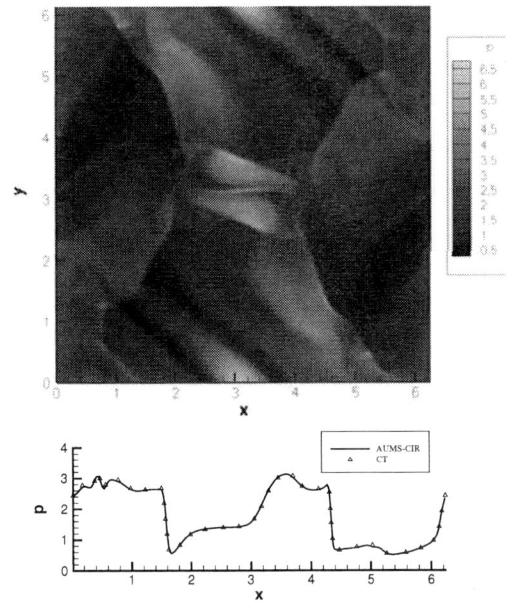

Figure 4: Top: evolution of thermal pressure of the Orszag-Tang vortex at $t = 3$ on a mesh of 200×200 elements. Bottom: comparison between the proposed method and a constrained transport solution [49] over a line at $y = 0.625\pi$ at $t = 3$.

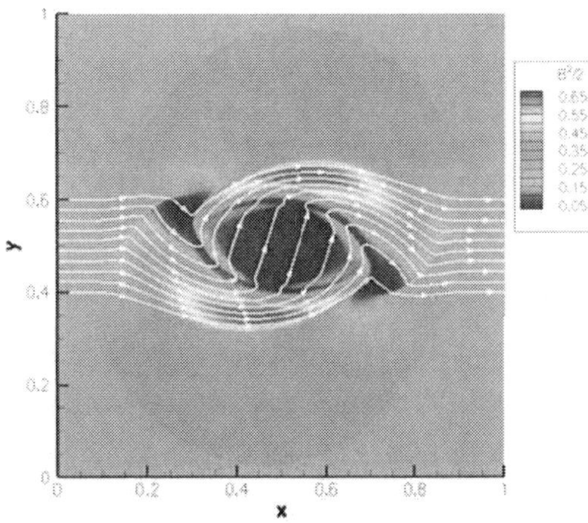

Figure 5: Magnetic pressure and magnetic field lines at $t = 0.2962$ of the MHD rotor on a mesh of 400×400.

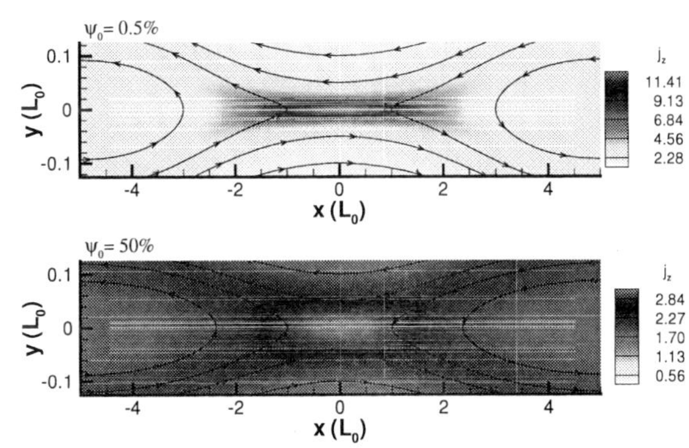

Figure 6: Current and magnetic field lines for a magnetic reconnection in chromospheric conditions. Two initial ionization levels are studied $\Psi_0 = 0.5\%$ (top) and $\Psi_0 = 50\%$.

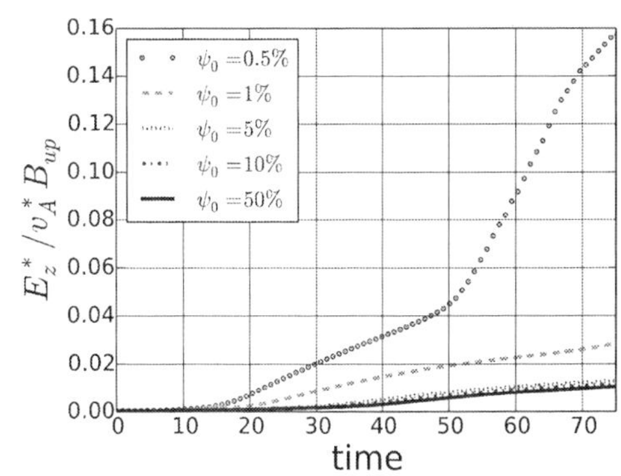

Figure 7: Reconnection rate under chromospheric conditions for different ionization levels.

IOP Conf. Series: Journal of Physics: Conf. Series **1031** (2018) 012015 doi:10.1088/1742-6596/1031/1/012015

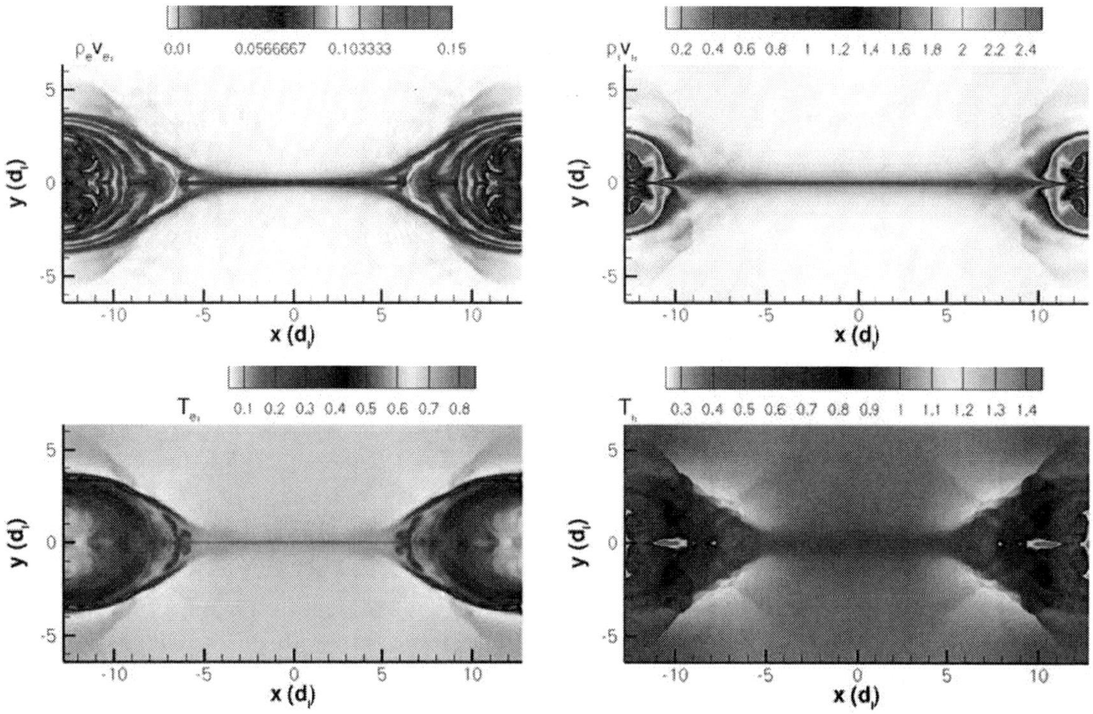

Figure 8: Results of the GEM challenge using an ideal two-fluid plasma model at $t = 25$ on a mesh of 512×512 cells. In the top panels the momentum of electrons (left) and ions (right) is shown. In the bottom panels we present the temperature of electrons (left) and ions (right).

4.4. Ideal two-fluid (electron+ion) plasma model

The ideal two-fluid plasma model considers ions and electrons as different fluids. From the numerical point of view, this system of equations is difficult to solve as the dynamics and scales of ions and electrons are very different because of their mass disparity. Additionally, even though the plasma remains quasi-neutral in lengths that are larger than the Debye length, small separation of the charges produces plasma waves that are very high frequency.

We simulate the Geospace Environmental Modeling (GEM) magnetic reconnection challenge [50] with a two-fluid model considering ions and electrons. The mass ratio is artificially reduced to 25 and the speed of light to $10v_A$ where v_A is the Alfvén speed. In Fig. 8, we show the momentum and temperature of electrons and ions. We observe the decoupling of the motion of ions and electrons that is responsible for the Hall effect. Additionally, in the temperature field of ions we note the formation of shocks as well as shock-shock interactions. Alternatively, the electrons do not present shocks as their speed of sound is much larger than this of ions.

The results presented in Fig. 8 qualitatively compare well to those of Fig. 14 of Hakim et al. [25]. A quantitative comparison of the reconnection rate that is obtained with the present numerical method is presented in Alvarez-Laguna et al. [34].

5. Conclusions

An innovative numerical method that solves for the multi-fluid plasma equations, including the transport, frictional, and chemical reactions terms, coupled to Maxwell's equations has been proposed. The numerical method features implicit time integration to tackle the stiffness of the system, a scheme that solves flows at all speeds regimes (from subsonic to supersonic), and a scheme for the electromagnetic field with a proper scaling for the numerical dissipation.

The numerical tool has been extensively verified and proves its versatility to represent a wide range of plasma regimes – from low-subsonic to supersonic, from reactive partially-ionized to fully-ionized, and from collisionally dominated to collisionless and magnetically dominated. The reusable implementation of the code allows for selecting the number of species and the closure models of choice. Hence, the same code can be used to simulate ideal and resistive MHD, reactive and collisional ion-neutral, and ion-electron models.

References

[1] J. P. Goedbloed, R. Keppens, and S. Poedts. *Advanced Magnetohydrodynamics*. April 2010.

[2] D. Biskamp. *Magnetic Reconnection in Plasmas*. Cambridge University Press, 2005.

[3] J. M. Dawson. Particle simulation of plasmas. *Reviews of Modern Physics*, 55:403–447, April 1983.

[4] E. Sonnendrücker, J. Roche, P. Bertrand, and A. Ghizzo. The Semi-Lagrangian Method for the Numerical Resolution of the Vlasov Equation. *Journal of Computational Physics*, 149:201–220, March 1999.

[5] Stefano Markidis, Giovanni Lapenta, and Rizwan-uddin. Multi-scale simulations of plasma with ipic3d. *Mathematics and Computers in Simulation*, 80(7):1509 – 1519, 2010. Multiscale modeling of moving interfaces in materials.

[6] Giovanni Lapenta. Particle simulations of space weather. *Journal of Computational Physics*, 231(3):795 – 821, 2012. Special Issue: Computational Plasma Physics.

[7] S. Ku, C.S. Chang, and P.H. Diamond. Full-f gyrokinetic particle simulation of centrally heated global itg turbulence from magnetic axis to edge pedestal top in a realistic tokamak geometry. *Nuclear Fusion*, 49(11):115021, 2009.

[8] F. Filbet and E. Sonnendrücker. Comparison of Eulerian Vlasov solvers. *Computer Physics Communications*, 150:247–266, February 2003.

[9] J. Martínez-Sykora, B. De Pontieu, V. Hansteen, and M. Carlsson. The role of partial ionization effects in the chromosphere. *Philosophical Transactions of the Royal Society of London Series A*, 373:20140268–20140268, April 2015.

[10] S. Briguglio, F. Zonca, and G. Vlad. Hybrid magnetohydrodynamic-particle simulation of linear and nonlinear evolution of alfvén modes in tokamaks. *Physics of Plasmas*, 5(9):3287–3301, 1998.

[11] Lars K.S. Daldorff, Gábor Tóth, Tamas I. Gombosi, Giovanni Lapenta, Jorge Amaya, Stefano Markidis, and Jeremiah U. Brackbill. Two-way coupling of a global hall magnetohydrodynamics model with a local implicit particle-in-cell model. *Journal of Computational Physics*, 268(Supplement C):236 – 254, 2014.

[12] Gábor Tóth, Xianzhe Jia, Stefano Markidis, Ivy Bo Peng, Yuxi Chen, Lars K. S. Daldorff, Valeriy M. Tenishev, Dmitry Borovikov, John D. Haiducek, Tamas I. Gombosi, Alex Glocer, and John C. Dorelli. Extended magnetohydrodynamics with embedded particle-in-cell simulation of ganymede's magnetosphere. *Journal of Geophysical Research: Space Physics*, 121(2):1273–1293, 2016. 2015JA021997.

[13] K.D. Makwana, R. Keppens, and G. Lapenta. Two-way coupling of magnetohydrodynamic simulations with embedded particle-in-cell simulations. *Computer Physics Communications*, 2017.

[14] C. Liu and K. Xu. A Unified Gas Kinetic Scheme for Multi-scale and Multi-component Plasma Transport. *ArXiv e-prints*, September 2016.

[15] R. J. Mason and C. W. Cranfill. Hybrid Two-Dimensional Electron Transport in Self-Consistent Electromagnetic Fields. *IEEE Transactions on Plasma Science*, 14:45–52, 1986.

[16] R. J. Mason. An electromagnetic field algorithm for 2D implicit plasma simulation. *Journal of Computational Physics*, 71:429–473, August 1987.

[17] J. Loverich and U. Shumlak. A discontinuous galerkin method for the full two-fluid plasma model. *Computer Physics Communications*, 169(1):251–255, July 2005.

[18] J. Loverich, A. Hakim, and U. Shumlak. A discontinuous galerkin method for ideal two-fluid plasma equations. *Communications in Computational Physics 9*, 240, 2011.

[19] B. Srinivasan and U. Shumlak. Analytical and computational study of the ideal full two-fluid plasma model and asymptotic approximations for Hall-magnetohydrodynamics. *Physics of Plasmas*, 18(9):092113, September 2011.

[20] E. M. Sousa and U. Shumlak. A blended continuous-discontinuous finite element method for solving the multi-fluid plasma model. *Journal of Computational Physics*, 326:56–75, December 2016.

[21] S. Baboolal. Finite-difference modeling of solitons induced by a density hump in a plasma multi-fluid. *Mathematics and Computers in Simulation*, 55(4):309 – 316, 2001. Nonlinear Waves: Computation and Theory.

[22] S. Baboolal and R. Bharuthram. Two-scale numerical solution of the electromagnetic two-fluid plasma-

maxwell equations: Shock and soliton simulation. *Mathematics and Computers in Simulation*, 76(1–3):3 – 7, 2007. Mathematical Modelling and Computational Methods in Applied Sciences and Engineering.

[23] Harish Kumar and Siddhartha Mishra. Entropy stable numerical schemes for two-fluid plasma equations. *Journal of Scientific Computing*, 52(2):401–425, 2012.

[24] U. Shumlak and J. Loverich. Approximate riemann solver for the two-fluid plasma model. *J. Comput. Phys.*, 187(2):620–638, May 2003.

[25] A. Hakim, J. Loverich, and U. Shumlak. A high resolution wave propagation scheme for ideal two-fluid plasma equations. *J. Comput. Phys.*, 219(1):418–442, November 2006.

[26] E. T. Meier and U. Shumlak. A general nonlinear fluid model for reacting plasma-neutral mixtures. *Physics of Plasmas*, 19(7):072508, July 2012.

[27] J. E. Leake, V. S. Lukin, M. G. Linton, and E. T. Meier. Multi-fluid Simulations of Chromospheric Magnetic Reconnection in a Weakly Ionized Reacting Plasma. *The Astrophysical Journal*, 760:109, December 2012.

[28] J. E. Leake, V. S. Lukin, and M. G. Linton. Magnetic reconnection in a weakly ionized plasma. *Physics of Plasmas*, 20(6):061202, June 2013.

[29] N. A. Murphy and V. S. Lukin. Asymmetric Magnetic Reconnection in Weakly Ionized Chromospheric Plasmas. *The Astrophysical Journal*, 805:134, June 2015.

[30] Y. G. Maneva, A. Alvarez Laguna, A. Lani, and S. Poedts. Multi-fluid Modeling of Magnetosonic Wave Propagation in the Solar Chromosphere: Effects of Impact Ionization and Radiative Recombination. *The Astrophysical Journal*, 836:197, February 2017.

[31] A. Alvarez Laguna, A. Lani, N. N. Mansour, H. Deconinck, and S. Poedts. Effect of radiation on chromospheric magnetic reconnection: Reactive and collisional multi-fluid simulations. *The Astrophysical Journal*, 842(2):117, 2017.

[32] V. Venkatakrishnan. On the accuracy of limiters and convergence to steady state solutions, 1993.

[33] A. Alvarez Laguna, A. Lani, H. Deconinck, N. N. Mansour, and S. Poedts. A fully-implicit finite-volume method for multi-fluid reactive and collisional magnetized plasmas on unstructured meshes. *Journal of Computational Physics*, 318:252–276, August 2016.

[34] A. Alvarez Laguna, Ozak N., A. Lani, H. Deconinck, N. N. Mansour, and S. Poedts. Fully-implicit finite volume method for the ideal two-fluid plasma model. *Computer Physics Communications*, Under review, 2017.

[35] C.-D. Munz, P. Ommes, and R. Schneider. A three-dimensional finite-volume solver for the Maxwell equations with divergence cleaning on unstructured meshes. *Computer Physics Communications*, 130:83–117, July 2000.

[36] R.J. LeVeque. *Finite Volume Methods for Hyperbolic Problems*. Cambridge Texts in Applied Mathematics. Cambridge University Press, 2002.

[37] R. Courant, E. Isaacson, and M. Rees. On the solution of nonlinear hyperbolic differential equations by finite differences. *Communication on Pure and Applied Mathematics*, 5:243–255, 1952.

[38] R. MacCormack. Numerical simulation of aerodynamic flow including induced magnetic and electric fields, 2008.

[39] R. MacCormack. Flow simulations within induced magnetic and electric fields, 2009.

[40] J. P. H. Goedbloed and S. Poedts. *Principles of Magnetohydrodynamics*. August 2004.

[41] T. Amano. Divergence-free approximate Riemann solver for the quasi-neutral two-fluid plasma model. *Journal of Computational Physics*, 299:863–886, October 2015.

[42] T. Barth. Aspects of unstructured grids and finitevolume solvers for the euler and navier-stokes equations, 1994.

[43] A. Lani, A. Alvarez Laguna, Nataly Ozak Munoz, Yana Maneva, Ray Vandenhoeck, Alexander Papen, Isaac Alonso, and Jesus Garicano Mena. COOLFluiD: Computational Object-Oriented Libraries for Fluid Dynamics, May 2017.

[44] D. S. Balsara. Second-Order-accurate Schemes for Magnetohydrodynamics with Divergence-free Reconstruction. *Astrophysical Journal Supplement Series*, 151:149–184, March 2004.

[45] D. S. Balsara. Divergence-free reconstruction of magnetic fields and WENO schemes for magnetohydrodynamics. *Journal of Computational Physics*, 228:5040–5056, August 2009.

[46] D. S. Balsara, T. Rumpf, M. Dumbser, and C.-D. Munz. Efficient, high accuracy ADER-WENO schemes for hydrodynamics and divergence-free magnetohydrodynamics. *Journal of Computational Physics*, 228:2480–2516, April 2009.

[47] S. A. Orszag and C.-M. Tang. Small-scale structure of two-dimensional magnetohydrodynamic turbulence. *Journal of Fluid Mechanics*, 90:129–143, January 1979.

[48] D. S. Balsara and D. S. Spicer. A Staggered Mesh Algorithm Using High Order Godunov Fluxes to Ensure Solenoidal Magnetic Fields in Magnetohydrodynamic Simulations. *Journal of Computational Physics*, 149:270–292, March 1999.

[49] Andrew J. Christlieb, James A. Rossmanith, and Qi Tang. Finite difference weighted essentially non-oscillatory schemes with constrained transport for ideal magnetohydrodynamics. *Journal of Computational Physics*, 268:302 – 325, 2014.

[50] J. Birn, J. F. Drake, M. A. Shay, B. N. Rogers, R. E. Denton, M. Hesse, M. Kuznetsova, Z. W. Ma, A. Bhattacharjee, A. Otto, and P. L. Pritchett. Geospace environmental modeling (gem) magnetic reconnection challenge. *J. Geophys. Res.*, 106(A3):3715–3720, 03 2001.

IOP Conf. Series: Journal of Physics: Conf. Series **1031** (2018) 012016 doi:10.1088/1742-6596/1031/1/012016

Quantification of Incertitude in Black Box Simulation Codes

A C Calder[1,2], M M Hoffman[3], D E Willcox[1], M P Katz[4], F D Swesty[1], and S Ferson[5]

[1] Department of Physics and Astronomy, Stony Brook University, Stony Brook, NY 11794-3800, USA
[2] Institute for Advanced Computational Science, Stony Brook University, Stony Brook, NY 11794-5250, USA
[3] National Radio Astronomy Observatory, Charlottesville VA, 22902, USA
[4] NVIDIA Corporation, 2788 San Tomas Expressway, Santa Clara, CA, 95050 USA
[5] Institute for Risk and Uncertainty, University of Liverpool L69 7ZF, UK

E-mail: alan.calder@stonybrook.edu

Abstract. We present early results from a study addressing the question of how one treats the propagation of incertitude, that is, epistemic uncertainty, in input parameters in astrophysical simulations. As an example, we look at the propagation of incertitude in control parameters for stellar winds in MESA stellar evolution simulations. We apply two methods of incertitude propagation, the Cauchy Deviates method and the Quadratic Response Surface method, to quantify the output uncertainty in the final white dwarf mass given a range of values for wind parameters. The methodology we apply is applicable to the problem of propagating input incertitudes through any simulation code treated as a "black box," i.e. a code for which the algorithmic details are either inaccessible or prohibitively complicated. We have made the tools developed for this study freely available to the community.

1. Introduction

Much of theoretical astrophysics relies on large-scale simulation for progress because phenomena of interest typically incorporate multiple physical processes interacting over a wide range of scales. Even study of the basic processes, e.g. plasma and fluid dynamics, requires simulations that can challenge extant computer architectures. Because of both the importance of simulation and the dynamic nature of state-of-the-art computer architectures, considerable resources go into the development and maintenance of simulation codes. Many of these advanced codes, built with person-years of development effort, are made publicly available and are used currently by many investigators in addition to the original developers.

Despite the capabilities of modern codes and architectures, uncertainty imposes limits on the results of simulations. This issue is often neglected in studies in part because of the difficulty in addressing it. As with verification and validation, uncertainty quantification is critical to credible computational science, particularly in astrophysics where most simulation results are predictions, that is, a description of the state of a system under conditions for which the computational model has not been validated [1, 2, 3].

There are two types of uncertainty to consider, aleatory uncertainty (variability) and epistemic uncertainty, also known as incertitude. Aleatory uncertainties occur when there is a variability or inherent randomness, and examples in stellar astrophysics include initial mass and composition among stars. Aleatory uncertainty is described by a (possibly unknown) distribution. In contrast, incertitude is a lack of knowledge about some aspect of the model and is a consequence of incomplete information, including a paucity or absence of data or an incomplete theoretical understanding.

The research presented here addresses the question of how to propagate incertitude in the input parameters though a complex simulation instrument while treating it as a "black box." Doing so avoids the possibly difficult process of analytically propagating uncertainty through the individual arithmetic operations involved in the calculation by rules of interval arithmetic, which may require more expertise than a typical code user has. The simulation instrument we employ in this study is the MESA stellar evolution code, and we treat it as a black box, applying it to the evolution of 1 $M_{\odot}$ ZAMS stars with minimal tuning. This paper presents initial results using relatively low resolution simulation data aimed at showcasing the method. The complete study is forthcoming [4].

2. Incertitude propagation techniques

The goal of incertitude propagation is to quantify uncertainty in the output of a simulation given ranges in input parameters representing their incertitudes. There are a variety of techniques for formally propagating incertitude via interval arithmetic, but most of these involve instrumenting or modifying a particular code or algorithm, which is not possible when addressing a black box code. In this study, we apply methods of incertitude propagation appropriate for black box codes, two simple sampling schemes and two more advanced techniques.

2.1. Sampling the Space

The simplest approach to exploring the bounds of a black box given incertitude in the input parameters is to sample the parameter space. Two obvious approaches are to uniformly sample it and to randomly sample it from a uniform distribution. We applied both of these sampling methods, and include the results from the random sample from a uniform distribution below.

If the black box is known to be a linear function, then sampling each uncertain parameter at its upper and lower bound will provide the bounds on the output. Such an approach is limited, however, by what is known as the "curse of dimensionality," which we discuss below. If the black box is not known to be a linear function, then both a uniform sample and a sample randomly drawn from a uniform distribution serve to characterize the output and can guide investigators in selecting more advanced methods of uncertainty quantification to apply. In principle, a large random sample would allow establishing bounds of the output, but in practice such an approach is limited by the possibility of pathological points in the underlying function and the curse of dimensionality.

2.2. The Cauchy Deviates Method

The Cauchy distribution (also referred to as the Lorentz distribution or the Breit-Wigner distribution) is described by the probability distribution

$$p(z) = \frac{\Delta}{\pi(z^2 + \Delta^2)} , \tag{1}$$

where Δ is the scale parameter of the distribution and z is the random variable. The method of Cauchy Deviates (CD) relies on properties of this distribution, particularly that a linear combination of independent Cauchy-distributed variables is itself Cauchy distributed. The method proceeds by associating the width of the uncertainty interval for each input variable with the scale parameter Δ in equation 1. Then, by sampling Cauchy-distributed inputs, one is able to obtain an interval for the output of the model, which, by the linear combination property, is also Cauchy distributed. One then extracts the scale parameter from the output distribution, Δ_y, which is associated with the width of the uncertainty just as with the input parameters. Details of this method may be found in [5] and complete details of our implementation may be found in [4].

It is important to note that in employing the Cauchy distribution *we make no assumption that the input parameters are Cauchy distributed.* This assertion may seem counter-intuitive. The method simply uses the fact that in a linear black-box, the input Cauchy distributions will be stretched and translated into an output Cauchy distribution. By comparing how the full-width half maxima (FWHM) of the input distributions, set by the scale parameters Δ_i for each of the n input distributions, are distorted into the FWHM for the output distribution, characterized by Δ_y, one can quantify how the black box translates and stretches input incertitude intervals into output intervals. The Cauchy Deviates method cannot predict the distribution of the model outputs as one does not know the distribution of the inputs. Thus the use of the input Cauchy distribution is merely a tool to quantitatively measure how much the black box function alters input intervals into output intervals.

The CD technique works for a linear functionality, providing a way to understand uncertainty propagation through a black box for an approximately linear function. As our study typified, astrophysical simulations are commonly nonlinear, and our approach to handling this nonlinear problem is to tile the input intervals into regions that are suspected to be nearly linear over that area and then apply the CD technique on each region.

It is possible to measure the uncertainty of this Δ_y and create confidence intervals, i.e., one can calculate the statistical uncertainty in our incertitude interval from the number of samples of the black box, N. One can relate the number of samples to the interval through

$$\sigma_{\Delta_y} = \Delta_y \cdot \sqrt{\frac{2}{N}}. \tag{2}$$

Using this relationship one can calculate the number of samples necessary to construct a confidence interval with a specific, statistical uncertainty, σ. One can then say at a chosen level of confidence that the true interval lies within our constructed Cauchy Deviates interval, plus or minus σ.

Because of this relationship between the confidence and the number of samples N, the CD method can constrain the uncertainty for a black box regardless of how large n is. This property of the method becomes more powerful as the number of uncertain inputs grows.

2.3. The Quadratic Response Surface Method

The Quadratic Response Surface Method (QRSM) addresses the problem of a nonlinear function by constructing a quadratic response surface that is fit to the output of the black box function that has been sampled over the domain of the input parameter space. The method thus avoids the need for tiling as is the case with the CD method. Also, despite being nonlinear, an astute choice of interval and fitting function allows determination of the final interval with a reasonable number of samples.

The QRSM method proceeds by first fitting a quadratic surface to the output of the black box function that has been sampled over the domain of the input parameter space. Once a quadratic surface has thus been obtained, the maximum and minimum of the surface can be found in order to obtain an estimate of the bounds on the output of the black box given the chosen ranges of input uncertainty. Finding the maximum and minimum of the quadratic response surface over the incertitude domain is essentially a constrained optimization problem, but finding the maximum and minimum of an n-dimensional quadratic function over the rectangular n-dimensional domain is challenging. We take the approach of [6] and use an ellipsoidal domain method, which allows for the use of a relatively simple mathematical algorithm for finding optima on an ellipsoidal domain.

As the input parameter space is rectangular, an obvious question that arises in moving to an elliptical domain is whether or not the elliptical domain should inscribe or circumscribe the rectangular domain. Figure 1 illustrates the two possibilities. The choice for the ellipse to

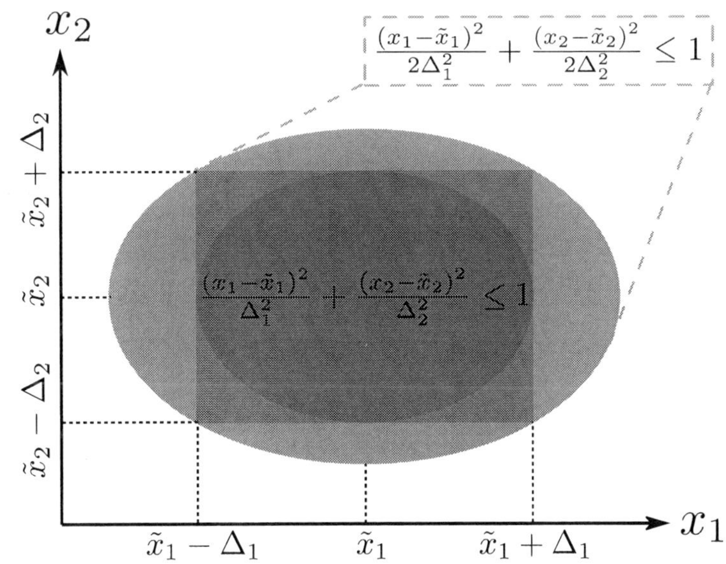

Figure 1. Ellipses constructed that either inscribe and circumscribe the original rectangular interval. We can construct ellipses from our square interval using the pictured equations. For a number of uncertain inputs $n > 2$, the ellipse is generalized to a multivariate ellipsoid.

inscribe or circumscribe the rectangle depends principally on whether or not it is safe to exclude the corner regions of the rectangular parameter domain. If one is confident that the jointly extreme combinations of the input variables do not occur in the modeled system, as might be tenable if one assumes their uncertainties are independent, then the inscribed ellipse is safe, as it is a subset of the rectangular uncertainty domain. If one is not sure that the maxima and minima will not occur on the corners of the input uncertainty domain, the safe path is to select the circumscribed ellipse and ensure that the optimization domain is at least as large as that within the rectangle. One caveat, here, follows from areas within the circumscribed ellipsoid being outside of the n-dimensional rectangle. These regions of parameter space may not be physically realizable, which could potentially lead to an unrealistic optimization solution, for which the black box function may be difficult to evaluate.

The ellipsoid serves as a constraint on the model, and there are many possible circumscribing ellipsoids for a given rectangle. We chose to minimize the volume of the n-dimensional circumscribing ellipsoids for a given n-dimensional rectangle and performed the fit accordingly. Ultimately, the goal is to find the maximum and minimum of an objective function, which for the QRSM method is a quadratic function given by

$$f(x_1, ..., x_n) = f_0 + \sum_{i=1}^{n} f_i \cdot x_i + \sum_{i=1}^{n} \sum_{j=1}^{n} f_{i,j} \cdot x_i \cdot x_j \tag{3}$$

that approximates the output of the black box model over the domain of the input parameters. Equation (3) describes a multivariate quadratic model, and one can obtain the coefficients f_0, f_i, and $f_{i,j}$ by obtaining a fit to the output data using multivariate quadratic regression analysis. The output of the black-box model is then approximately bounded by the maximum and minimum of equation (3) over the ellipsoidal regions in the input parameter domain.

3. Case Study: the MESA Simulation Instrument

The instrument we treat as our black box code for this study is MESA (Modules for Experiments in Stellar Astrophysics) [7, 8, 9]. The MESA code is open source and is widely used in astrophysical research. It is an excellent example of a community code that is the culmination of years of development. As might be expected from incorporating multiple physical effects on multiple scales, the algorithms in MESA are complicated and defy the relatively easy analysis needed to analytically propagate incertitude in input parameters.

We applied the two advanced incertitude propagation techniques to the problem of evolving stars with MESA. Specifically, we performed suites of simulations evolving zero age main sequence (ZAMS) stars through their lives until they become C/O white dwarfs. The simulations were based on a setup available in the MESA test suite, `1M_pre_ms_to_wd`, with models of 1 $M_\odot$ with a solar metallicity, $Z_\odot = 0.02$, and the stars were evolved to a luminosity of $0.1 L_\odot$, the endpoint of stellar evolution.

Stellar winds are the mechanism of mass loss during the evolution of stars, but a first-principles understanding of these remains elusive. Instead, descriptions are parameterized by stellar properties such as luminosity, mass, and radius. MESA implements a number of prescriptions for mass loss during evolution on the red giant branch (RGB) and asymptotic giant branch (AGB) [10, 11, 12, 13]. For simplicity in this study, we chose models of the winds during these phases that do not assume a temperature dependence.

The two epistemically uncertain parameters for these winds are the Reimers and Blöckers scaling factors, η_R and η_B, corresponding to RGB winds and AGB winds respectively. These parameters describe the RGB mass loss rate [11] and AGB mass loss rate [12], as

$$\dot{M}_R = 4 \cdot 10^{-13}\, \eta_R\, \frac{LR}{M}\, [M_\odot\ \mathrm{yr}^{-1}] \tag{4}$$

and

$$\dot{M}_B = 1.932 \cdot 10^{-21}\, \eta_B M^{-3.1} \cdot L^{3.7} R\, [M_\odot\ \mathrm{yr}^{-1}]\,. \tag{5}$$

Here L, M, and R are the luminosity, mass, and radius of the star in solar units. The scaling factors, η_R and η_B, describe the strength of the winds and MESA expresses these through the controls `Reimers_scaling_factor` and `Blocker_scaling_factor`, respectively.

4. Results

In this section, we describe performing suites of simulations with MESA and applying the methods of uncertainty quantification. The results for each suite are a range of white dwarf masses presenting the bounds on the mass given the incertitude of our input.

We began our study with a regularly sampled uniform grid with η_B sampled between 0.0 and 0.1, and η_R was sampled between 0.4 and 0.9. These values were chosen following an earlier study of $1.0 M_\odot$ ZAMS mass stars with similar evolution tracks in [14] and [15]. The results are shown in figure 2. Points at $\eta_B = 0.0$ failed to complete for lower values of η_R, as seen in the figure. After this initial study, the domain was slightly adjusted so that η_B was sampled regularly between 0.01 and 0.1 and η_R was sampled regularly between 0.3 and 0.9. In addition to an initial exploration, this suite of regularly sampled simulations also served as a baseline with which to compare the CD and and QRSM methods. We note that any robust method

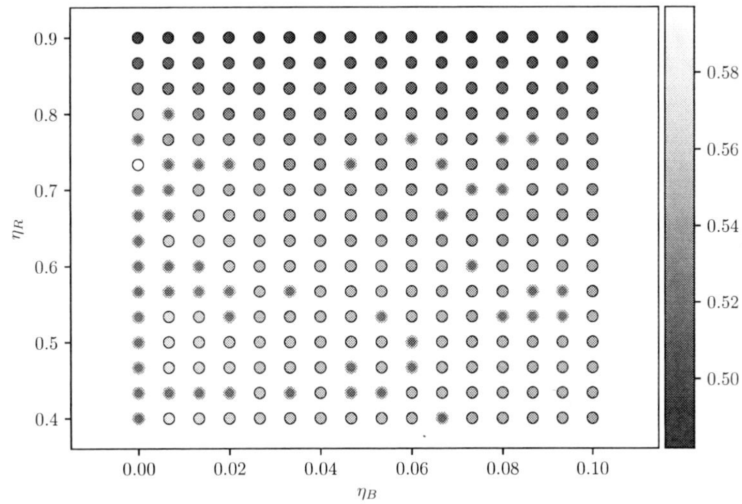

Figure 2. Initial regular grid of 256 $1.0M_{\odot}$ ZAMS models. The axes show η_R vs. η_B, and the colormap indicated the final CO WD star masses. Models for which the simulations did not complete are highlighted in green.

for incertitude propagation must be able to accommodate the realistic situation that simulation codes are unable to carry out simulations for all values of input parameters for a variety of reasons. In this regard, MESA is a typical proxy for many other astrophysical simulation codes.

We next performed a suite of 200 simulations randomly sampled from a uniform distribution. In this case, results from simulations that did not complete are omitted. This suite better illustrated the parameter space, particularly the obvious nonlinearity. Figure 3 depicts the results of this suite of simulations. For this suite, the minimum sampled mass is 0.4902 $M_{\odot}$, and the maximum sampled mass is 0.5984 $M_{\odot}$.

The obvious nonlinearity observed in the first two suites of simulations indicated modification of the CD method would be required. As explained by [5], a way to deal with non-linearity is to subdivide the parameter space into linear sections and for each calculate a set of Cauchy-distributed inputs within the intervals. When bounds are found for each section, the maximum of the maxima and the minimum of the minima produce the bounds for the full space. Figure 4 illustrates the resultant tiles. For the suites of simulations that went into the CD study, we found an interval 0.5395 $\pm$ 0.0542 $M_{\odot}$ with a standard deviation of 0.0077 $M_{\odot}$, giving the range as [0.4776, 0.6014] [0.4699, 0.6091] [0.4622, 0.6168] $M_{\odot}$ with 68.3%, 99.5%, and 99.7% confidence, respectively.

The objective function found for our QRSM study is given by

$$\begin{aligned} M_{\mathrm{WD}} = f(\eta_B, \eta_R) = \; & 0.5596 - 1.1323\ \eta_B + 0.1378\ \eta_R \\ & + 5.7688\ \eta_B^2 - 0.2060\ \eta_R^2 \\ & + 0.4646\ \eta_B \cdot \eta_R\ , \end{aligned} \tag{6}$$

where the coefficients f_0, f_i, and $f_{i,j}$ in equation 3 have been replaced by the actual values obtained by fitting (see figure 5 below). For the purpose of obtaining the objective function, the sampling was done over the actual incertitude domain instead of on either the inner or outer ellipse. The computation of the output incertitude range based on the objective function is, however, found using the elliptical constraints. Figure 5 illustrates the quadratic regression

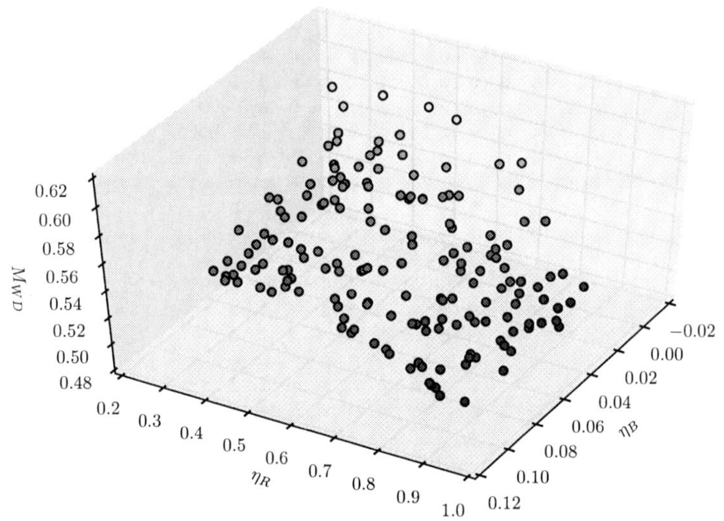

Figure 3. A suite of 200 models with values of η_R and η_B chosen at random from a uniform distribution between $\eta_R = [0.3, 0.9]$ and $\eta_B = [0.01, 0.1]$.

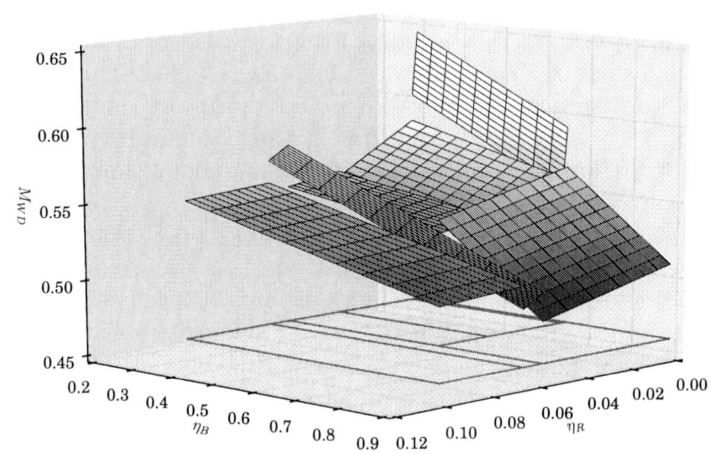

Figure 4. Parameter space broken up into 6 linear planes for application of the CD approach to uncertainty quantification. Linear tiles are determined via a region-growing algorithm that fits planes to samples of the parameter space subject to a constraint of the coefficient of determination for each plane fit.

fit. For our QRSM study, we found ranges [0.4979,0.5606] $M_\odot$ and [0.4811,0.6119] $M_\odot$ for the inscribed and circumscribed ellipses, respectively.

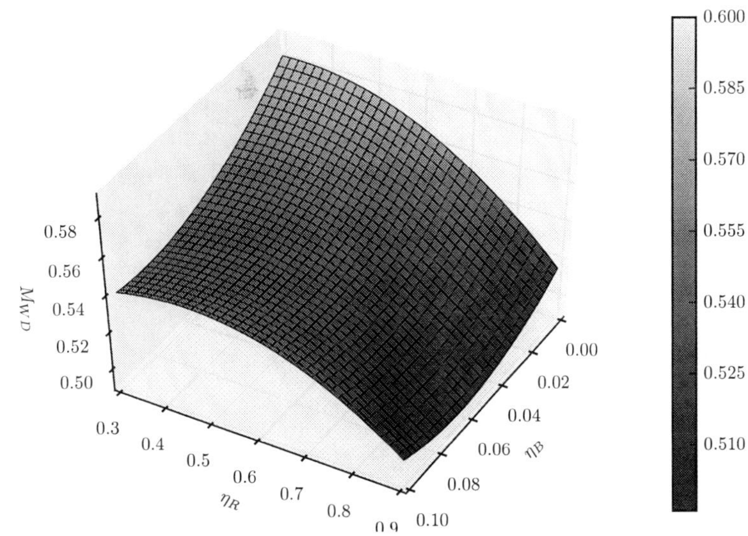

Figure 5. Surface plot of the multivariate quadratic regression fit. Note that the maximum and minimum are on the edges—there are no local extrema within the space.

5. Discussion and Conclusions

The purpose of this manuscript is to present the methodology for a study of incertitude propagation in black box simulation codes, and the results presented here, ranges of white dwarf masses, are not definitive. We presented this study as a look at work in progress and defer final calculated ranges to the complete work [4]. As may be seen in the results from the first uniform grid of points in the input parameter domain, many of the MESA simulations did not converge. While it is possible to produce more robust simulations with MESA by "tuning" the input to improve convergence, we specifically did not do this so as to keep our black box completely consistent for all simulations. The complete study will review the issue of convergence of the simulations. Our conclusion is that these methods, CD and QRSM, allow the formal propagation of the input incertitude into incertitude intervals of the simulation outputs in a straight forward and computationally tractable way.

In this case study, we utilized only two variables with incertitude. The full advantage of the advanced methods that we demonstrated, CD and QRSM, does not become apparent until applied to higher dimensional incertitude problems. One might be tempted to conclude that since the case study problem can be broken up into linear sub regions that one could easily evaluate the model at the extrema of the incertitude ranges. While this could be easily done for the two-dimensional case, in a situation with n uncertain input parameters, this approach would require 2^n model evaluations to estimate the output uncertainty of the model. This is the so-called curse of dimensionality. In contrast, the advanced methods require far fewer model evaluations in order to estimate the model output uncertainty.

Another comparison we wish to make is between the CD and QRSM methods. The CD method has an advantage in that one has some knowledge of how many model sample points are needed in order to specify the desired confidence in the estimate of the model output uncertainty. In contrast, with the QRSM method there is no statistical knowledge of how many sample points are needed in order assess the precision with which the model output incertitude is known. The QRSM method, however, does posses two advantages over the CD method. First, the QRSM method makes no assumption of linearity and does not require subdividing the input parameter incertitude intervals into quasi-linear sub-intervals, thereby avoiding computationally costly model evaluations needed to evaluate the degree of linearity. Second, the QRSM method potentially can require far fewer model evaluations in that as few as $n(n-1)/2+2n+1$ are needed to determine the coefficients of the ellipsoid given in equation 3. In a situation where n is large, this property could result in a significant advantage realized through the use of the QRSM method. One can only gain confidence in the estimates of the model output uncertainty, however, by carrying out a convergence test where the number of sample points is increased. The minimum number of points yields an exact output uncertainty only in the case of a perfectly quadratic model.

Finally, we wish to point out a limitation on the interpretation of the output model response. This response cannot be physically interpreted as a probability distribution because the distribution of input parameters is unknown by definition of incertitude. All that can be determined in the case of input incertitudes is a model output uncertainty. Were the distributions of the input uncertainties known, the output response could have an interpretation as a probability distribution. The methods described here, however, are not suitable to address this problem. Research into uncertainty quantification for the problem of known distributions of input uncertainties, however, is underway.

The source code for the tools applied to this study, Star Simulation Techniques for Research in Uncertainty Quantification, is freely available at `https://github.com/StarSTRUQ`. The site includes the software for generating the Cauchy deviates samples, performing the QRSM analysis, and the tiling, which subdivides a sampled domain into planar tiles constrained by the planar fit.

Acknowledgments

This work was supported in part by the US Department of Energy under grant DE-FG02-87ER40317 and by the US National Science Foundation under grant AST-1211563. The research described here originated with co-author Melissa Hoffman's thesis at Stony Brook University. The simulations performed for this study were performed with the MESA stellar evolution code and the authors thank its developers and community for making this resource available. The authors would like to thank Stony Brook Research Computing and Cyberinfrastructure, and the Institute for Advanced Computational Science at Stony Brook University for access to the high-performance LIred and SeaWulf computing systems, the latter of which was made possible by a $1.4M National Science Foundation grant (#1531492). The authors also thank Tracy M. Calder for previewing the manuscript. This research has made use of NASA's Astrophysics Data System Bibliographic Services.

References

[1] Roache P 1998 *Verification and Validation in Computational Science and Engineering* (Hermosa) ISBN 9780913478080 URL `https://books.google.com/books?id=ENR1QgAACAAJ`

[2] Oberkampf W and Roy C 2010 *Verification and Validation in Scientific Computing* (Cambridge University Press) ISBN 9781139491761 URL `https://books.google.com/books?id=7d26zLEJ1FUC`

[3] AIAA 1998 *AIAA Guide for the Verification and Validation of Computational Fluid Dynamics Simulations* AIAA G: American Institute of Aeronautics and Astronautics (American Institute of Aeronautics and Astronautics) ISBN 9781563472855 URL `https://books.google.com/books?id=OUAHAAAACAAJ`

[4] Hoffman M M, Willcox D E, Katz M P, Ferson S and Swesty F D 2017 *in preparation, to be submitted to ApJ*

[5] Kreinovich V and Ferson S 2004 *Reliability Engineering and System Safety* **85** 267–279 ISSN 0951-8320

[6] Kreinovich V, Hajagos J G, Tucker W T, Ginzburg L R and Ferson S 2008 Propagating uncertainty through a quadratic response surface model Tech. Rep. SAND2008-5983 Sandia National Laboratory

[7] Paxton B, Bildsten L, Dotter A, Herwig F, Lesaffre P and Timmes F 2011 *ApJS* **192** 3 (*Preprint* `1009.1622`)

[8] Paxton B, Cantiello M, Arras P, Bildsten L, Brown E F, Dotter A, Mankovich C, Montgomery M H, Stello D, Timmes F X and Townsend R 2013 *ApJS* **208** 4 (*Preprint* `1301.0319`)

[9] Paxton B, Marchant P, Schwab J, Bauer E B, Bildsten L, Cantiello M, Dessart L, Farmer R, Hu H, Langer N, Townsend R H D, Townsley D M and Timmes F X 2015 *ApJS* **220** 15 (*Preprint* 1506.03146)
[10] Reimers D 1975 *Memoires of the Societe Royale des Sciences de Liege* **8** 369–382
[11] Reimers D 1975 *Circumstellar envelopes and mass loss of red giant stars* pp 229–256
[12] Blöcker T 1995 *A&A* **297** 727
[13] de Jager C, Nieuwenhuijzen H and van der Hucht K A 1988 *A&AS* **72** 259–289
[14] Karakas A I and Lugaro M 2016 *ApJ* **825** 26 (*Preprint* 1604.02178)
[15] Pignatari M, Herwig F, Hirschi R, Bennett M, Rockefeller G, Fryer C, Timmes F X, Ritter C, Heger A, Jones S, Battino U, Dotter A, Trappitsch R, Diehl S, Frischknecht U, Hungerford A, Magkotsios G, Travaglio C and Young P 2016 *ApJS* **225** 24 (*Preprint* 1307.6961)

IOP Conf. Series: Journal of Physics: Conf. Series **1031** (2018) 012017 doi:10.1088/1742-6596/1031/1/012017

Towards an exascale code for GRMHD on dynamical spacetimes

Sven Köppel

Frankfurt Institute for Advanced Studies (FIAS), Ruth-Moufang-Strasse 1, 60438 Frankfurt am Main, Germany; Institute for theoretical Physics (ITP), Goethe University Frankfurt, Max-von-Laue-Strasse 1, 60438 Frankfurt am Main, Germany

E-mail: koeppel@fias.uni-frankfurt.de

Abstract. We present the first results on the construction of an exascale hyperbolic PDE engine (ExaHyPE), a code for the next generation of supercomputers with the objective to evolve dynamical spacetimes of black holes, neutron stars and binaries. We solve a novel first order formulation of Einstein field equations in the conformal and constraint damping Z4 formulation (CCZ4) coupled to ideal general relativistic magnetohydrodynamics (GRMHD), using divergence-cleaning. We adopt a novel communication-avoiding one-step ADER-DG scheme with an a-posteriori subcell finite volume limiter on adaptive spacetrees. Despite being only at its first stages, the code passes a number of tests in special and general relativity.

1. Introduction

Gravitational waves ejected by the merger of compact objects have been recently detected for both black hole-black hole mergers [1] as well as neutron star-neutron star mergers [2]. Both phenomena take place in the strong field regime of general relativity, described by Einstein field equations and so far only accessible by numerical relativity. The general relativistic fluid description of matter allows to predict precisely the dynamics of single and binary neutron stars. Furthermode, electromagnetism plays an important role for the realistic description of the fate of the binary system. General relativistic (magneto)hydrodynamics—in short GR(M)HD—is a successful theory to describe these pheomena, also underlying the gravitational wave banks used for the recently detected NS-NS merger.

The coupled numerical evolution of Einsteins and the GRMHD equations is a comparatively young field of research, as the first successful evolution of a BH-BH system dates back only one decade [3]. Since then, many GRHD and GRMHD codes for solving coalescing neutron stars have been developed (for instance [4–11], see also [12, 13] for a review).

In the time of gravitational wave multi-messenger astronomy, the quantiative and qualitative need for accurate numerical simulations is essential. In the same time, computational science is entering the era of *exascale* computing, i.e. the first supercomputer to archive 10^{18} arithmetic operations per second is estimated to be built in the next three years. It is excepted that in order to reasonably exploit these resources, the next generation of numerical codes must satisfy tight constraints in locality, communication, storage, cache and energy efficiency [14].

In this work, we present ingredients for such a code which solves generic nonlinear hyperbolic first order balance laws,

$$\partial_t Q_k + \partial_i F_k^i(Q) + B_k^{ij}(Q)\partial_i Q_j = S_k(Q)\,, \tag{1}$$

with a state vector $\vec{Q}$, conserved fluxes $\vec{F}^i$, a nonconservative part described by the matrix $\vec{B}^{ij}$ and algebraic source terms $\vec{S}$. We present particular PDEs, defined by $\{\vec{Q}, \vec{F}^i, \vec{B}^{ij}, \vec{S}\}$, as well as first results with actual implementations. The paper is structured as follows: In Section 2 we present the PDE system, in Section 3 we present the numerical scheme to solve this system, in Section 4 we present the grid representation and communication, in Section 5 we present first benchmarks and results while in Section 6 we give a summary and outlook of the next steps.

We work in geometric units (speed of light $c = 1$ and gravitational constant $G = 1$). Greek indices run from 0 to 3, Latin indices run from 1 to 3 and Einstein sum convention is applied over repeated incides. The signature of the metric tensor is assumed to be $(-,+,+,+)$.

2. The Einstein-Maxwell-Euler system

Three theories which describe a wide range of astrophysical phenomenae are

$$\text{Einstein equations} \qquad R g_{\mu\nu} + R_{\mu\nu} = 8\pi G\, T_{\mu\nu}\,, \tag{2}$$

$$\text{Maxwell equations} \qquad \nabla^*_\mu F^{\mu\nu} = 0 \quad \text{and} \quad \nabla^\mu F_{\mu\nu} = 0\,, \tag{3}$$

$$\text{Euler equations} \qquad \nabla^\mu T_{\mu\nu} = 0 \quad \text{and} \quad \nabla_\mu(\rho u^\mu) = 0\,. \tag{4}$$

with Ricci scalar R, metric $g_{\mu\nu}$, energy momentum tensor $T_{\mu\nu}$, Faraday tensor $F^{\mu\nu}$, dual Faraday tensor ${}^*F_{\mu\nu}$, rest mass density ρ and fluid velocity u^μ. In the following, 3+1 initial value formulations in the language of (1) are presented.

2.1. *A first order formulation of Einstein equations: FO-CCZ4*

The formulation of Einstein equations used in this work builds upon successors of the original ADM Cauchy initial value formulation of Einsteins equations [15] which introduces the 16 dynamical fields lapse α, shift β^i, 3-metric γ_{ij} and 3-extrinsic curvature K_{ij}. The ADM equations are not hyperbolic [16–18]. Two successors of the ADM equations are the BSSNOK formulation [19, 20] and the class of Z4 formulations [21, 22]. Both approaches were unified in the conformal and constraint-damping CCZ4 formulation [23, 24] which was recently rewritten in a first order formulation in space and time [25] which we subsequently refer to as FO-CCZ4. The full derivation and discussion of these equations including an eigenstructure analysis can be found in [25] where also strong hyperboliciy for the system in certain gauges is proven.

In order to briefly outline the equations, we introduce the conformal factor $\phi = (\det \gamma_{ij})^{-1/6}$ and subsequently $\tilde{\gamma}_{ij} = \phi^2 \gamma_{ij}$. We also define a conformal and trace-free extrinsic curvature tensor $\tilde{A}_{ij} = \phi^2(K_{ij} - 1/3\, K\gamma_{ij})$ where it's trace $K = K_{ij}\gamma^{ij}$ has been seperated. Furthermore, we define the Christoffel variable contraction $\tilde{\Gamma}^i = \tilde{\gamma}^{ij}\tilde{\Gamma}^i_{jk}$ as well as $\hat{\Gamma}^i = \tilde{\Gamma}^i + 2\tilde{\gamma}^{ij}Z_j$. Recall that $Z_\mu = (\Theta, Z_i)$ is the vector field which measures the distance to an analytical solution of Einsteins equations with constraint $Z_\mu = 0$. For the first-order formulation, we introduce the auxilliary vectors $A_i = \partial_i \ln\alpha$ and $P_i = \partial_i \ln\phi$ as well as tensors $B_k^i = \partial_k \beta^i$ and $D_{kij} = 1/2\partial_k\tilde{\gamma}_{kj}$. Based on these quantities, the FO-CCZ4 system is given in Table 1 where system parameters $(\tilde{\tau}, s, f, e, \kappa_{1,2,3}, c, \eta, \mu)$ and slicing condition $g(\alpha)$ are highlighted in red whereas matter contributions (cf. Section 2.2) are highlighted in blue. The state vector Q_i collects 59 evolved variables. For details on the imposed constraint equations, applied Bona-Massó gauge conditions and meaning of the parameters we refer again to [25].

Table 1. FOCCZ4 system $\partial_t Q + B(Q)\nabla Q = S(Q)$

	Q_i	NCP_{Q_i}: Nonconservative product $B\nabla Q$	S_{Q_i}: Algebraic source S
ODE-ADM	$\ln\alpha$	0	$\beta^k A_k - \alpha g(\alpha)(K - K_0 - 2\Theta c)$
	β^i	0	$s\beta^k B^i_k + s\, f\, b^i$
	$\tilde\gamma_{ij}$	0	$\beta^k 2D_{kij} + \tilde\gamma_{ki}B^k_j + \tilde\gamma_{kj}B^k_i - 2/3\tilde\gamma_{ij}B^k_k - 2\alpha\left(\tilde A_{ij} - 1/3\,\tilde\gamma_{ij}\mathrm{tr}\tilde A\right) - 1/\tau(\tilde\gamma - 1)\,\tilde\gamma_{ij}$
	$\ln\phi$	0	$\beta^k P_k + 1/3\left(\alpha K - B^k_k\right)$
SO-CCZ4	$\tilde A_{ij}$	$-\beta^k\partial_k\tilde A_{ij} + 1/3\,\tilde\gamma_{ij}\left(-\nabla^k\nabla_k\alpha + \alpha R + 2\alpha\nabla_k Z^k\right)$ $-\phi^2\left(-\nabla_i\nabla_j\alpha + \alpha R_{ij} + \alpha\nabla_i Z_j + \alpha\nabla_j Z_i\right)$	$\tilde A_{ki}B^k_j + \tilde A_{kj}B^k_i - 2/3\,\tilde A_{ij}B^k_k - \phi^4 8\pi\left(S_{ij} - 1/3\,\tau\tilde g_{ij}\right)$ $+\alpha\tilde A_{ij}(K - 2\Theta c) - 2\alpha\tilde A_{il}\tilde\gamma^{lm}\tilde A_{mj} - 1/\tau\,\tilde\gamma_{ij}\,\mathrm{tr}\tilde A$
	K	$-\beta^k\partial_k K + \nabla^i\nabla_i\alpha - \alpha(R + 2\nabla_i Z^i)$	$\alpha K(K - 2\Theta c) - 3\alpha\kappa_1(1 + \kappa_2)\Theta + 4\pi(S - 3\tau)$
	Θ	$-\beta^k\partial_k\Theta - 1/2\,\alpha c^2(R + 2\nabla_i Z^i)$	$1/2\,\alpha c^2(2/3K^2 - \tilde A_{ij}\tilde A^{ij}) - \alpha\Theta K c - Z^i\alpha A_i - \alpha\kappa_1(2 + \kappa_2)\Theta - 8\pi\alpha\tau$
	$\hat\Gamma^i$	$-\beta^k\partial_k\hat\Gamma^i + 4/3\,\alpha\tilde\gamma^{ij}\partial_j K - 2\alpha\tilde\gamma^{ki}\partial_k\Theta$ $-s\tilde\gamma^{kl}\partial_{(k}B^i_{l)} - s/3\,\tilde\gamma^{ik}\partial_{(k}B^l_{l)} - s2\alpha\tilde\gamma^{ik}\tilde\gamma^{nm}\partial_k\tilde A_{nm}$	$2/3\tilde\Gamma^i B^k_k - \tilde\Gamma^k B^i_k + 2\alpha(\tilde\Gamma^i_{jk}\tilde A^{jk} - 3\tilde A^{ij}P_j) - 2\alpha\tilde\gamma^{ki}\left(\Theta A_k + 2/3 K Z_k\right) - 16\pi\alpha\tilde\gamma^{ij}S_j$ $-2\alpha\tilde A^{ij}A_j - 4s\alpha\tilde\gamma^{ik}D^{nm}_k\tilde A_{nm} + 2\kappa_3\left(2/3\,\tilde\gamma^{ij}Z_j B^k_k - \tilde\gamma^{jk}Z_j B^i_k\right) - 2\alpha\kappa_1\tilde\gamma^{ij}Z_j$
	b^i	$-s\beta^k\partial_k b^i$	$s(\partial_t\hat\Gamma^i - \beta^k\partial_k\hat\Gamma^i - \eta b^i)$
FO-CCZ4	A_k	$-\beta^l\partial_l A_k + \alpha g(\alpha)\left(\partial_k K - \partial_k K_0 - 2c\partial_k\Theta\right)$ $+s\,\alpha\, g(\alpha)\tilde\gamma^{nm}\partial_k\tilde A_{nm}$	$-s\,\alpha\, g(\alpha)\partial_k\tilde\gamma^{nm}\tilde A_{nm}$ $-\alpha A_k\left(K - K_0 - 2\Theta c\right)\left(g(\alpha) + \alpha g'(\alpha)\right) + B^l_k\, A_l$
	B^i_k	$-s\beta^l\partial_l B^i_k - s\big(f\partial_k b^i - \mu\,\tilde\gamma^{ij}\left(\partial_k P_j - \partial_j P_k\right)$ $+\mu\,\tilde\gamma^{ij}\tilde\gamma^{nl}\left(\partial_k D_{ljn} - \partial_l D_{kjn}\right)\big)$	$B^l_k\, B^i_l$
	D_{kij}	$-\beta^l\partial_l D_{kij} - s/2\,\tilde\gamma_{mi}\partial_{(k}B^m_{j)} - s/2\,\tilde\gamma_{mj}\partial_{(k}B^m_{i)}$ $+s/3\,\tilde\gamma_{ij}\partial_{(k}B^m_{m)} + \alpha\partial_k\tilde A_{ij} - 1/3\,\alpha\tilde\gamma_{ij}\tilde\gamma^{nm}\partial_k\tilde A_{nm}$	$B^l_k D_{lij} + B^l_j D_{kli} + B^l_i D_{klj} - 2/3\,B^l_l D_{kij} + 1/3\,\alpha\tilde\gamma_{ij}\partial_k\tilde\gamma^{nm}\tilde A_{nm}$ $-\alpha A_k(\tilde A_{ij} - 1/3\,\tilde\gamma_{ij}\mathrm{tr}\tilde A)$
	P_k	$\beta^l\partial_l P_k - 1/3\,\alpha\partial_k K + 1/3\,\partial_{(k}B^i_{i)} - s/3\,\alpha\tilde\gamma^{nm}\partial_k\tilde A_{nm}$	$1/3\,\alpha A_k K + B^l_k P_l + s/3\,\alpha\partial_k\tilde\gamma^{nm}\tilde A_{nm}$

2.2. The GRMHD system

The GRMHD system describes general relativistic hydrodynamics coupled to the magnetohydrodynamic approximation of Maxwell theory where magnetic field lines are advected with the fluid and the electric field is given by $\vec E = -\vec v \times \vec B$. Thus, the Maxwell equations reduce to an evolution equation for the magnetic field as well as the constraint equation $\nabla \cdot \vec B = 0$. Violations are treated with a cleaning approach, i.e. we evolve an additional scalar field ϕ to transport constraint violations (no damping is applied). The PDE system is given in Table 2, where $D = W\rho$ is the conserved density which is related to the rest mass density ρ by the Lorentz factor $W = (1 - v_i v^i)^{-1/2}$ with fluid velocity v^i. The conserved momentum is given by $S_j = \rho h W^2 v_j$. We also evolve the rescaled energy density $\tau = E - D = \rho h W^2 - p - \rho W$ where E is the energy density measured by the Eulerian observer and p is the fluid pressure. B^i represents the magnetic field and $w^i = \alpha v^i - \beta^i$ the advection velocity relative to the coordinates, also refered to as transport velocity. The GRMHD 3-energy-momentum tensor is given by $S^{ij} = S^i v^j + (p - T^k_k/2) - T^{ij}$ with $T^{ij} = B^i B^j/W^2 + (B^i v^j)(B^k v_k)$ the Maxwell 3-energy momentum tensor. The primitive recovery follows [26–28]. Note that all fields are rescaled to a tensor density by $\sqrt{g} \equiv \sqrt{\det g_{ij}}$ and the actual PDE therefore reads

Table 2. GRMHD system components

Q_i	Conserved Flux F^i	Nonconservative product $B\nabla Q$
D	$w^i D$	0
S_j	$\alpha W^i_j - \beta^i S_j$	$E\partial_j\alpha - \frac{\alpha}{2}S^{lm}\partial_j\gamma_{lm} - S_k\partial_j\beta^k$
τ	$\alpha(S^i - v^i D) - \beta^i\tau$	$S^i\partial_i\alpha - \alpha S^{ij}K_{ij}$
B^j	$w^i B^j - v^j B^i - B^i\beta^j$	$B^i\partial_i\beta^j + \alpha\gamma^{ij}\partial_i\phi$
ϕ	$\phi\beta^i - \alpha B^i$	$\phi\partial_i\beta^i + \frac{1}{2}\phi\gamma^{ij}\beta^k\partial_k\gamma_{ij} - B^i\partial_i\alpha$

$$\partial_t\sqrt{g}Q_i + \nabla_j\sqrt{g}F^j(Q) + \sqrt{g}B\nabla Q = 0\,. \tag{5}$$

For a comprehensive discussion of this system in context of our ADER-DG approach (Section 3), we refer to our publication [29].

3. Communication-avoiding ADER-DG scheme with a subcell limiter

We adopt a path-conservative arbitrary high order derivative discontinous Galerkin (ADER-DG) scheme with a finite volume (FV) limiting approach, adaptive mesh refinement (AMR) and local timestepping. This scheme was already applied on several nonrelativistic problems [11, 30–32] and recently formulated for GRMHD [29].

The computational (spatial) domain Ω is covered by Cartesian nonoverlapping *cells* Ω_i (also refered to as *patches* or *elements*). The analytic DG solution $u_h(x, t^n) = \sum_l \hat{u}^n_{il} \Phi_l(x)$ is written in an orthogonal basis $\Phi_l(x)$ in each cell. We choose the Euler-Legendre polynomial basis at order N and thus store $N + 1$ degrees of freedom (DOF) on a non-uniform nodal *subgrid*. The PDE system (1) is multiplied by these spatial test functions and written in weak integral form as

$$0 = \int_{t^n}^{t^{n+1}} \mathrm{d}t \int_{\Omega_i} \mathrm{d}x \left(\frac{\partial Q}{\partial t} + \frac{\partial F^i(Q)}{\partial x^i} + B^{ij} \frac{\partial Q_j}{\partial x^i} - S(Q) \right) \Phi_k \,. \tag{6}$$

3.1. Space-time predictor

In order to archieve element-locality, the scheme (6) is seperated in an element-local space-time *predictor* and a *corrector* step where boundary face values are communicated,

$$0 = \left(\hat{u}^{n+1}_{kl} - \hat{u}^n_{kl} \right) M_{kl} + \int_{t^n}^{t^{n+1}} \mathrm{d}t \left\{ \int_{\partial\Omega_i} \vec{n} \mathrm{d}S \; \Phi_k \mathcal{J}(q_h^-, q_h^+) + \int_{\Omega_i \backslash \partial\Omega_i} \mathrm{d}x \; \Phi_k \left[B(q_h) \nabla q_h - S \right] \right\}. \tag{7}$$

Here, the q_h is the element-local predictor solution and $M_{kl} = \int_{\Omega_i} \mathrm{d}x \, \Phi_k \Phi_l$ the diagonal element mass matrix. The surface integral collects the conservative and nonconservative jump terms $\mathcal{J}$ between cells and is subject of an approximate Riemann solver, we apply a path-conservative HLLEM method [33] which takes the nonconservative terms into account.

The predictor solution $q_h(x, t)$ is approximated in time from a known solution without considering the interaction with the neighbouring cells. We thus solve a *local* Cauchy problem inside the cell,

$$\int_{t^n}^{t^{n+1}} \mathrm{d}t \int_{\Omega_i} \mathrm{d}x \, \Theta_k(x,t) \left(S(q_h) - B^{ij} \nabla q_h - \nabla F(q_h) \right) = \int_{t^n}^{t^{n+1}} \mathrm{d}t \int_{\Omega_i} \mathrm{d}x \, \Theta_k(x,t) \frac{\partial q_h}{\partial t} \tag{8}$$

$$= \int_{\Omega_i} \mathrm{d}x \Theta_k(x, t^{n+1}) \left(q_h(x, t^{n+1}) - u_h(x, t^n) \right) - \int_{t^n}^{t^{n+1}} \mathrm{d}t \int_{\Omega_i} \mathrm{d}x \frac{\partial \Theta_k}{\partial t} q_h(x,t) \,, \tag{9}$$

which was rewritten by partial integration in order to archieve a fixed-point problem which can easily be solved with fast converging algorithms [34, 35].

3.2. Subcell limiter

In case of discontinuities in the solution, the ADER-DG scheme can produce spurious oscillations (Gibbs phenomena) or even unphysical solutions. We encounter these problems by switching to a robust finite volume scheme on a regular subgrid. This switch is either done *a priori* on a purely geometrical criterion (e.g. static puncture position) or *a posteriori*, *after* computing a DG timestep. In the later case, several criteria are applied to ensure the correctness of the solution: Mathematical admissibility criteria (the occurance of floating point errors), physical admissibility criteria (e.g. $D > 0$, $\vec{v}^2 < 1$, $\rho > 0$ in the GRMHD equations or $\alpha < \alpha_0$ in CCZ4) as well as an heuristic discrete maximum principle (DMP) which checks whether all cell solutions are below the maxima and above the minima of the adjacent cells.

By choosing a subgrid size of $M = 2N + 1$ cells per axis, where N is the order of the DG polynomial, the number of subcells is maximized without breaking the CFL condition which would make the limited cells evolve with smaller timesteps than untroubled DG cells. We can not only preserve all $N + 1$ DOF per axis of a troubled DG cell, the limiter also acts as a natrual mesh refinement on its own.

In principle, any FV scheme can be used. We implemented an ADER-TVD [29] and ADER-WENO scheme [25] but also, for communication avoidance, a 1st order Godunov scheme and a 2nd order MUSCL-Hancock scheme [36]. The limiter is extensively described in the references [25, 29, 36]. In short, it completely replaces the DG scheme in the troubled cell. In order to properly communicate the boundary values, a DOF prolongation of DG→FV is also neccessary in the adjacent cells. After each timestep, the restriction FV→DG takes place to restore the DG polynomial.

4. Exascale architecture

It is assumed that many assumptions of contemporary high performance computing no longer hold at the exascale. For instance, due to the heterogenity and complex, error-prone and fault-tolerant computer architectures, equal work distribution (i.e. naive Cartesian slicing which minimizes the allocation surface) is no longer assumed to lead to balanced computing time. Similarly, massive parallel scalability will become even more important, and thus the communication overhead has to be reduced at any price [14].

4.1. Seperation of physics and machinery: Loose of control

Cactus and the Einstein Toolkit [37, 38] are examples of mature frameworks for PDE solvers. While they focus on abstracting the distributed memory parallelization (MPI), the advent of manycore processors requires hybrid codes (shared memory parallelization, e.g. OpenMP or TBB) and diffused the clear seperation of physics and computational infrastructure [39, 40]. In contrast, the novel ExaHyPE code forces the users to step back from implementing random access schemes as it prescribes both the data layout, programming workflow as well as the major algorithmic steps [36, 41, 42]. Primarily, the physicist as a user *declares* the PDE system without describing its control flow.

4.2. Spacetime AMR by Peano

We use the Peano framework [43] which implements tree-structured Cartesian meshes with adaptive mesh refinement (AMR). The name indicates that the Peano space filling curve is employed for storage locality [44]. The code also supports Hybrid (MPI and TBB) parallelization with dynamic load balancing.

Peano implements the inversion of control paradigm where users give up control of the program workflow in order to profit from high-level task restructuring. Peano defines the program workflow in terms of two coupled finite state machines, one for the grid traversal and one for the major steps of the PDE solver. Peano is one of the building blocks of the exascale hyperbolic PDE engine ExaHyPE [41]. In ExaHyPE, the concrete numerical scheme (e.g. the particular Riemann solver used) is exchangable and delegated to modular *kernels* which can be tailored to a specific PDE.

5. Benchmarks and Results

This section collects seperated results from vacuum spacetime [25], GRMHD in fixed background Cowling approximation [29], and dedicated nonrelativistic tests with the limiter in the Peano framework [36]. All figures are taken from the respective publications. The evolved fluids follow an ideal gas equation of state.

5.1. Nonrelativistic benchmarks

Figure 1 visualizes the Limiter functionality on an Eulerian explosion problem (spherical Sod shock tube [45, 46]). Figure 2 shows a similar but non-spherical Sod shock tube. Both figures highlight the transition steps between limited (troubled) cells and unlimited cells. Figure 3 shows the 2D SRMHD rotor problem [47] in a poor resolution, cf. [31, fig. 4] for a high resolution. Since the limiter acts only per-cell, a minimum resolution is neccessary to properly work.

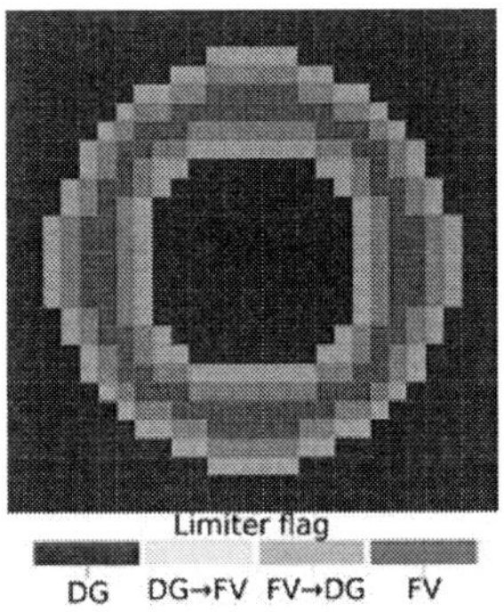

Figure 1. Radial explosion problem with clearly visible cell status.

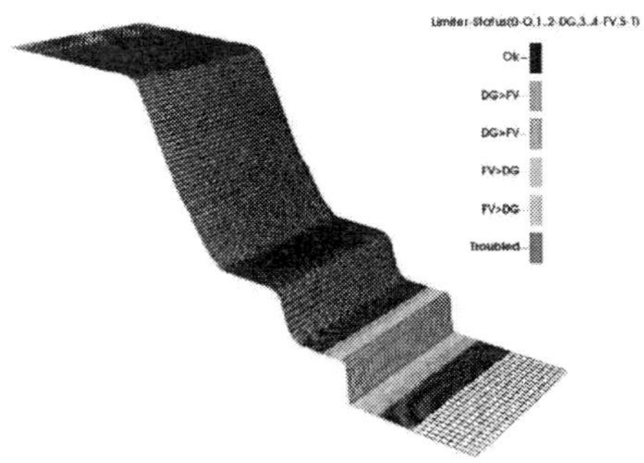

Figure 2. 2D Sod shock tube. The rarefraction wave and shock front are computed with ADER-DG, only the contact discontinuity is limited.

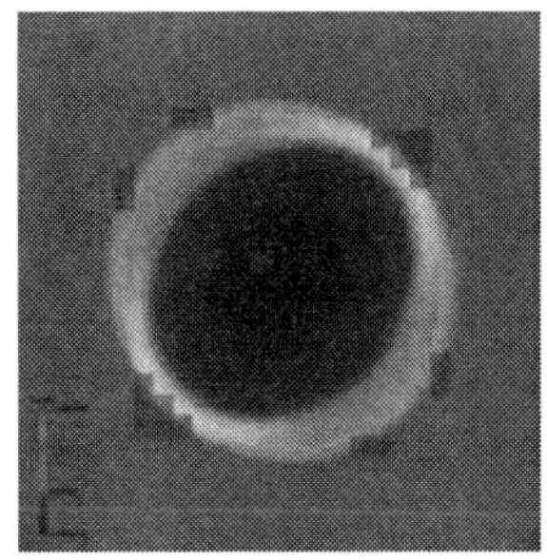

Figure 3. MHD Rotor, poorly resolved: Coincidental limiting due to heuristics.

5.2. Cowling GRHD benchmarks

Figure 4 shows the Limiter active in a 2D thick torus [48] solution after $t = 1000M$. Figure 5 shows the grid around a TOV [49] neutron star surface. Both figures show the color encoded rest mass density profile. The limiter always activates at the sharp crossover from matter to vacuum (atmosphere).

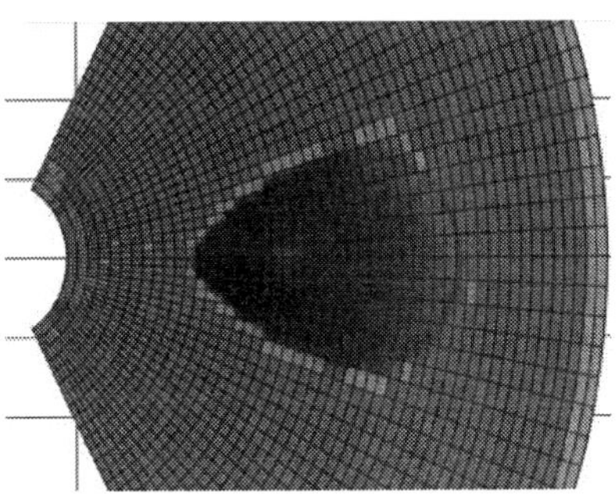

Figure 4. 2D thick torus in polar coordinates. The red cells indicate limited cells.

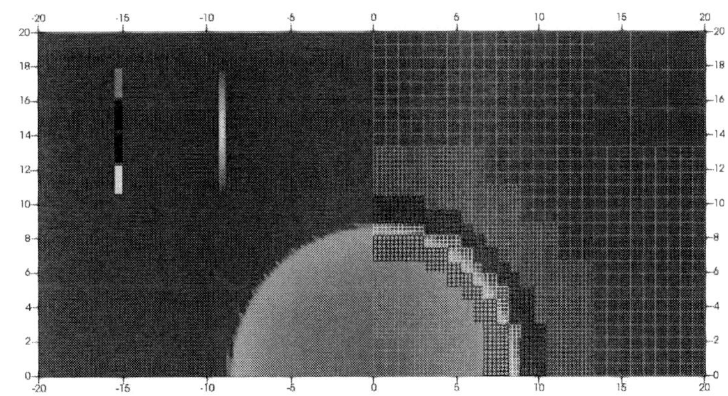

Figure 5. 3D cut of a stable neutron star. Green cells indicate limited cells on a 3-level AMR grid.

5.3. Vacuum spacetime benchmarks

Figure 6 shows the Hamiltonian and momentum constraint violations during the evolution of a static Schwarzschild black hole spacetime with the FO-CCZ4 system. Limiting is applied only in the vicinity of the puncture. Figure 7 shows a smooth gauge wave after 160 crossing times which is recovered with 100 cells at third order ($p = 3$) already very good.

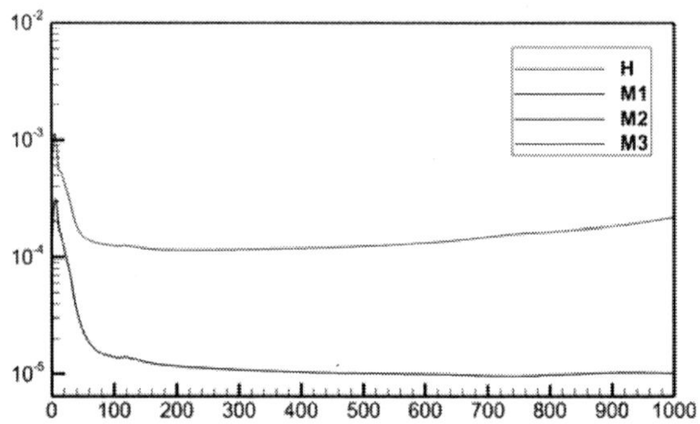

Figure 6. Errors in the black hole spacetime evolution up to $t = 1000M$.

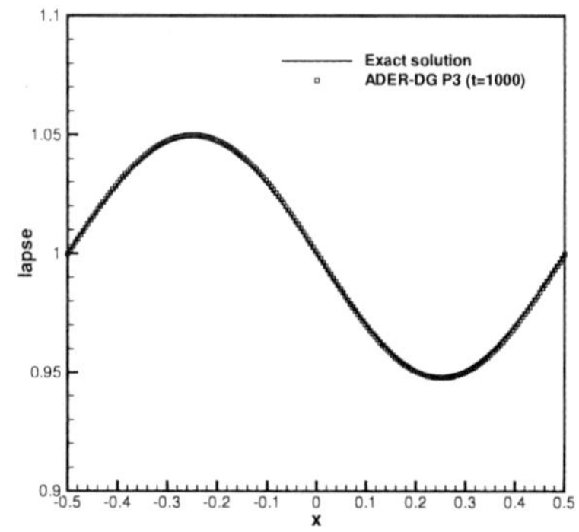

Figure 7. Large amplitude gauge wave

6. Summary

We have shown first results on a new code which is the unification of three different scientific branches: The development of a new formulation of Einstein Equations, the application of sophisticated high-order schemes and the implementation in a pioneering high performance grid framework.

The proposed novel first order CCZ4 formulation was proven to be strongly hyperbolic. Due to the special treatment of the nonconservative terms as Borel measures, the system can be evolved very accurately. The dynamical coupling of CCZ4 and GRMHD were sketched and are subject to ongoing investigations.

The ADER-DG scheme provides high order time integration and high order space discretization. The local space-time predictor is communication avoiding and allows to integrate with only one data transfer/grid traversal per time step [36]. The goal is to maximize the arithmetic intensity ("science per watt"). The subcell limiter preserves the degrees of freedom by resolving physical oscillations with adaptive mesh refinement. It is shock capturing as it allows to evolve shock fronts stably with any finite volume scheme.

The Peano framework provides a modern AMR framework where users give up control in order to obtain scalability and maintainablity. Peano promises scaling to millions of cores. All together, the three ingredients (i) CCZ4+GRMHD (ii) the communication avoiding ADER-DG and (iii) Peano, were merged to a single code called ExaHyPE from which first results have been shown in this work and more are expected soon.

Acknowledgments

S.K. thanks the members of the ExaHyPE consortium, especially Dominic Charrier, Francesco Fambri and Michael Dumbser, for contributing plots and results to this proceeding. The author also thanks Tobias Weinzierl, Michael Bader and especially Luciano Rezzolla for discussion and support. The author also thanks the Helmholtz Graduate School HGS-HiRE for FAIR and the Frankfurt International Graduate School for Science for support. This project has received funding from the European Unions Horizon 2020 research and innovation programme under grant agreement No 671698 (ExaHyPE).

References

[1] Abbott B P *et al.* (LIGO Scientific Collaboration and Virgo Collaboration) 2016 *Phys. Rev. Lett.* **116**(6) 061102 URL https://link.aps.org/doi/10.1103/PhysRevLett.116.061102
[2] Abbott B P *et al.* (LIGO Scientific Collaboration and Virgo Collaboration) 2017 *Phys. Rev. Lett.* **119**(16) 161101 URL https://link.aps.org/doi/10.1103/PhysRevLett.119.161101
[3] Pretorius F 2005 *Phys. Rev. Lett.* **95** 121101 (*Preprint* gr-qc/0507014)
[4] Baiotti L, Hawke I, Montero P J, Löffler F, Rezzolla L, Stergioulas N, Font J A and Seidel E 2005 *Phys. Rev. D* **71** 024035 (*Preprint* gr-qc/0403029)
[5] Duez M D, Liu Y T, Shapiro S L and Stephens B C 2005 *Phys. Rev. D* **72** 024028 astro-ph/0503420
[6] Anninos P, Fragile P C and Salmonson J D 2005 *Astrophys. J.* **635** 723
[7] Antón L, Zanotti O, Miralles J A, Martí J M, Ibáñez J M, Font J A and Pons J A 2006 *Astrophys. J.* **637** 296 (*Preprint* astro-ph/0506063)
[8] Giacomazzo B and Rezzolla L 2007 *Classical Quantum Gravity* **24** S235
[9] Radice D and Rezzolla L 2012 *Astron. Astrophys.* **547** A26 (*Preprint* 1206.6502)
[10] Radice D, Rezzolla L and Galeazzi F 2013 (*Preprint* arXiv:1306.6052)
[11] Zanotti O and Dumbser M 2015 *Computer Physics Communications* **188** 110–127
[12] Font J A 2008 *Living Rev. Relativ.* **6** 4 URL http://www.livingreviews.org/lrr-2008-7
[13] Martí J M and Müller E 2015 *Living Reviews in Computational Astrophysics* **1**
[14] Kogge P, Bergman K, Borkar S *et al.* 2008 Exascale computing study: Technology challenges in achieving exascale systems dARPA report URL http://www.cse.nd.edu/Reports/2008/TR-2008-13.pdf
[15] Arnowitt R, Deser S and Misner C W 1962 *Gravitation: An introduction to current research* ed Witten L (New York: John Wiley) pp 227–265 (*Preprint* gr-qc/0405109)
[16] Baumgarte T W and Shapiro S L 2010 *Numerical Relativity: Solving Einstein's Equations on the Computer* (Cambridge University Press, Cambridge UK)
[17] Gourgoulhon E 2012 *3+1 Formalism in General Relativity* (*Lecture Notes in Physics, Berlin Springer Verlag* vol 846)
[18] Rezzolla L and Zanotti O 2013 *Relativistic Hydrodynamics* (Oxford University Press, Oxford UK)
[19] Shibata M and Nakamura T 1995 *Phys. Rev. D* **52** 5428
[20] Baumgarte T W and Shapiro S L 1998 *Phys. Rev. D* **59** 024007
[21] Bona C, Ledvinka T, Palenzuela C and Zacek M 2003 *Phys. Rev. D* **67** 104005
[22] Alic D, Bona C and Bona-Casas C 2009 *Phys. Rev. D* **79** 044026
[23] Alic D, Bona-Casas C, Bona C, Rezzolla L and Palenzuela C 2012 *Phys. Rev. D* **85**(6) 064040
[24] Alic D, Kastaun W and Rezzolla L 2013 (*Preprint* 1307.7391)
[25] Dumbser M, Guercilena F, Köppel S, Rezzolla L and Zanotti O 2017 *submitted to Physical Review D* (*Preprint* 1707.09910)
[26] Noble S C, Gammie C F, McKinney J C and Del Zanna L 2006 *Astrop. J.* **641** 626–637
[27] Del Zanna L, Zanotti O, Bucciantini N and Londrillo P 2007 *Astronomy and Astrophysics* **473** 11–30
[28] Porth O, Olivares H, Mizuno Y, Younsi Z, Rezzolla L, Moscibrodzka M, Falcke H and Kramer M 2017 *Computational Astrophysics and Cosmology* **4** 1
[29] Fambri F, Dumbser M, Köppel S, Rezzolla L and Zanotti O 2018 *Monthly Notices of the Royal Astronomical Society* **477** 4543–4564 URL http://dx.doi.org/10.1093/mnras/sty734
[30] Zanotti O, Fambri F, Dumbser M and Hidalgo A 2015 *Computers and Fluids* **118** 204 – 224 ISSN 0045-7930
[31] Zanotti O, Fambri F and Dumbser M 2015 *Mon. Not. R. Astron. Soc.* **452** 3010–3029
[32] Fambri F, Dumbser M and Zanotti O 2017 *Computer Physics Communications* ISSN 0010-4655 URL http://www.sciencedirect.com/science/article/pii/S0010465517302448
[33] Dumbser M and Balsara D S 2016 *Journal of Computational Physics* **304** 275–319 ISSN 0021-9991
[34] Dumbser M, Balsara D S, Toro E F and Munz C D 2008 *Journal of Computational Physics* **227** 8209–8253
[35] Dumbser M, Castro M, Parés C and Toro E 2009 *Computers and Fluids* **38** 1731–1748
[36] Charrier D E and Weinzierl T 2018 Stop talking to me – a communication-avoiding ader-dg realisation *in preparation*
[37] Löffler F, Faber J, Bentivegna E, Bode T, Diener P, Haas R, Hinder I, Mundim B C, Ott C D, Schnetter E, Allen G, Campanelli M and Laguna P 2011 *ArXiv e-prints* (*Preprint* 1111.3344)
[38] Goodale T, Allen G, Lanfermann G, Massó J, Radke T, Seidel E and Shalf J 2003 *Vector and Parallel Processing – VECPAR'2002, 5th International Conference, Lecture Notes in Computer Science* (Berlin: Springer)

[39] Löffler F, Brandt S R, Allen G and Schnetter E 2013 *CoRR* **abs/1309.1812** URL http://arxiv.org/abs/1309.1812
[40] Schnetter E, Blazewicz M, Brandt S R, Koppelman D M and Löffler F 2014 *CoRR* **abs/1410.1764** URL http://arxiv.org/abs/1410.1764
[41] Charrier D E, Dumbser M, Duru K, Fambri F, Gabriel A, Gallard J M, Köppel S, Osorio A, Rannabauer L, Rezzolla L, Tavelli M and Weinzierl T 2017 Exahype, an exascale hyperbolic pde engine http://www.exahype.eu
[42] Schwarz A, Charrier D E, Guera F, Gallard J M, Samfass P, Köppel S, Weinzierl T and Varduhn V 2017 The ExaHyPE guidebook http://dev.exahype.eu/guidebook.pdf
[43] Weinzierl T 2015 *CoRR* **abs/1506.04496** URL http://arxiv.org/abs/1506.04496
[44] Bader M 2013 *Space-Filling Curves: An Introduction with Applications in Scientific Computing* (Springer)
[45] Sod G A 1978 *Journal of Computational Physics* **27** 1–31
[46] Toro E 1999 *Riemann Solvers and Numerical Methods for Fluid Dynamics* 2nd ed (Springer)
[47] Balsara D S and Spicer D 1999 *Journal of Computational Physics* **148** 133–148
[48] Font J A and Daigne F 2002 *Mon. Not. R. Astron. Soc.* **334** 383–400
[49] Tolman R C 1939 *Phys. Rev.* **55** 364

IOP Conf. Series: Journal of Physics: Conf. Series **1031** (2018) 012018 doi:10.1088/1742-6596/1031/1/012018

Adaptive Kinetic-Fluid Models for Expanding Plasmas

Robert Arslanbekov [a] **and Vladimir Kolobov** [a,b,1]

[a] CFD Research Corporation, Huntsville, AL 35806, USA
[b] The University of Alabama in Huntsville, AL 35899, USA

E-mail: vladimir.kolobov@cfdrc.com

Abstract. We describe gas dynamic and plasma effects for transient plasma jets under different conditions. Our Unified Flow Solver (UFS) is evaluated for simulations of mixed continuum-rarefied flows, transient gas expansion dynamics and collisionless plasma expansion. We identify further advances that are required for hybrid fluid-kinetic simulations of laser-ablation and EEE-induced plasma jets.

1. Introduction

Expanding plasmas appear in various fields of space science and industrial technologies. Two types of expanding plasmas can be distinguished. The first type includes dense plasma expanding into vacuum or an ambient gas. These types of problems are characterized by a slow time scale, which is associated with adiabatic electron motion and by ion acceleration in plasma-induced electric fields. The second type includes plasmas formed inside gas plumes during propagation of fast ionization waves. These plasma jets are characterized by a fast time scale associated with ionization processes induced by electrons in a neutral gas; the ion motion at this time scale can be neglected. In particular, such plasma jets have been produced in atmospheric air inside plumes of rare gases, and are currently being studied for numerous applications [1]. The disparity of time scales for electron and ion transport allows substantial simplifications in modelling both types of plasma jets.

Here, we focus our attention on the first type of jets. An expansion of evaporated solid/liquid material into vacuum or an ambient background gas has been widely studied for pulsed laser ablation [2]. During the gas plume expansion, the evaporated material is accelerated as its initial internal energy is converted into kinetic energy. The resulting expansion velocity contains information about the properties of the hot dense gas before expansion including its ionization degree.

Micro-explosions also result in the formation of plasma jets expanding from a solid surface. During a jet expansion, dense non-ideal plasma turns into ideal plasma as the plasma density gradually decreases and inter-particle collisions are reduced. Plasma flares expanding into vacuum from arc-induced micro-explosions (ectons) on the cathode surface have been previously analyzed using both fluid and kinetic models [3,4,5]. The importance of hybrid fluid-kinetic approach to simulations of Explosive Electron Emission (EEE) and associated plasma expansion has been emphasized in Ref. [4]. However, no analysis of ambient gas effects on the dynamics of EEE-induced plasma jets has been reported so far.

In this paper, we describe simulations of gas dynamic and plasma effects for transient jets. We apply our Unified Flow Solver (UFS) [6] for mixed continuum-rarefied flows, investigate gas plume

[1] To whom any correspondence should be addressed.

dynamics and collisionless plasma expansion. We identify further advances that are required for hybrid fluid-kinetic simulations of laser-ablation and EEE-induced plasma jets.

2. Simulations of Gas Plume Dynamics

Plume expansion produced by pulsed laser ablation in vacuum has been previously investigated experimentally and computationally by Direct Simulation Monte Carlo (DSMC) [7] for low-fluency regimes, when the plume ionization and plasma effects are negligible. DSMC simulations have demonstrated that hyperthermal kinetic energies, which are observed for ablated metals and graphite, can be explained through purely gas-dynamic considerations. The increase of particle kinetic energy is due to two effects: hydrodynamic plume acceleration into the forward direction and kinetic selection of fast particles in the on-axis region. The problem is characterized by two dimensionless parameters: $b = R/(v_T \tau)$ and $Kn = \lambda_0 / R$. The first one is the ratio of the spot radius R to the distance travelled by the evaporated particles with thermal velocity v_T during the duration of the laser pulse τ. The second parameter is the Knudsen number, i.e. ratio of the particle mean free path to the spot radius, which defines initial collisionality degree of the plume [8]. For plume expansion into a steady ambient gas, additional parameters are the density ratio of the background particles and evaporated particles, $\eta = n_b / n_0$, and the temperature ratio, $\eta_T = T_b / T_0$.

We have first performed simulations of plume expansion into same background gas for different values of n_0 / n_b using Boltzmann and Euler solvers in UFS. For simplicity, we used 2D-2V Boltzmann solver with the simplest BGK model for collisions. The base conditions, $\tau \sim 10$ ns, $R \sim 10$ μm, $v_T \sim 10^3$ m/s, correspond to $b \sim 1$. We used dimensionless variables in our calculations with velocity normalized to v_T, and density normalized to n_0. The computational domain consists of a unit box: $0 < x < 1$, $-0.5 < y < 0.5$. The boundary condition at the surface ($x = 0$), corresponds to an evaporating spot with dimensionless radius $R = 0.05$. Inside the spot, the dimensionless density, velocity and temperature are assumed to be equal to one during the power pulse, and then drop instantly to the ambient gas values, η and η_T. The temperature ratio $\eta_T = 0.1$ was used in all simulations reported below.

Figure 1 shows an example of instantaneous spatial distributions of gas density, mean velocity and temperature at $t/\tau = 1.2$ for $\eta = 0.01$ and $Kn = 0.2$. The mesh was dynamically adapted based on gradients of gas density and temperature. The velocity mesh for the Boltzmann solver was selected as $-6 < \xi_x, \xi_y < 6$, with 24 cells in each velocity direction.

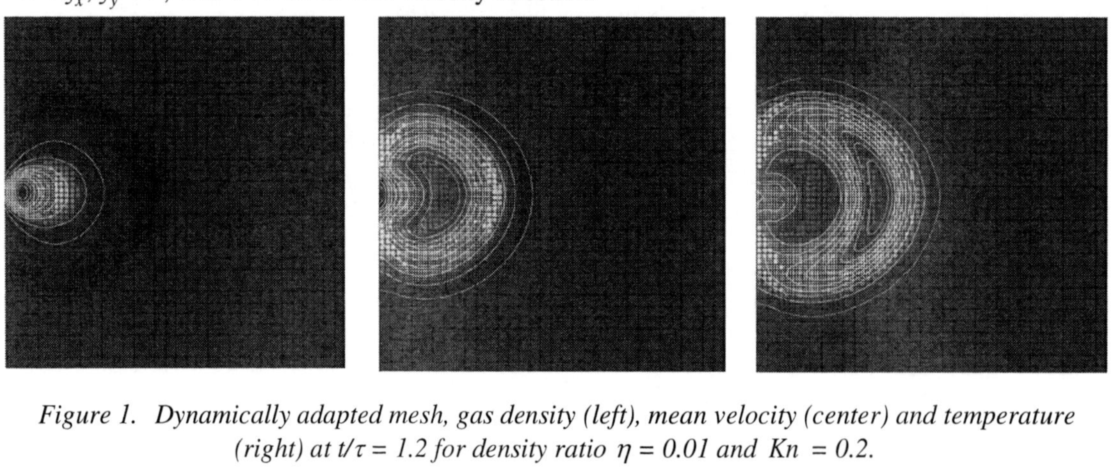

Figure 1. Dynamically adapted mesh, gas density (left), mean velocity (center) and temperature (right) at $t/\tau = 1.2$ for density ratio $\eta = 0.01$ and $Kn = 0.2$.

Simulations have been performed for different η and Kn. Figure 2 shows a time dependence of the gas density on the axis at $L = 10R$ for $\eta = 0.1$, 0.01, and 0.001. Solid lines correspond to $Kn = 0.2$, dashed line is for $Kn = 0.01$. The calculated density variations correspond to the time-of-flight (TOF) distributions measured by density-sensitive detectors in experiments. It is seen that the shock wave propagates faster at lower background densities (lower η), and a decrease of Kn number results in a sharper front of the TOF function.

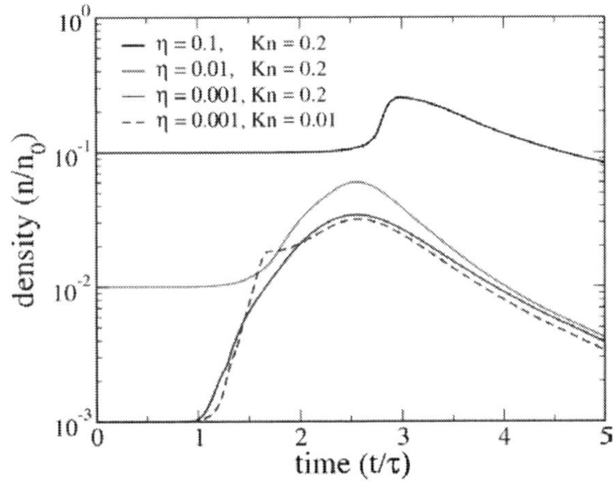

Figure 2. Time of flight function for different initial density ratios at L=10R.

3. Expansion of Evaporated Material into Vacuum or into Background Gas

3.1. 1D Simulations of Cu Expansion into Argon with a Navier-Stokes Solver

Expansion of evaporated metal plumes into vacuum or a background gas has been studied in relation to laser ablation for thin film deposition (see, e.g., Ref. [9,10] and references cited therein). We have first solved this problem with a multispecies Navier-Stokes (NS) solver. Following Ref. [10], we first considered a 1D expansion problem with a background Ar gas pressure of 200 mTorr and a temperature of 300 K at rest. The mean free path, λ, for Ar atoms under these initial is ~10 μm. At the beginning, we created a region of high density and high pressure of Cu atoms with density of 10^{19} cm^{-3} and temperature 7000 K. The density ratio is thus ~1500, and the pressure ratio ~35000. The size of this initial Cu core is 50 μm (~5 initial Ar atom mean free paths and ~7000 Cu atom mean free paths). The computation domain of 15 cm (~15000 initial Ar atom mean free paths) was utilized. Because of such a large variation of the problem spatial scales, as well as moving strong shocks and discontinuities, a dynamically adapted grid was used with adaptation on species mass fractions, density, pressure and Mach number (each with a different threshold value).

The flow field profiles at two time instances are shown in Figure 3. One can see that shortly after the expansion starts, a density peak is formed due to gas compression (the snowplowing effect, see, e.g., Ref.[10]). This compression takes place at the Cu plume front. Further downstream from this front, there develops a moving shock propagating with a velocity of ~3000 m/s. At this time, the Cu plume expands over ~5 mm (or ~500 initial mean free paths of Ar atoms). At a later time of 108 μs, the shock expands over 12 cm, while the Cu plume decelerates to much smaller speeds, hardly moving in the preceding 100 μs. The shock propagates into the background gas and the density drops by 3 orders of magnitude. By comparing these results with those presented in Ref.[10] for a 1D plume expansion we observed close agreement considering that different metal and background species were simulated and a laser source was used (see below). In particular, at the expansion time of 100 μs, a negative flow velocity of ~200 m/s was predicted in Ref. [10] (which is an indication of a counter flow flux of Ar atoms), while NS predicts a negative velocity of 190 m/s at this time, see Figure 3, right.

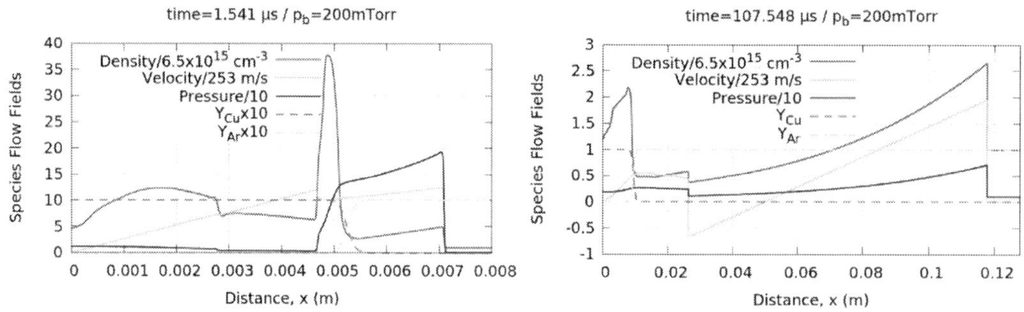

Figure 3. Species flow fields obtained by NS solver at times of ~1.5 μs (left) and ~107 μs (right).

It is well known that the presence of background gas radically changes the dynamics of metal plume expansion [10]. To study this effect, we have also conducted simulations for the background pressures of 20 mTorr and 2 Torr. The time evolutions of the peak velocities and temperatures are compared in Figure 4 for three background gas pressures. One can see that the plume propagation peak velocity and temperature decrease with increasing gas pressure, as has been previously observed in numerical and experimental studies. In addition to the free expansion simulations, we have carried out simulations by assuming source terms for mass density and energy, which are present during the action of a short laser pulse according to Ref. [10]. To simplify the analysis, we approximated the initial flow solutions from Ref. [9] obtained with source terms and carried out the subsequent computations. As one can see in Figure 4, the plume velocities of ~5000 m/s are predicted, while the previous free expansion simulations with the step-function initial conditions show smaller velocities by a factor of ~2. These results are in very good agreement with Ref. [10], where also a factor of 2 differences have been predicted. Also, we note the NS solver correctly reproduces typical velocities of the initial metal plume propagation, which are of the order of 10^4 m/s, as has been observed experimentally and numerically, see Ref. [9]. One can also see that the predicted peak temperatures with the sources (reaching ~25000 K) are about a factor of 2 higher compared to the free expansion results. These findings emphasize the importance of metal content initial conditions for accurate predictions of plume expansion into vacuum or low background pressure environment.

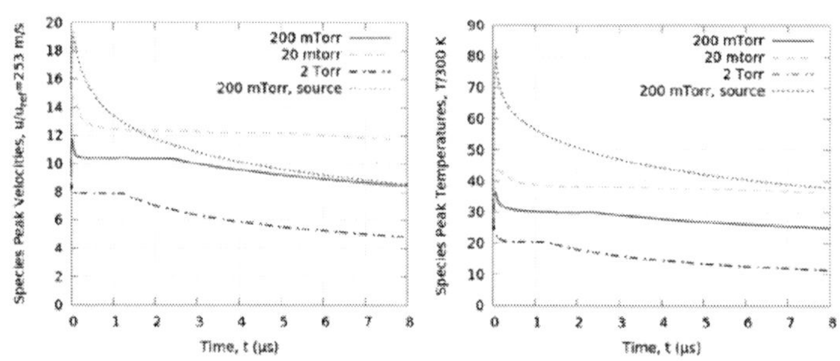

Figure 4. Temporal evolutions of species mean velocities and temperatures for 20 mTorr, 200 mTorr and 2 Torr for the free expansion case. Also shown are evolutions in case with a laser source.

3.2. 1D Simulations of Cu Plume Expansion into Argon with a Boltzmann Solver

We next study the problem of high-density Cu plume expansion into a low-pressure Ar gas by the multispecies Boltzmann kinetic solver. The same initial conditions are used as in the previously described simulations with the NS solver. The results for Ar pressure of 200 mTorr are shown in Figure 5 at different time instances. One can see that the velocity of the Cu plume front quickly increases to high values of ~16 × u_{ref} (= 16×253 m/s) and compression of the background gas starts to occur as the Ar density in the vicinity of the Cu plume front increases several with respect to the

IOP Conf. Series: Journal of Physics: Conf. Series **1031** (2018) 012018 doi:10.1088/1742-6596/1031/1/012018

background gas density. This corresponds to the snow-plow mechanism discussed in the literature.[9,10] At later times of ~1 μs, a compression wave in Ar gas develops forming a supersonic shock wave propagating over the background gas. The most interesting feature of the kinetic simulations is the fact that over the initial time of about 1–2 μs the Cu atom mean velocity and temperature differ significantly from those of the Ar gas. Indeed, Cu atoms attain velocities almost double of that of the Ar atoms, while Cu atoms temperatures are more than double of temperature of Ar atoms, reaching values of ~60 × T_{ref} (= 300 K), or ~18000 K. At later times (t >1–2 μs), the macro-parameters of Ar and Cu species start to coincide with each other. At these times, energetic Cu atoms have travelled a distance of over several 100's of mean free paths and become in equilibrium with the background gas.

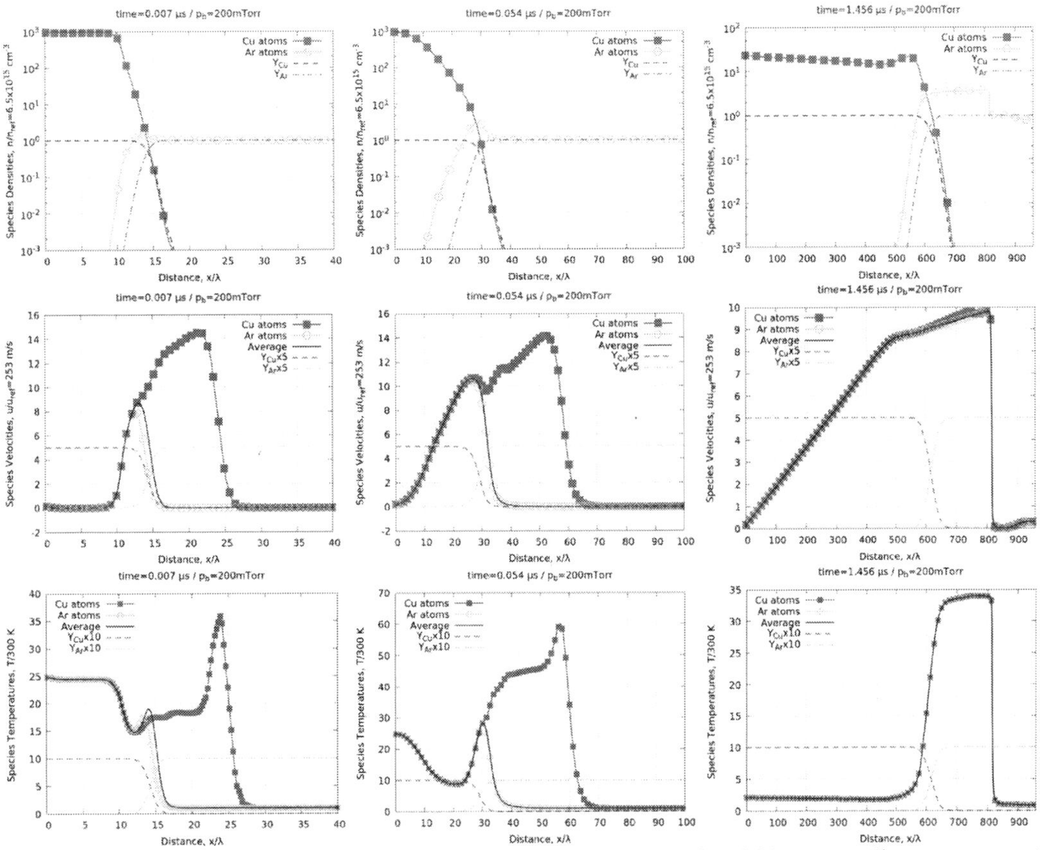

Figure 5. Species macroparameters obtained by Boltzmann solver for 200 mTorr. Shown are species densities (top row), mean velocities (middle row) and mean temperatures (bottom row) at 3 times of 0.007, 0.054 and 1.46 μs. Also shown are Cu and Ar mass fractions as well as mixture average quantities. Spatial scale is normalized to λ =10 μm.

At the initial times after the Cu plume release, a small amount of highly energetic Cu atoms run away from the dense plume region and find themselves in a low-density background gas. The fact that the Cu atoms are heavier than the Ar atoms makes it even more difficult for them to dissipate energy in collisions with the background species and reach an equilibrium with the background gas. This mechanism is clearly evidenced in Figure 6 where we plot a typical instantaneous Cu atom velocity distribution function (VDF) are two locations: one inside the high-density Cu plume (explosion core) region and one in front of the propagating plume (at a distance of ~10–20 λ). The VDF inside the core is close to Maxwellian due to frequent collisions among Cu atoms. In front of the plume boundary (low-density region), a highly non-equilibrium VDF forms, which is characterized by two distant peaks in velocity space. One peak corresponds to the low-energy part of the VDF replicating that in the high-density core, and the other peak corresponds to the highly energetic Cu atoms escaped from the high-density plume zone and thus becoming almost collisionless. The VDF of such a form is characterized by high mean velocities and temperatures as observed in Figure 5 and Figure 6. The number density of the highly energetic atoms is almost 5 orders of magnitude lower than that in the Cu plume region as seen in Figure 6. This makes it difficult to accurately capture such a VDF with statistical DSMC methods making the DVM Boltzmann solver advantageous for modeling transient kinetic processes associated with the formation of energetic metal species during the metal plume expansion. The energetic metal atoms can propagate over long distances from the initial explosive core and can be detected experimentally. These neutral atoms can be easily ionized (e.g., in charge-exchange collisions) with surrounding plasma and become ions. The energy and angular distributions of ions can be a good diagnostic tool for the physical processes occurring in the small explosive zones far from detector locations.

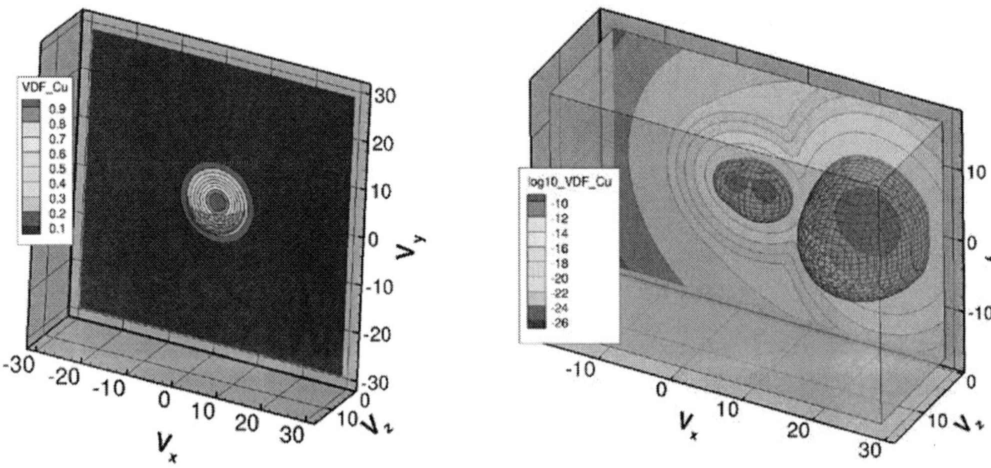

Figure 6. Typical Cu atoms VDFs in 3D velocity space at two locations: explosion core (left, linear scale) and run-away zone (right, log scale). The velocities are normalized to u_{ref} = 253 m/s.

To study the influence of background gas on the dynamics of metal plume expansion, we have varied the background gas pressure. For the background gas pressures of 20 mTorr, the mean free path is λ = 100 μm and for 2 Torr λ = 1 μm. The results for the species temperatures for 20 mTorr are presented in Figure 7. One can see that for the lower background pressure, there is even stronger non-equilibrium, which manifests itself, among other things, in very high Cu atom temperatures reaching values of 120 times the background temperature. These high temperature regions extend over 100 mean free paths from the Cu propagation front. And because of the longer mean free path in the lower pressure conditions, the actual non-equilibrium zone is larger by a factor of at least several 10's than for 200 mTorr. The temporal evolutions of the species mean velocities and temperatures shown in Figure 8 indicate that the initial stages of the plume expansion are characterized by both high velocities (up to ~6500 m/s) and high temperatures (up to ~36,000 K).

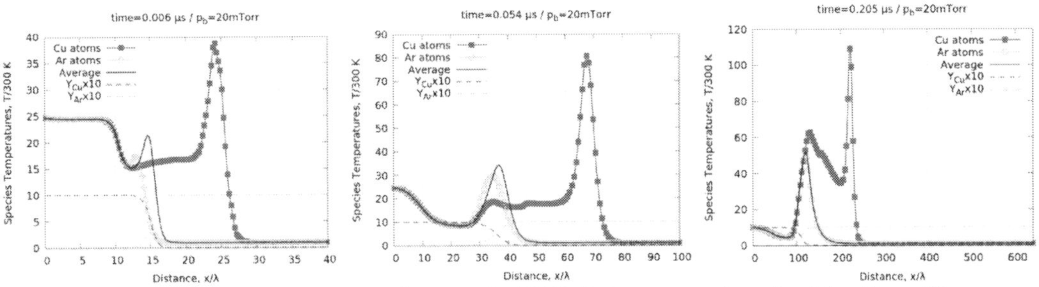

Figure 7. Species macroparameters obtained by UFS-Boltzmann solver for 20 mTorr. Shown are species mean temperatures together with mixture average temperatures and species mass fractions at 3 times instances of 0.006, 0.054 and 0.2 μs. Spatial scale is normalized to λ =100 μm.

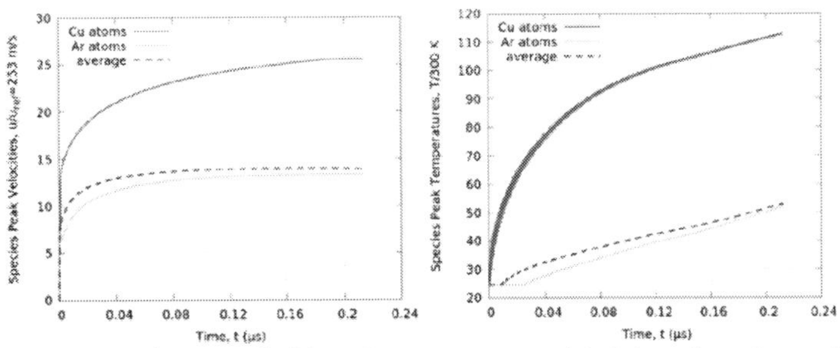

Figure 8. Species peak mean velocities (left) and temperatures (right) as functions of time obtained by UFS-Boltzmann for 20 mTorr. Shown also are mixture average quantities.

Splitting metal plume species into slow and fast components has been previously observed both experimentally and computationally using simplified models (see, e.g., Refs. [11,12]). It has been observed experimentally [12] that the plasma plume splitting occurs over time periods of up to several hundreds of ns and strongly influenced by the background gas pressure. Both these experimental observations are in excellent agreement with our predictions.

3.3. 1D Simulations of Cu Plume Expansion into Argon with a Hybrid Solver

The kinetic simulations of the high-density metal plume expansion into low-pressure Ar gas have clearly demonstrated that the initial stage of this process is highly non-equilibrium. The non-equilibrium zone extended up to 100 mean free paths in front of the metal expansion boundary. For simulation domains extending over several tens of thousands of mean free paths (see Figure 3), kinetic simulations are rather costly. Furthermore, the non-equilibrium zones typically exist over a short period of time during initial metal plume expansion when highly energetic metal atoms are generated. For later dynamics of the plume expansion, fluid models may be adequate.

Using UFS methodology, we have conducted hybrid kinetic-fluid simulations of the plume expansion into background gas. Details of the cell-by-cell selection of the kinetic and fluid solvers and coupling these solvers at kinetic-fluid interfaces have been previously described in [13,14]. A slightly modified continuum breakdown criterion based on the mean free path was used in present simulations, and the computational grid was adapted in the same way as in the previous simulations. By utilizing the dynamic load balancing capabilities for efficient partitioning of the kinetic and fluid cells, we obtained very good load balance across all computational cores utilized. The results for Ar pressure of 200 mTorr are presented in Figure 9. One can see that at the time instances shown, kinetic regions of about 100 and 200 mean free paths in length are required. By comparing with the full Boltzmann solutions, we could observe very good agreement, and a gain factor of up to 10 in terms of CPU time and required RAM have been achieved for a non-optimized kinetic breakdown criterion. Optimization of this criterion is expected to result in even more significant gain factors.

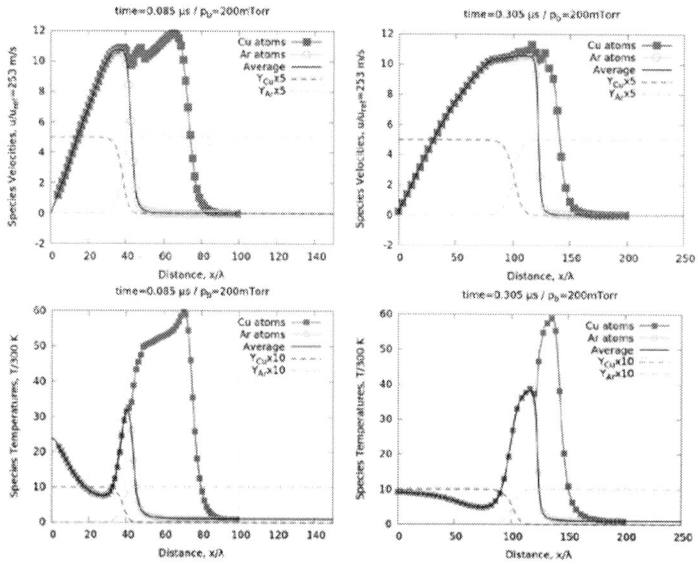

Figure 9. Macroparameters obtained by hybrid solver for 200 mTorr: species mean velocities (top row) and mean temperatures (bottom row), at 3 times of 0.09 and 0.3 μs. Also shown are Cu and Ar mass fractions as well as mixture average quantities. Spatial scale is normalized to λ =10 μm.

3.4. 2D Axi-Symmetric Simulations of Cu Plume Expansion with the Navier-Stokes Solver

We finally demonstrate the UFS capabilities for the problem of Cu plume expansion into a low-pressure Ar in multi-dimensional settings. We present here NS simulations of a spherical expansion of an initial high-pressure Cu core of 50 μm (5 Ar mean free paths) in radius with domain size of 80 λ. Snapshots of the Mach number and adapted grid are shown in Figure 10. We observe that due to the "snow-plow" mechanism, a supersonic shock wave propagates over background gas slowing down to subsonic velocities. Dynamicaly adapted grid allows resolving the large spatial scale disparity between the initial plume zone and the large expansion zone, as well as moving shocks and discontinuities.

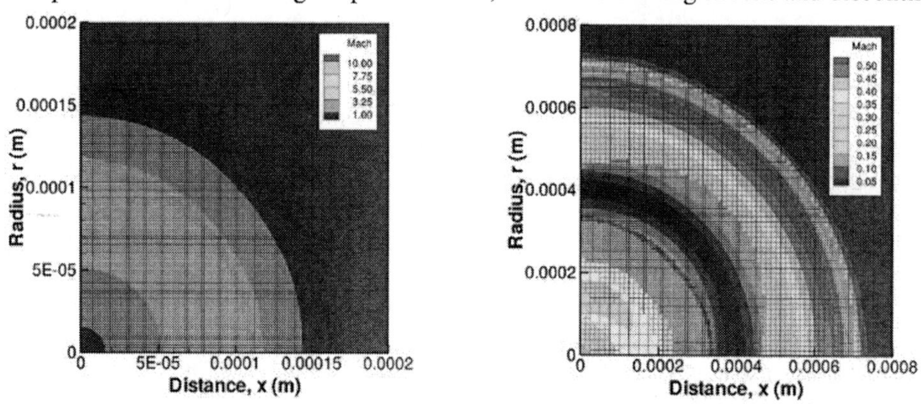

Figure 10. 2D-axi NS solution for problem of metal plume free expansion for 200 mTorr. Shown are Mach numbers at t = 0.07 μs (left) and 0.7 μs (right) and dynamically adapted mesh.

Furthermore, we have conducted simulations for metal atom injection over short periods of time in order to observe metal content propagation in 2D/3D setting. For these simulations, a high-pressure Cu source was set at the solid boundary in a shape of a cylinder (sphere in 3D). The opposite boundary is assumed to be a wall with zero velocity and room temperature boundary conditions. For a 25 ns source (a short laser pulse), we have observed that the metal content exhibits strongly non-spherical, forward-directed pattern. Such patterns have been observed experimentally in many works (e.g., Ref. [12]). The dynamic AMR allows resolving the plume dynamics over a wide range of scales making the simulations very efficient. The initially strict CFL time step limitation was lifted for later stages and computations proceeded with larger time steps and muck coarser grids.

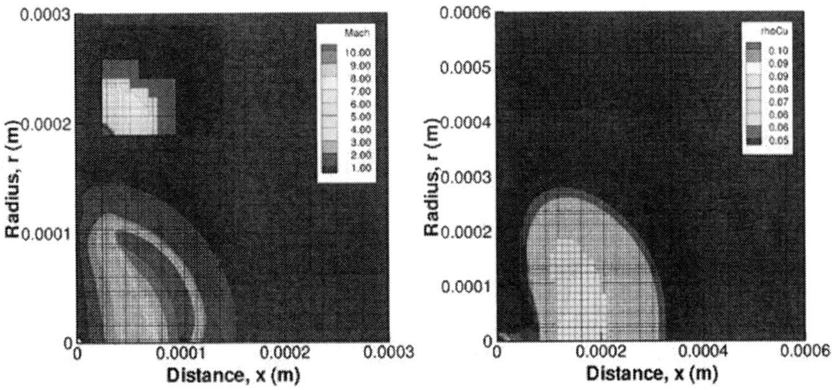

Figure 11. 2D-axi NS solution for problem of metal plume expansion from a solid object for 200 mTorr. Shown are Mach number at t = 0.08 μs (left; insert shows close up view on initial grid with refinement levels from 4 to 9 around solid object) and Cu mass density at t = 0.52 μs (right). AMR grids at both time instances are also shown.

4. Plasma Expansion into Vacuum

Plasma expansion into vacuum has been studied in a large number of works since publications by Gurevich et al. [15], due to high relevance to laboratory devices and applications as well as space science. Dense plasma expands into a vacuum due to thermal electron pressure; some ions can be accelerated to very high energies as a result of this process [16]. The electrons can be assumed to remain in equilibrium with the electric field since the expansion occurs on a time scale long compared with electron time scale [17].

In a series of recent works (see Ref. [18] and references cited therein), two regimes of plasma expansion have been distinguished. In the first regime, when electron motion is collisional, the process is largely similar to the conventional gas-dynamic expansion. In this regime, the electrons could remain Maxwellian during the expansion. In the second, collisionless regime, the assumption of Maxwellian electrons is not justified. A kinetic analysis of this problem in Ref. [18] has led to a conclusion that the deviation from the Maxwellian distribution fundamentally alters the ion acceleration with hot non-thermal electrons playing a critical role in the process [19].

4.1. Implementation of fluid model for collisionless plasma expansion in UFS

We have adapted UFS for simulations of collisionless plasma expansion. We assumed that electrons stay in equilibrium with the electrostatic potential, while the ion motion is described by two equations for density and mean velocity. The major difference between our model and other works is using the Euler framework. In a Lagrangian formulation,[21] the computational grid is moving with the ion front using analytically computed boundary conditions at the ion front. Such a formulation can provide more accurate results for 1d problems but is not practical for more general geometries (such as 2D and 3D). In our implementation, the electron density was updated at each time step using the Boltzmann relationship. We have implemented a Poisson-Boltzmann solver by extending the multi-grid Poisson solver in UFS. As a first step, we applied relaxation on electron density and potential during multi-grid and algebraic solver iterations. More sophisticated techniques can be used in future work, such as those involving non-linear Newton iterations with a defined Jacobian matrix, as well as coupled with electron and ion evolution strategies (see, e.g., Ref.[21]).

IOP Conf. Series: Journal of Physics: Conf. Series **1031** (2018) 012018 doi:10.1088/1742-6596/1031/1/012018

Results of 1D simulations of collisionless plasma expansion are shown in Figure 12. The electric field peaks at the locations of ion front decrease in magnitude as the ion front expands. A close up view on the ion front region is also shown in Figure 12 (bottom, right). One can see that the electric field has a pronounced peak at the ion front location, as expected for a plasma double layer where the ion density drops to low values within a short distance forming a space charge region. There are no peaks on the ion profiles at the plasma fronts, as in some predictions obtained by Lagrangian codes by other groups.[21] This can be explained by the fact that in our setup a finite computational domain (320 Debye lengths) was used and the initial ion profile was not an ideal step function due to the finite grid resolution. In Refs.[20,21], the issue of the initial ion front was discussed in detail: some works did find peaks of ion density, while others [[22]] did not find such peaks. The authors of Ref.[20] conclude that from an experimenter`s point of view, a finite width of the edge will always be present in laser-produced plasmas. We expect that the inclusion of collisions and multi-dimensional effects will relax the strict numerical constrains placed on solvers for modeling plasma expansion into vacuum.

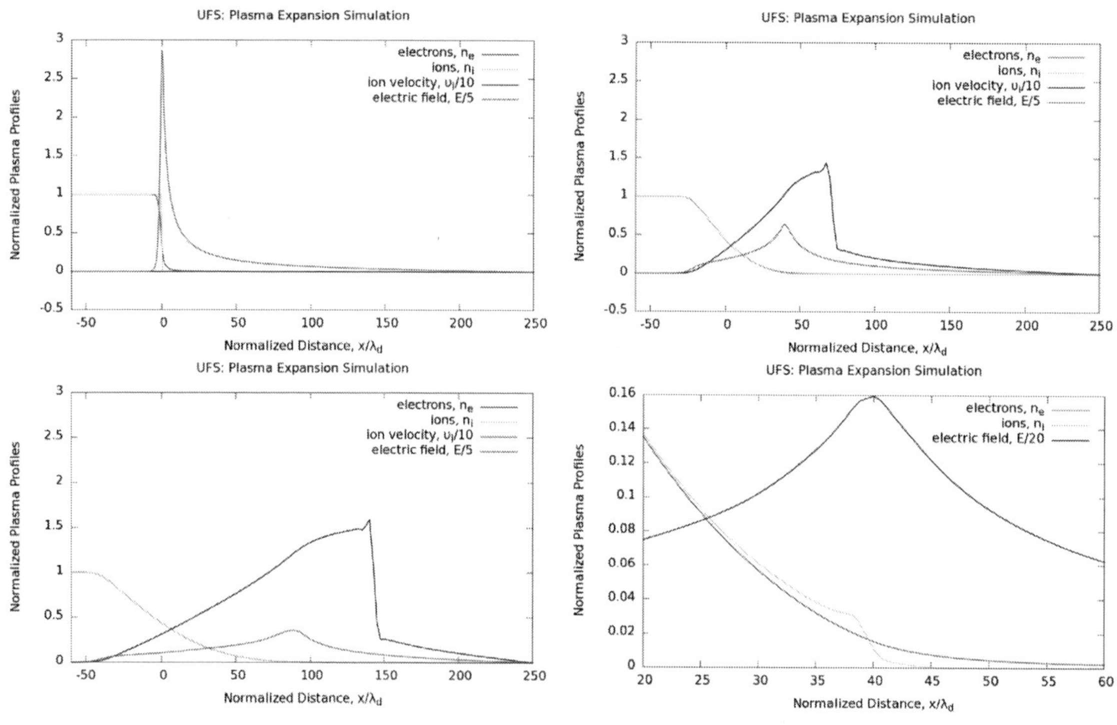

Figure 12. Electron and ion densities, ion velocity and electric field at normalized times 0 and 5 (top row) and 10 (bottom, left) for simulation of plasma expansion into vacuum. Bottom, right shows close up view of the ion front region at normalized time 5.

An important quantity to look at is the ion velocity at the location of the ion front. One can see in Figure 12 that the normalized velocities increase with time and reach values of around 1.5–2 at the normalized times between 5 and 10. These values, as well as the ion velocity time dependence, are consistent with those obtained in the literature; the predicted ion front velocities are slightly lower.

4.2. 3D Simulations of Collisionless Plasma Expansion

For 3D simulations of collisionless plasma expansion, the initial plasma density profile was assumed to be of a spherical shape with a radius of $20\lambda_d$. The Poisson-Boltzmann solver was used to compute self-consistent electron density and electrostatic potential distributions. The domain size was set to 240

IOP Conf. Series: Journal of Physics: Conf. Series **1031** (2018) 012018 doi:10.1088/1742-6596/1031/1/012018

$\lambda_d \times 240\lambda_d \times 240\lambda_d$. Grid adaptation strategy was based on gradients of electron and ion densities, and the space charge. Time evolution of ion density is shown in Figure 13 at several time instances. One can see that at early stages plasma expansion (t < 1) is close to spherical. However, at t > 1, a flat-shaped ion front forms, which moves with a slowly increasing speed until t < 3–4. For times t > 5, the flat-shaped ion front starts to collapse inward while undergoing self-focusing and forming a well-defined bunch at later times. The dynamically adapting grid allows capturing lower density ion bunch at later times when its amplitude drops by a factor of 20–30. Resolving the low-density ion density at long distances from the explosion is crucial for successful comparison with the experimentally measured ion distributions.

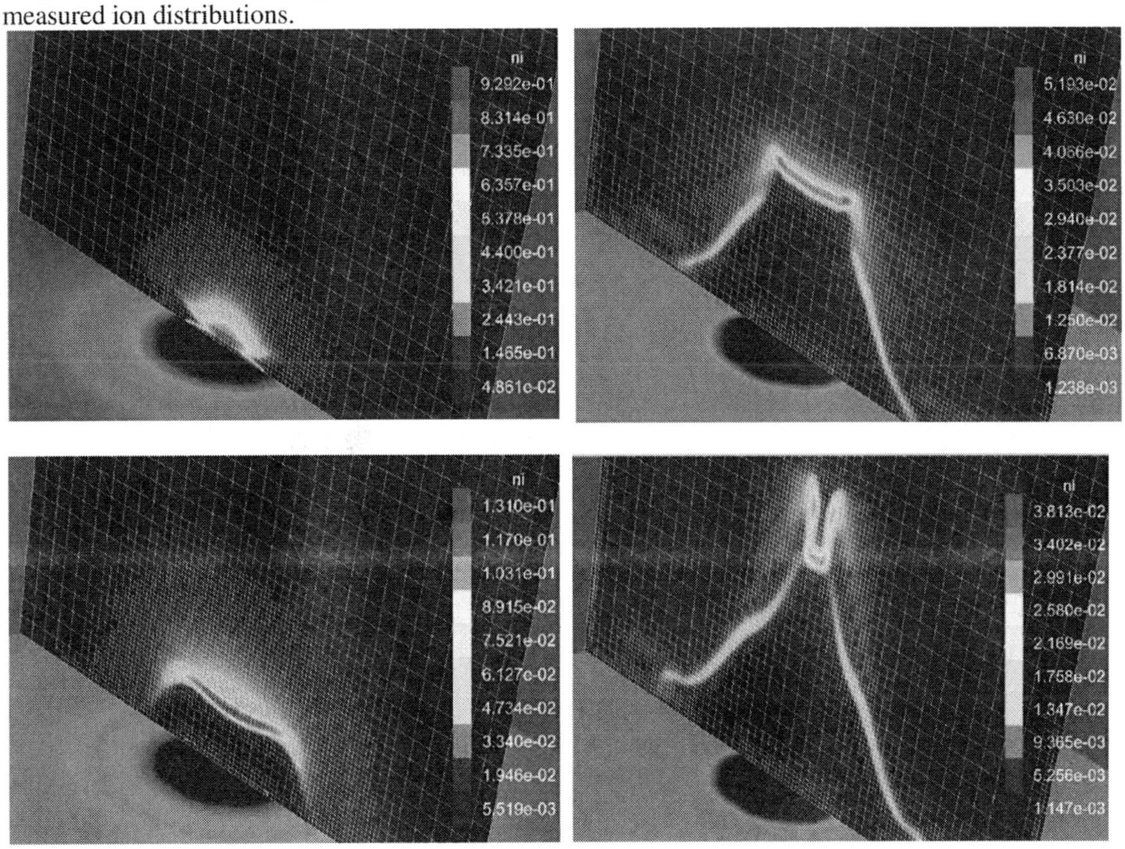

Figure 13. Contours of ion density at four time instances: 1, 3 (left column, top to bottom), 5 and 6 (right column, top to bottom), together with adapted grid for 3D simulation of plasma expansion into vacuum.

5. Conclusions

We have investigated gas dynamic and plasma effects for expanding plasma jets. The Unified Flow Solver (UFS) has been adapted and enhanced for simulations of mixed continuum-rarefied flows, transient gas plume dynamics and collisionless plasma expansion. The enhancements allowed us to efficiently reconstruct a 3D picture of plasma expansion into vacuum. We have observed an interesting phenomenon of ion bunching in our 3D simulations. Since a simplified model of plasma expansion has been used, more accurate predictions and comparison with experiments will require implementations of additional physics. We have demonstrated that the UFS framework is capable simulating different aspects of plasma expansion problems. We identified further advances that are required for hybrid fluid-kinetic simulations of laser-ablation and EEE-induced plasma jets.

Acknowledgments
This work was partially supported by the DOE SBIR Project DE-SC0015746, by the US Department of Energy Office of Fusion Energy Science Contract DE-SC0001939, and by the NSF EPSCoR project OIA-1655280. Thanks are due to Anna Frolova for help with the code development.

References

1 J Winter, R Brandenburg and K-D Weltmann, Atmospheric pressure plasma jets: an overview of devices and new directions, Plasma Sources Sci. Technol. **24** (2015) 064001 (19pp)
2 Q Ma, V Motto-Ros, X Bai, and J Yu, Experimental investigation of the structure and the dynamics of nanosecond laser-induced plasma in 1-atm argon ambient gas, Appl. Phys. Lett. **103** (2013) 204101
3 D L. Shmelev, S A. Barengolts, Modeling of Cathode Plasma Flare Expansion, IEEE Trans Plasma Sci. **41** (2013) 1964
4 S V Anishchenko and A A Gurinovich, Modeling of high-current devices with explosive electron emission, Computational Science & Discovery **7** (2014) 015007
5 E. V. Nefedtsev and A. V. Batrakov, Numerical Simulations of Stable Explosive-Emission Center Plasma Expansion in Vacuum, J. Exper. Theoretical Physics **121** (2015) 706.
6 S. Zabelok, R. Arslanbekov, V. Kolobov, Adaptive kinetic-fluid solvers for heterogeneous computing architectures, J. Comput. Phys. **303** (2015) 455
7 A. A. Morozov, A. B. Evtushenko, and A. V. Bulgakov, Gas-dynamic acceleration of laser-ablation plumes: Hyperthermal particle energies under thermal vaporization, Applied Physics Letters **106** (2015) 054107
8 A A Morozov, Analytical formula for interpretation of time-of-flight distributions for neutral particles under pulsed laser evaporation in vacuum, J. Phys. D: Appl. Phys. **48** (2015) 195501
9 J.-L. Meunier, Pressure limits for the vacuum arc deposition technique, IEEE Trans. Plasma Sci. **18** (1990) 904
10 J. N. Leboeuf, K. R. Chen, J. M. Donato, D. B. Geohegan, C. L. Liu, A. A. Puretzky, and R. F. Wood, Modeling of plume dynamics in laser ablation processes for thin film deposition of materials, Physics of Plasmas **3** (1996) 2203
11 R. F. Wood, K. R. Chen, J. N. Leboeuf, A. A. Puretzky, and D. B. Geohegan, Dynamics of Plume Propagation and Splitting during Pulsed-Laser Ablation, Phys. Rev. Lett. **79** (1997) 1571
12 S. Mahmood, et al., On the plume splitting of pulsed laser ablated Fe and Al plasmas, Phys. Plasmas **17** (2010) 103105
13 V.I. Kolobov and R.R. Arslanbekov, Towards Adaptive Kinetic-Fluid Simulations of Weakly Ionized Plasmas, J. Comput. Phys. **231** (2012) 839
14 V.I.Kolobov, R.R. Arslanbekov, V.V.Aristov, A.A.Frolova, S.A.Zabelok, Unified solver for rarefied and continuum flows with adaptive mesh and algorithm refinement, J. Comput. Phys. **223** (2007) 589.
15 A.V. Gurevich, L.V.Pariĭskaya, and L.P. Pitaevskiĭ, Self-similar motion of rarefied plasma. Sov. Phys. JETP **22** (1966) 449
16 M. Widner, I. Alexeff, and W. D. Jones, Plasma Expansion into a Vacuum, Physics of Fluids **14** (1971) 795
17 J. E. Crow, P. L. Auer and J. E. Allen, The expansion of a plasma into a vacuum, J. Plasma Physics **14** (1975) 65
18 A. V. Arefiev and B. N. Breizman, Collisionless plasma expansion into vacuum: Two new twists on an old problem, Phys. Plasmas **16** (2009) 055707
19 D. Bendaceur-Doumaz, D. Bara, and M. Djebli, Self-similar two-electron temperature plasma expansion into vacuum, Laser and Particle Beams **33** (2015) 723.
20 J.E.Allen and M. Perego, On the ion front of a freely expanding plasma, Physics of Plasmas **21** (2014) 034504
21 M. Perego, P. D. Howell, M. D. Gunzburger, J. R. Ockendon, and J. E. Allen, The expansion of a collisionless plasma into a plasma of lower density, Phys. Plasmas **20** (2013) 052101
22 P. Mora, Plasma Expansion into a Vacuum, Phys. Rev. Lett. **90** (2003) 185002

Two-way coupled MHD-PIC simulations of magnetic reconnection in magnetic island coalescence

Kirit Makwana, Rony Keppens, Giovanni Lapenta

Center for Mathematical Plasma Astrophysics, Department of Mathematics, Celestijnenlaan 200B, KU Leuven, Heverlee 3001, Belgium

E-mail: kirit.makwana@kuleuven.be

Abstract. We present simulations of magnetic reconnection with a newly developed coupled MHD-PIC code. In this work a global magnetohydrodynamic (MHD) simulation receives kinetic feedback within an embedded region that is modeled by a kinetic particle-in-cell (PIC) code. The PIC code receives initial and boundary conditions from the MHD simulation, while the MHD solution is updated with the PIC state. We briefly describe this coupling mechanism. This method is suitable for simulating magnetic reconnection problems, as we show with the example of reconnection in the coalescence of magnetic islands. We compare the MHD, Hall-MHD, fully PIC and coupled MHD-PIC simulations of the magnetic island coalescence. We find that the kinetic simulations are very different from the MHD and Hall-MHD results, while the coupled MHD-PIC simulations can remedy this discrepancy while saving computing time. The diffusion region is well resolved in the kinetic simulations, which is also captured by the coupled MHD-PIC model. The coupled simulation also reproduces the kinetic Hall magnetic fields correctly. We calculate the reconnection rates and find differences between the MHD and kinetic results. We find that the coupled MHD-PIC code can reasonably reproduce the kinetic reconnection rate when a larger PIC feedback region is used, while still saving significant computing time.

1. Introduction

Simulating plasmas is a very challenging task because plasma processes are multi-scale as well as multi-physics in nature. For example, turbulence can be driven at large scales by fluid processes, like solar flares or some magnetohydrodynamic (MHD) instabilities. The nonlinear interactions at the high Reynolds and Lundquist numbers in space and astrophysical plasmas give rise to energy transfer to multiple scales, going right down to the ion and electron scales, which are several orders of magnitude smaller [1, 2] than the driving scales. In solar corona MHD processes operate at $10^9 m$ while electron scales lie at $10^{-1} m$ [3]. Similarly in Earth's magnetosphere the solar wind driver operates at $10^9 m$ while electron scales lie at $10^3 m$ [4]. Simulating this range of scales is not possible even with the computing facilities of foreseeable future. However, an even greater challenge is that while the plasma can be treated as a fluid at the largest scales, the kinetic, particulate nature of plasma becomes critical at the small scales. MHD models are fluid models which work at scales larger than the ion gyroradius or skin depth [5] and are computationally ideal for global simulations. However, they do not capture the kinetic physics that results from the particulate nature of plasma. The Maxwell-Vlasov equations are typically used to describe a plasma kinetically and many advanced particle-in-cell (PIC) techniques have been developed to simulate them. However, these are significantly more computationally demanding then MHD as they simulate a 6D phase space compared to the 3D physical space of fluid models. Moreover, they need to resolve the finer kinetic length and time scales. Hence, implicit PIC methods have been developed to allow larger time-steps and grid-spacing [6]. Despite all this, it is still very difficult to simulate global space and astrophysical plasmas kinetically for physically relevant times.

To overcome some of these problems, we present simulations with a coupling of MHD and PIC in physical space. We simulate the well-known magnetic island coalescence problem with a coupled MHD-PIC code where the global simulation is done with MHD, while a domain around the reconnection region is simulated with PIC. We briefly describe the details of this two-way coupling in Sec. 2. Sec. 3 describes the problem setup of magnetic island coalescence which is a useful problem to study flux-rope merging in tokamaks as well as solar corona [7, 8]. We also do a comparison of the results produced by MHD, Hall-MHD, coupled MHD-PIC, and fully PIC simulations. We find that the coupled simulation saves computing time compared to a full PIC simulation while capturing the relevant kinetic physics effects which are not reproduced in MHD and Hall-MHD. In Sec. 4 we show the reconnection rates obtained in these simulations and how we can get the correct reconnection rates from the coupled MHD-PIC approach. The results are summarized and discussed in Sec. 5.

2. Two-way coupling of MHD with PIC

The coupling is implemented between the codes MPI-AMRVAC [9] and iPIC3D [10, 11]. MPI-AMRVAC solves a variety of hyperbolic advection problems, including MHD, with conservative, shock-capturing finite volume and finite difference schemes. It offers versatility in terms of geometry of the grid, dimensionality, the spatial and temporal discretization, and computer platform [12, 13, 14]. It has the resistive MHD and Hall-MHD modules which are critical in coupling with a kinetic code. It also has a grid-block based adaptive mesh refinement scheme. In this work we have used the TVDLF spatial discretization with a CADA3 limiter function [15], along with a three-step RK time stepping scheme. The $\nabla \cdot \mathbf{B} = 0$ condition is maintained by a generalized lagrangian multiplier (GLM) scheme [16]. iPIC3D is a semi-implicit moment method for PIC simulations of plasmas. It solves a linearized version of the implicit Ampère's law to advance the electric field. Then the magnetic field is obtained from the induction equation and the particles are also moved implicitly. As a result of this semi-implicit scheme, iPIC3D can take larger grid spacing ($10\times - 50\times$) and larger time steps ($5\times - 10\times$) compared to explicit PIC codes.

This coupling is feasible because of the complimentary characteristics of the two codes. MPI-AMRVAC has an adaptive mesh which we can utilize to zoom in with a high resolution around the area where kinetic effects are expected to be important. As a result the grid spacing of the two codes can be made similar. The time-step of MPI-AMRVAC is determined by the CFL condition where $\Delta t < \Delta x/v_f$, where v_f is the characteristic speed of the fastest wave, typically the fast magnetosonic wave. On the other hand, the grid spacing of an explicit PIC code is limited by the Debye length $\Delta x < \lambda_D$ and the time step is restricted by the speed of light, $\Delta t < \Delta x/c$. These scales are very different from the scales of an MHD simulation model. However, with the semi-implicit approach of iPIC3D these constraints are relaxed. In iPIC3D the grid spacing is limited by the electron skin depth $\Delta x < d_e$ and the time step is limited by the electron thermal velocity $\Delta t < \Delta x/u_{th,e}$. As a result the grid spacing of the MHD and PIC are not very different and with mesh refinement the MPI-AMRVAC grid size can be brought within the iPIC3D grid spacing. Assuming the same grid spacing in MHD and PIC, the ratio of the time steps in the two schemes is

$$\frac{\Delta t_{\text{iPIC3D}}}{\Delta t_{\text{AMRVAC}}} \sim \sqrt{\left(1 + \frac{T_i}{T_e} + \frac{1}{\beta_e}\right)\left(\frac{m_e}{m_i}\right)} \qquad (1)$$

Thus by cleverly adjusting the electron beta β_e, electron temperature T_e, and the ion-to-electron mass ratio, we can have similar time steps in MPI-AMRVAC and iPIC3D, making their coupling feasible.

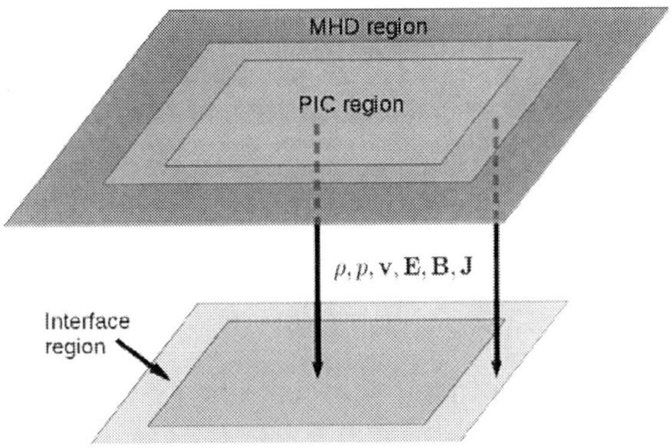

Figure 1. Cartoon showing the coupling scheme in 2D. The blue region at the upper level represents a global MHD simulation. The green region in the middle (reproduced on the lower level) is the domain within the global simulation which is simulated with PIC. Yellow is the interface region between MHD and PIC. The MHD quantities like density (ρ), pressure (p), velocity ($\mathbf{v}$) and magnetic field ($\mathbf{B}$) are exchanged between the two codes for coupling.

The concept of coupling is shown in Fig. 1. This is very similar to the coupling implemented between the MHD code BATS-R-US and iPIC3D [17, 18]. The global domain is simulated with MPI-AMRVAC. At some time step, a region within the global domain is selected by the user to be simulated with iPIC3D. In MHD to PIC coupling, information from MPI-AMRVAC is used to provide initial conditions to iPIC3D to start the simulation. The primitive variables of MHD namely density (ρ), pressure (p), velocity ($\mathbf{v}$), and magnetic field ($\mathbf{B}$) are supplied to iPIC3D. Using this the electromagnetic fields are initialized. The particles are initialized with a Maxwellian distribution using the thermal velocity derived from MHD pressure and drift velocity derived from MHD velocity and current density. After this MPI-AMRVAC is advanced by one time step and the time averaged state at the interface between MHD and PIC is supplied to iPIC3D as the boundary conditions for its time step. After iPIC3D advances its state to the next time step, moments of the distribution function are gathered to obtain the MHD primitive variables. In PIC to MHD coupling this information from iPIC3D is used to update the MPI-AMRVAC solution. The MHD solution is overwritten by a weighted average of PIC solution and the MHD solution in the PIC domain. The weight function varies smoothly across the interface region such that the updated MHD solution smoothly changes from purely MHD at the edge to purely PIC in the interior. This is the two-way coupling method. A more detailed explanation with several tests is provided in Ref. [19].

This coupling optimizes the parallelization in an efficient way. In Fig. 2 we show the parallelization strategy of this coupling. MPI-AMRVAC is an adaptive mesh code which utilizes the distribution of MPI-AMRVAC grid blocks on different processors according to the Morton curve, shown as an example in Fig. 2. There is a region of refined grid blocks in the center out of which the dashed rectangle is selected to be simulated with PIC. The grid blocks are arranged in a complex manner on different processors, shown by 4 different colors for 4 different processors. On the right we zoom into the iPIC3D region. The PIC simulation is distributed and run on the same set of processors as MPI-AMRVAC (represented by the same colors). This is done by running the iPIC3D as a library of MPI-AMRVAC. As a result of this each processor handles some part of the MHD domain as well as some part of the PIC domain. This means that each processor is continuously advancing the MHD and PIC simulations, alternately, maximizing the use of all processors. However, the physical domains of MHD and PIC simulations processed by a processor can be totally different. As can be seen, for ex., the red processor processes very different physical regions in the MHD and PIC parts. Therefore there are more communication costs in this approach. The communication is currently handled via master processors of MHD

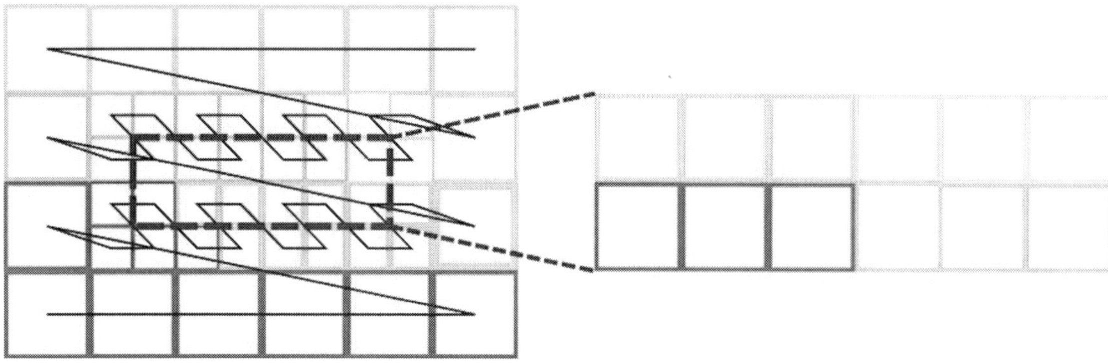

Figure 2. This figure shows the parallelization strategy. The left figure represents a MPI-AMRVAC grid with one level of refined mesh in the center. The squares represent grid blocks. These blocks are distributed on various processors according to the Morton Z-order curve shown by the thin black lines. The grid blocks are distributed on 4 processors - shown by the 4 colors green, red, blue, and yellow. The PIC region is identified by the dashed blue rectangle. The right figure zooms into this PIC region and shows the distribution of the PIC grid on the different processors.

and PIC which gather/distribute information and communicate it to the other program via MPI. We now apply this method to simulate the coalescence of magnetic islands.

3. Numerical setup of magnetic island coalescence

We setup the problem of magnetic island coalescence given by the Fadeev equilibrium [20]. This has the magnetic vector potential given as [21, 22]

$$A_z = B_0 \lambda \ln[\cosh(y/\lambda) + \epsilon \cos(x/\lambda)] \tag{2}$$

where $\epsilon = 0.4$. The density is given by

$$n = n_b + \frac{n_0(1-\epsilon^2)}{[\cosh(y/\lambda) + \epsilon \cos(x/\lambda)]^2} \tag{3}$$

The asymptotic magnetic field is kept at $B_0 = 0.1$. The background density is taken as $n_b = 0.3n_0$ where the ion mass is normalized to unity and n_0 is also unity. The velocity is zero and the pressure is given by $p = nk_B(T_{i0} + T_{e0})$ where $T_{i0} + T_{e0} = B_0^2/(2\mu_0 n_0 k_B)$ to maintain the equilibrium. The initial condition is perturbed by a sinusoidal perturbation of the form given in Ref. [23] to initiate the coalescence of the two islands.

Fig. 3 shows the state of the MHD solution with the coupled PIC feedback after $0.5\tau_A$, where τ_A is the Alfvén time. The black box in Fig. 3 represents the domain where PIC feedback is supplied. The PIC domain is inside $-16 \leq x \leq 16$ and $-16 \leq y \leq 16$. This coupling is done with MHD. Initially the magnetic islands start out with a separation of $2\pi\lambda = 30d_i$ which is at the edge of the PIC box. After $t = 0.5\tau_A$ they have smoothly moved inside the PIC box. Across all the 8 primitive variables of MHD we find that the coalescing islands smoothly merge across the MHD-PIC interface. In Fig. 3(e) we see that the inflow velocity v_x varies continuously across the MHD-PIC interface. We use a smoothing function of width $2d_i$ to average the MHD and PIC quantities at the interface [19].

The simulations are carried out in a box of size $L_x = L_y = 60d_i$, where d_i is the ion skin depth. The parameter λ which is the island width is set by the relation $L_x = 4\pi\lambda$. The grid

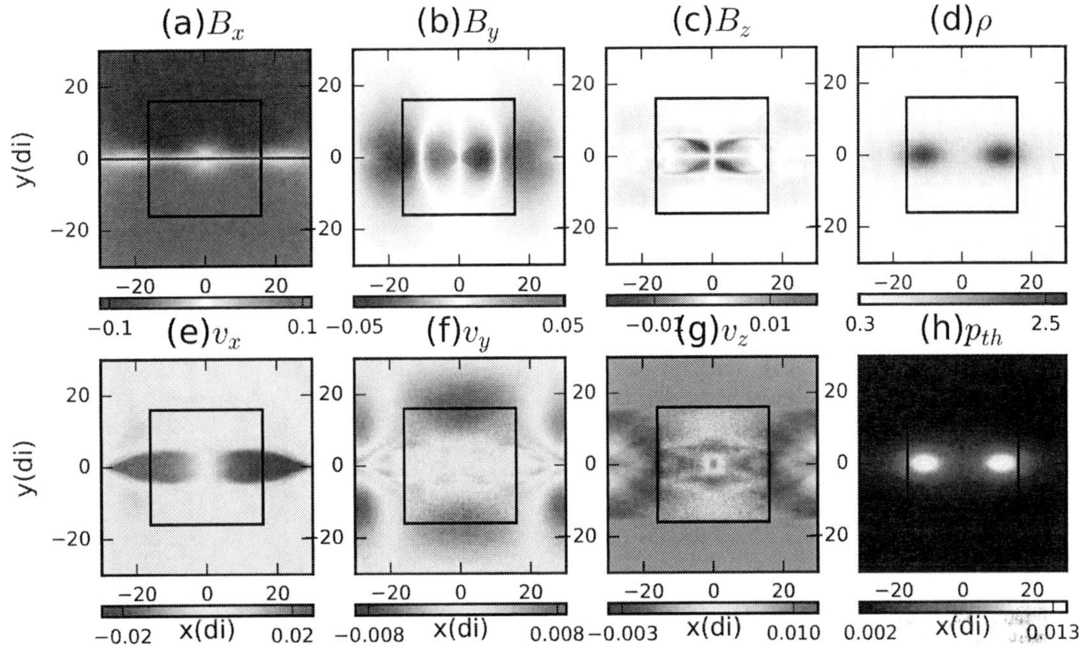

Figure 3. We plot the 8 MHD quantities (a)B_x, (b)B_y, (c)B_z, (d) ρ, (e)v_x, (f)v_y, (g)v_z, and (h)p in the coupled MHD-PIC simulations. This is the state after $t = 0.5\tau_A$. The black box shows the embedded PIC region within the MHD simulation.

is taken of $(600, 600)$ cells giving a resolution of $\Delta x = \Delta y = 0.1d_i$. We simulate this setup with the MHD module of MPI-AMRVAC, with the Hall-MHD module, with the coupled MHD-PIC approach and with a fully iPIC3D simulation. For the MHD simulation the resistivity is taken to be $\eta = 0.003$ in code units which is required at this resolution in order to have a resolved current sheet forming between the two coalescing islands. For lower resistivity values we find that the thickness of the current sheet decreases as the resolution is increased, implying that it is not resolved. With $\eta = 0.003$ the Lundquist number is $S_\lambda = \lambda v_A/\eta \approx 160$. In the coupled MHD-PIC simulations, the resistivity in the MHD is kept to zero. The Hall-MHD module adds the extra Hall-term $(\eta_h/\rho)\mathbf{J} \times \mathbf{B}$ in the Ohm's law equation [9]. In units where the length is normalized to a reference ion skin depth and velocity is normalized to speed of light, the Hall parameter η_h is equal to the reference Alfvén velocity normalized to c which is simply unity. In the coupled MHD-PIC simulations, the resistivity (η) and the Hall parameter (η_h) in the MHD part are kept zero. For the MHD and Hall-MHD simulations the boundary conditions are symmetric for the density, pressure, tangential velocity and normal magnetic field component, while antisymmetric for the normal velocity and tangential magnetic field along the x-boundaries. It is continuous for the density and pressure, symmetric for tangential velocity and tangential magnetic field, while antisymmetric for the normal velocity and magnetic field in the y-boundaries. For the PIC simulations (both stand-alone as well as coupled to MHD) the mass ratio is set at $m_i/m_e = 25$ with the temperature ratio $T_{i0}/T_{e0} = 5$. For the fully PIC simulations where the entire simulation domain is treated with PIC, the boundary conditions are perfectly conducting for fields along all boundaries and perfect mirror for particles along the x-boundaries while along the y-boundaries the outgoing particles are deleted and reintroduced with a random thermal velocity taken from the Maxwellian distribution. For the coupled MHD-

PIC simulations the MHD domain has boundary conditions as described above for MHD and Hall-MHD while the PIC domain receives boundary conditions from the MHD state [19]. The time step for the MHD simulations with these parameters is kept at $0.2\omega_{pi}^{-1}$ which satisfies the CFL condition, and the same time step is used in the PIC simulations as well as in the coupled MHD-PIC simulations. The time step for Hall-MHD is significantly reduced to around $0.008\omega_{pi}^{-1}$ due to the dispersive whistler waves.

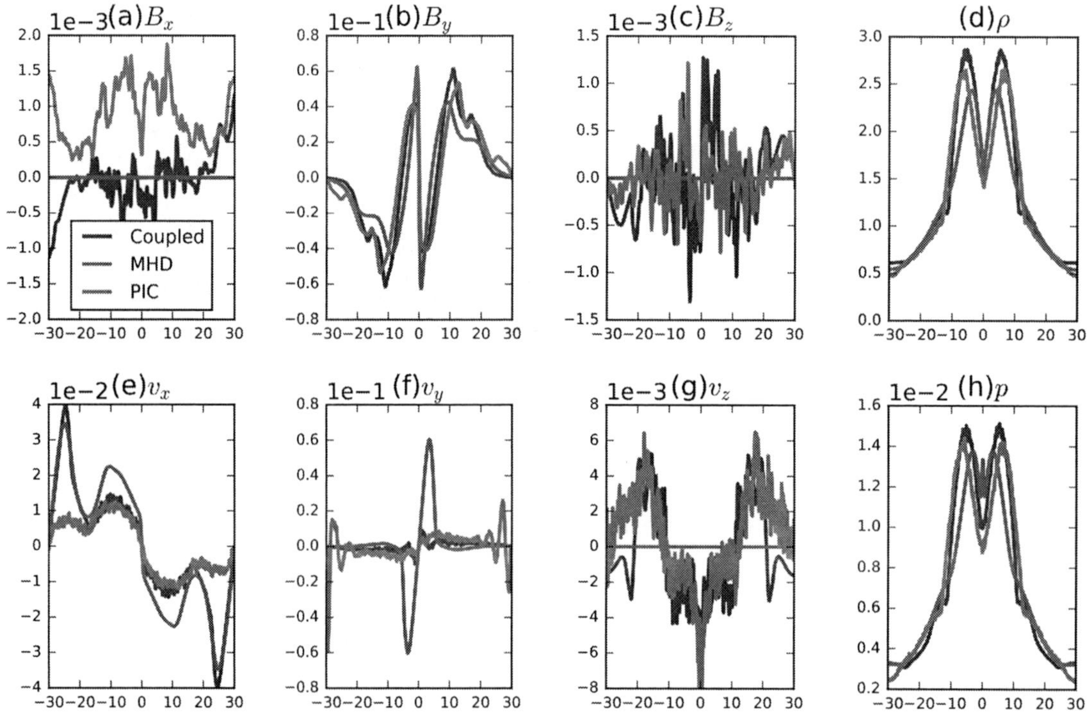

Figure 4. We take a trace of the 8 MHD quantities along the x-axis of Fig. 3 (except for the (f)v_y figure which is a trace along the y-axis of Fig. 3(f)). Therefore the x axis represents the co-ordinate along the x axis in all figures, except in (f) where it shows the y co-ordinate.The green line is taken from the MHD simulation, the blue line is from the coupled MHD-PIC simulation, while the red line is taken from the fully PIC simulation. This is the state of the simulations after 1.0 Alfvén time.

The solutions along the X and Y axes are shown in Fig. 4 at 1 Alfvén time. This is the time around which the reconnection rate peaks. Three simulations are shown in this figure, the green line is the MHD simulation, the blue line is from the coupled MHD-PIC simulation, while the red line is from the fully PIC simulation. The magnetic field B_x is close to zero along the x axis because that is where it flips sign, therefore it is noisy in all the three simulations. The vertical field B_y agrees well between the coupled and PIC simulations, but is stronger and more concentrated in the MHD simulation. The out-of-plane magnetic field is zero in MHD, but non-zero in the coupled and PIC simulations, which agree with each other. The density and pressure match well across the three simulations, with a closer match between the coupled and PIC simulations. There is a spike in the pressure at the central current sheet for the MHD which is not present in the coupled and PIC simulations. The inflow velocity v_x matches well between the coupled and PIC simulations inside the PIC feedback region $-16 \leq x \leq 16$. However, outside of it the coupled simulation result follows the MHD result. The trace for the outflow velocity v_y

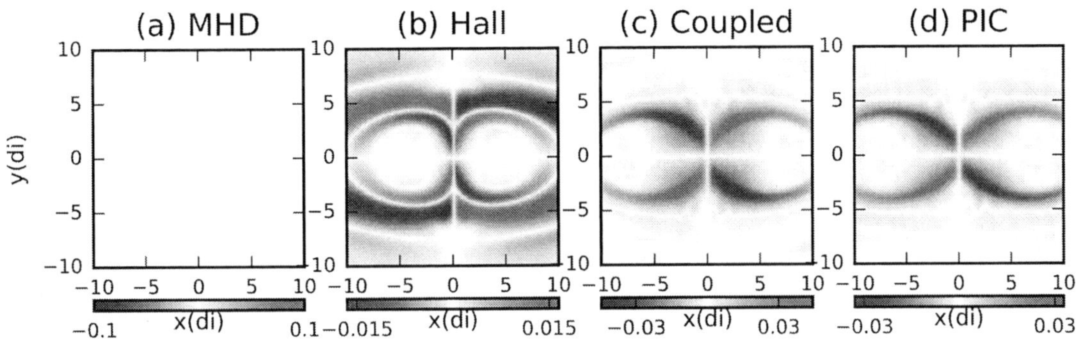

Figure 5. Comparison of out-of-plane magnetic field in the four different simulations, (a) MHD, (b) Hall-MHD, (c) Coupled MHD-PIC, and (d) fully PIC. The plot is in a box of size $(20d_i, 20d_i)$ around the central current sheet.

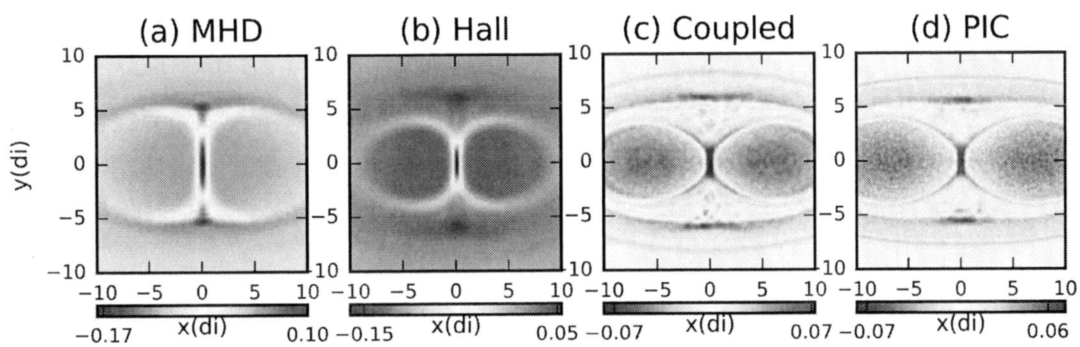

Figure 6. Comparison of out-of-plane current density in the four different simulations. This is shown at 1 Alfvén time.

is taken along the y-axis unlike the other panels, because there is no interesting behavior along the x axis. Here the PIC and coupled solution match well except at the boundaries where the PIC solution develops some numerical effects. We see that the outflow is much stronger and very different in the MHD simulation than in the coupled and PIC simulation. If we look at the 2D profile of the outflow for MHD, we find that this is due to the fact that the outflow in MHD is much more collimated while it is much more spread out in PIC (Fig. 3(f)). The out-of-plane velocity v_z is again zero in the MHD description but shows a strong negative value near the central region in both coupled and PIC simulations that agree quite well with each other. Outside the PIC feedback, the coupled solution diverges from the PIC solution.

In Fig. 5 we show the out-of-plane Hall magnetic field around the diffusion region for the 4 different simulations at 1 Alfvén time. MHD does not develop a Hall field whereas Hall-MHD produces a quadrupolar structure around the central current sheet. However, this structure is different from that produced in the coupled MHD-PIC and fully PIC simulations. The coupled and PIC simulations match quite well showing a thicker region with prominent B_z field. The strength of B_z is also much stronger in the PIC and coupled simulations compared to Hall-MHD. The current density for the four different simulations is showed in Fig. 6. The current sheet that forms in MHD looks very different than the current sheet in PIC and coupled simulations. The Hall-MHD current sheet has a lower aspect ratio compared to MHD and is closer to a point-like geometry. We have found that to obtain a correct X-point like Hall-MHD structure, we need to go to significantly lower resistivity, higher resolution, and also larger system size. However, both MHD and Hall-MHD show sheets that are thinner than the ion skin-depth [22]. On the other hand, the coupled and PIC simulations show well-resolved current sheets. Their thickness is close to 1 ion skin depth d_i and does not change with resolution. This is similar to results from simulations of turbulence as well [24, 25]. Also their structure is closer to an X-point geometry.

4. Reconnection rates

The reconnection rate of the merging islands is an important property of the underlying physics of coalescence. To calculate the reconnection rate, the magnetic flux along the x axis is defined as $\psi(x) = \int_0^x B(y)\mathrm{d}x$. This is calculated by integrating along the x-axis starting from the X-point which lies at the origin. The reconnection rate (E_R) is calculated as [21]

$$E_R = \frac{1}{v_{Am}B_m}\left|\frac{\partial(\psi_X - \psi_O)}{\partial t}\right|, \tag{4}$$

where ψ_X and ψ_O are the fluxes at X and O points respectively. The O-point is identified as the center of the island where the flux is maximum. For comparison with Ref. [21] it is normalized to the Alfvén velocity v_{Am} and magnetic field B_m at the point between the islands where the magnetic field is maximum, which is around 0.035 in our setup.

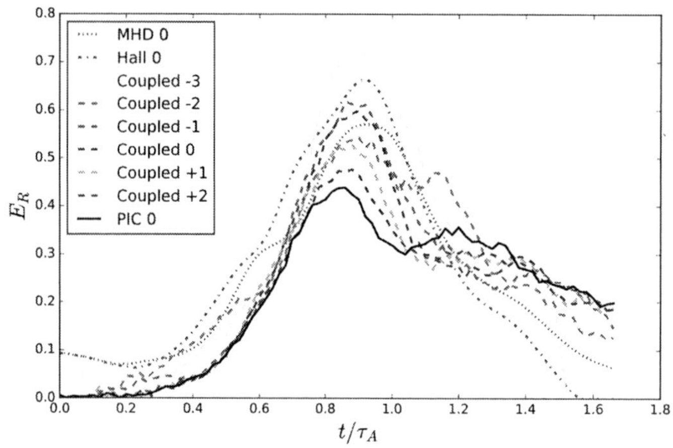

Figure 7. Reconnection rate as a function of time in different simulations. The 4 simulations described above are labelled with 0. The simulation parameters are given in Table. 1. The coupled MHD-PIC simulations match with the fully PIC simulations for larger embedded PIC domains.

We calculate the reconnection rates for several different simulations and they are plotted in Fig. 7 as a function of time. The 4 simulations described above are labeled as MHD 0, Hall 0, Coupled 0, and PIC 0. In addition, there are several other simulations performed with the coupled MHD-PIC code where different sizes of the embedded PIC region were used. For ex., in Coupled 0 the embedded PIC domain was within $-16 \leq x \leq 16$, $-16 \leq y \leq 16$. In Coupled -1, a slightly smaller PIC region is used, lying within $-12 \leq x \leq 12$, $-12 \leq y \leq 12$. These sizes are specified in Table. 1 for the different simulations. We find that the MHD 0 and Hall 0 simulations show similar time evolution of the reconnection rate, with the Hall simulation giving a higher peak reconnection rate than MHD. They peak later than the PIC simulation, closer to 1 Alfvén time and after $t = 1.2\tau_A$ their reconnection rate drops significantly compared to PIC 0. The Coupled and PIC simulations agree in their reconnection rates up to $0.7\tau_A$. After that the coupled simulations start to show a higher reconnection rate. We see that as the domain of PIC feedback is reduced, the reconnection rate shoots higher, overestimating the actual PIC rate. As

Table 1. Important reconnection parameters. PIC domain shows the size of PIC feedback region in case of coupled MHD-PIC simulations in units of d_i. τ_{PR} is the time (in Alfvén units) of peak reconnection rate. Peak E_R is the peak reconnection rate and Averge E_R is the average reconnection rate from $0.5\tau_A$ to $1.5\tau_A$. The last column gives the percentage error in the average E_R compared to fully PIC result.

Simulation	PIC domain	τ_{PR}	Peak E_R	Average E_R	% error in avg E_R
MHD 0	NA	0.92	0.571	0.266	26%
Hall 0	NA	0.94	0.579	0.268	27%
PIC 0	NA	0.86	0.438	0.211	0%
Coupled -3	$(8, 8)$	0.86	0.734	0.282	34%
Coupled -2	$(16, 16)$	0.9	0.543	0.260	23%
Coupled -1	$(24, 24)$	0.86	0.618	0.244	16%
Coupled 0	$(32, 32)$	0.9	0.599	0.235	11%
Coupled +1	$(40, 40)$	0.88	0.527	0.223	6%
Coupled +2	(48,48)	0.88	0.476	0.213	1%

the PIC feedback region is increased, the reconnection rate approaches closer to the PIC result. Coupled simulations -1, 0, +1, and +2 show reasonable similarity with the PIC result.

In Table. 1 we summarize the reconnection characteristics of the different simulations. The peak reconnection rate is reported and also the time at which it occurs τ_{PR}. The reconnection rate is averaged from time $0.5\tau_A$ to $1.5\tau_A$ and is also reported. The peak reconnection rate of 0.438 in PIC 0 simulation matches well with the result of Ref. [21]. The MHD, Hall MHD, as well as coupled MHD-PIC simulations overestimate the peak and average reconnection rates compared to the PIC result. However, the match between coupled and PIC simulations gets better as the size of PIC feedback region increases. The result of the Coupled -1 simulation is reasonably close enough to the PIC result with an error of less than 20% in the average reconnection rate. However, the Coupled -1 simulation took only 26% of the computing time of the PIC 0 simulation, leading to significant computational savings.

5. Conclusions

We have simulated the magnetic island coalescence problem with different simulation methods. This is a subtle problem where neither MHD nor Hall-MHD are able to resolve the reconnection dynamics properly, compared to a fully kinetic code. The MHD and Hall-MHD simulations produce very thin current sheets at Lundquist numbers greater than 10^3 which need a very high resolution to resolve. On the other hand, the kinetic simulations show a well resolved current sheet which does not become thinner than 1 ion skin depth. The kinetic simulations show a X-point like geometry, while the MHD produces a Sweet-Parker geometry and Hall-MHD shows a lower aspect ratio current sheet. We also simulate this with a new coupled code where a PIC simulation is embedded in a smaller region within a global MHD simulation. This code coupling works reasonably well for this problem. It reproduces the correct current sheet structure and magnetic field structure as a fully PIC simulation, while taking significantly less computing time than a fully PIC simulation. The out-of-plane Hall magnetic field is actually much broader and stronger in the kinetic simulation than in Hall-MHD. For the out-of-plane velocity and the inflow velocity, the coupled method reproduces the kinetic solution well inside the PIC feedback region, but there is some divergence outside of it.

The reconnection rates are also calculated. We find that the MHD and Hall-MHD show a different reconnection rate time evolution compared to a fully PIC solution. The peak and average reconnection rates are overestimated in the MHD and Hall-MHD simulations. Coupled MHD-PIC simulations also overestimate the reconnection rate when the size of PIC feedback region is small. As the size of PIC feedback increases, we recover the PIC reconnection rate, but we still save significant computing time. This problem is a test case for flux tube merging in solar corona and tokamaks. It will be useful to apply this for larger system sizes to obtain scaling results of the reconnection process with system size. This will also help us in studying this process at large, realistic Lundquist numbers.

Acknowledgments

This research was supported by the Interuniversity Attraction Poles Programme (initiated by the Belgian Science Policy Office, IAP P7/08 CHARM) and by the KU Leuven GOA/2015-014. Simulations used the VSC (Flemish Supercomputer Center) funded by the Hercules foundation and the Flemish government.

References

[1] Kiyani K H, Osman K T and Chapman S C 2015 *Phil. Trans. R. Soc. A* **373** 20140155
[2] Alexandrova O, Saur J, Lacombe C, Mangeney A, Mitchell J, Schwartz S J and Robert P 2009 *Phys. Rev. Lett.* **103** 165003
[3] Aschwanden M J 2004 *Physics of the Solar Corona: An Introduction* (Springer) ISBN 3-540-22321-5
[4] Kivelson M G and Russell C T 1995 *Introduction to Space Physics* (Cambridge University Press) ISBN 0521457149
[5] Schekochihin A A, Cowley S C, Dorland W, Hammett G W, Howes G G, Quataert E and Tatsuno T 2009 *The Astrophysical Journal Supplement Series* **182** 310
[6] Mason R J 1981 *Journal of Computational Physics* **41** 233
[7] Browning P K, Cardnell S, Evans M, Lucini F A, Lukin V S, McClements K G and Stanier A 2016 *Plasma Phys. Control. Fusion* **58** 014041
[8] Priest E R, Heyvaerts J F and Title A M 2002 *The Astrophysical Journal* **576** 533
[9] Porth O, Xia C, Hendrix T, Moschou S P and Keppens R 2014 *The Astrophysical Journal Supplement Series* **214** 4
[10] Markidis S, Lapenta G and Rizwan-uddin 2010 *Mathematics and Computers in Simulation* **80** 7
[11] Lapenta G 2012 *Journal of Computational Physics* **231** 795
[12] Xia C, Teunissen J, Mellah I E, Chané E and Keppens R 2017 *Submitted to ApJS*
[13] Keppens R, Nool M, Tóth G and Goedbloed J P 2003 *Comput. Phys. Comm.* **153** 317
[14] Tóth G 1997 *J. Comput. Phys.* **138** 981
[15] Čada M and Torrilhon M 2009 *Journal of Computational Physics* **228** 4118
[16] Dedner A, Kemm F, Kröner D, Munz C D, Schnitzer T and Wesenberg M 2002 *Journal of Computational Physics* **175** 645
[17] Daldorff L K S, Tóth G, Gombosi T I, Lapenta G, Amaya J, Markidis S and Brackbill J U 2014 *Journal of Computational Physics* **268** 236
[18] Tóth G, Jia X, Markidis S, Peng I B, Chen Y, Daldorff L K S, Tenishev V M, Borovikov D, Haiducek J D, Gombosi T I, Glocer A and Dorelli J C 2016 *J. Geophys. Res. Space Physics* **121** 1273
[19] Makwana K D, Keppens R and Lapenta G 2017 *Computer Physics Communications* **221** 81
[20] Fadeev V M, Kvartskhava I F and Komarov N N 1965 *Nucl. Fusion* **5** 202
[21] Stanier A, Daughton W, Chacón L, Karimabadi H, Ng J, Huang Y M, Hakim A and Bhattacharjee A 2015 *Phys. Rev. Lett.* **115** 175004
[22] Ng J, Huang Y M, Hakim A, Bhattacharjee A, Stanier A , Daughton W, Wang L and Germaschewski K 2015 *Phys. Plasmas* **22** 112104
[23] Daughton W, Roytershteyn V, Albright B J, Karimabadi H, Yin L and Browers K J 2009 *Phys. Rev. Lett.* **103** 065004
[24] Makwana K D, Zhdankin V, Li H, Daughton W and Cattaneo F 2015 *Phys. Plasmas* **22** 042902
[25] Makwana K D, Li H, Guo F and Li X 2017 *J. Phys.: Conf. Ser.* **837** 012004

IOP Conf. Series: Journal of Physics: Conf. Series **1031** (2018) 012020 doi:10.1088/1742-6596/1031/1/012020

Fast magnetic reconnection: The *ideal* tearing instability in classic, Hall, and relativistic plasmas.

E. Papini[1], S. Landi[1,2] and L. Del Zanna[1,2,3]

[1]Dipartimento di Fisica e Astronomia, Università degli Studi di Firenze, Italy.
[2]INAF - Osservatorio Astrofisico di Arcetri, Firenze, Italy.
[3]INFN - Sezione di Firenze, Italy.

E-mail: papini@arcetri.astro.it

Abstract. Magnetic reconnection is believed to be the driver of many explosive phenomena in Astrophysics, from solar to gamma-ray flares in magnetars and in the Crab nebula. However, reconnection rates from classic MHD models are far too slow to explain such observations. Recently, it was realized that when a current sheet gets sufficiently thin, the reconnection rate of the tearing instability becomes "ideal", in the sense that the current sheet destabilizes on the "macroscopic" Alfvénic timescales, regardless of the Lundquist number of the plasma. Here we present 2D compressible MHD simulations in the classical, Hall, and relativistic regimes. In particular, the onset of secondary tearing instabilities is investigated within Hall-MHD for the first time. In the frame of relativistic MHD, we summarize the main results from Del Zanna et al. [1]: the relativistic tearing instability is found to be extremely fast, with reconnection rates of the order of the inverse of the light crossing time, as required to explain the high-energy explosive phenomena.

1. Introduction: Fast magnetic reconnection and the *ideal* tearing instability

Magnetic reconnection is believed to be the driver of many explosive phenomena observed in laboratory and astrophysical plasmas, such as sawtooth crashes in tokamaks and solar flares, which involve the conversion of magnetic energy into heat and particle acceleration in the surrounding plasma. Reconnection may also explain several high-energy events observed in extreme environments, for instance the gamma-ray flares observed in the Crab nebula and the soft gamma repeaters observed in magnetars, to name a few. All these phenomena are impulsive, characterized by a violent release of energy, and happen on very short timescales. However, classic models of reconnection predict characteristic times which, for highly conducting plasmas, are too slow to explain the rapid evolution of the observed events. In the following we briefly review these models and highlight recent breakthroughs from fast magnetic reconnection models in classic and relativistic plasmas.

In the frame of resistive magnetohydrodynamics (MHD) the evolution of the magnetic field $\mathbf{B}$ is governed by the induction equation

$$\partial_t \mathbf{B} = \nabla \times (\mathbf{v} \times \mathbf{B}) + \eta \nabla^2 \mathbf{B} \tag{1}$$

where $\mathbf{v}$ is the plasma flow velocity and η is a (constant) magnetic diffusivity. We remind that magnetic reconnection can occur only in presence of a non-zero resistivity (i.e. a finite value

IOP Conf. Series: Journal of Physics: Conf. Series **1031** (2018) 012020 doi:10.1088/1742-6596/1031/1/012020

of S). From the above equation it is possible to identify two characteristic timescales, $\tau_A = L/c_A$ and $\tau_D = L^2/\eta$, which describe the fluid advection and the Ohmic diffusion of magnetic fields respectively. Here c_A is the Alfvén speed and L is a characteristic macroscopic length. The ratio of the two timescales defines the Lundquist number $S = Lc_A/\eta$. For astrophysical plasmas, S is very high ($\sim 10^{12}$), thus reconnection is allowed only in current sheets with strong magnetic field gradients.

Early models of magnetic reconnection concern two-dimensional stationary equilibria. In particular, the Sweet-Parker model [2, 3] (hereafter SP) of steady incompressible reconnection driven by velocity inflows v_{in} predicts, for a current sheet of length L and thickness a, a reconnection time

$$\tau_{SP} = \tau_A(c_A/v_{\rm in}) = \tau_A(L/a) = \tau_A S^{1/2} \tag{2}$$

which increases with S, thus far too slow to explain the observed events. Note that the SP current sheet must have a fixed aspect ratio L/a scaling as $S^{1/2}$. Concurrently, Furth et al. [4] found that an infinitely long currentsheet is locally prone to the linear *tearing* instability, which leads to the formation of multiple X-points and magnetic islands (plasmoids) during the reconnection process. The e-folding growth time τ_t of the fastest reconnecting mode, calculated by using the current sheet thickness a as characteristic length scale, reads

$$\tau_t \simeq \tau_A^a S_a^{1/2}\,/\,0.6, \quad S_a = ac_A/\eta, \quad \tau_A^a = a/c_A, \tag{3}$$

again too slow and increasing with the Lundquist number S_a. However, by considering the onset of the tearing instability in extremely thin current sheets and renormalizing all quantities in Eq. (3) with respect to the global scales L, one obtains linear growth rates $\gamma = 1/\tau_t$ which are very fast and increase with S for thin enough current sheets. In particular, recent theoretical and numerical works (see, e.g., [5, 6, 7, 8, 9]) have shown that, for high $S \gtrsim 10^4$ and for a current sheet with a Sweet-Parker aspect ratio $L/a \sim S^{1/2}$, reconnection is indeed very fast, γ increasing with S as $\gamma\tau_A \sim S^{1/4}$, and leads to the formation of a chain of plasmoids.

However, the existence of this plasmoid instability brings to a paradox, since it predicts an infinite reconnection rate as $S \to \infty$, even though reconnection is impossible in ideal MHD. This paradox can be avoided if we proove that SP current sheets cannot possibly form. In fact, the plasmoid instability is so fast that a current sheet could be disrupted by the tearing instability before reaching the desired aspect ratio of SP. To assess this possibility, it is convenient to analyze the mechanisms at the base of current sheet formation.

We can identify four mechanisms at play into the dynamic evolution of a forming current sheet: the diffusion of the bulk magnetic field, the pile-up of plasma toward the current sheet center due to the presence of inflows with velocity v_{in}, the expulsion of plasma along the current sheet outflows, and finally the tearing instability, which tend to disrupt the current sheet. Accordingly, the characteristic time scales associated to the above mechanisms are

$$\tau_D = a^2/\eta, \quad \tau_{up} = a/v_{in} = \tau_A(a/L)(c_A/v_{in}), \quad \tau_A = L/c_A, \quad \tau_t = \tau_A S^{1/2}\,(a/L)^{3/2} \tag{4}$$

where τ_t has been renormalized with respect to L and we assumed the outflow to be alfvénic, that is a low limit for the expulsion time. The tearing instability is always faster than diffusion in the cases where $S > L/a$, which is true for SP current sheets, but also for sheets with a smaller aspect ratio. The presence of an outflow has a stabilizing effect, since plasmoids can be expelled before they grow enough to destabilize the current sheet. The plasma pile-up is the main formation mechanism for the current sheet. Hence the current sheet disrupt if the conditions

$$\tau_t < \tau_A \iff a/L < S^{-1/3}, \quad \text{and} \quad \tau_t < \tau_{up} \iff a/L < S^{-1}\,(c_A/v_{in})^2 \tag{5}$$

are satisfied. The first condition fixes the critical aspect ratio at which the current sheet is disrupted

$$L/a = S^{1/3}, \tag{6}$$

which already implies that a SP current sheet will never form, since it would be destroyed by the tearing instability before reaching the ratio $L/a = S^{1/2}$. The further matching of the second condition (5) sets the scaling for the inflows, namely $v_{in}/c_A = S^{-1/3}$. A current sheet with a critical aspect ratio has another important property: the growth rate of the fastest tearing mode becomes *ideal*, in the sense that is independent on the Lundquist number and of the order of the macroscopic Alfvén time, i.e., $\tau_t \sim \tau_A$. The ideal tearing instability was studied theoretically by [10], who calculated the tearing eigenmodes in a current sheet of ideal aspect ratio $L/a = S^{1/3}$ and found that the linear growth rate of the fastest reconnecting mode asymptotically tends to $\gamma\tau_A \simeq 0.67$, in agreement with Eq. (3). The results have been confirmed and extended to the nonlinear regime by [11] and [12] using numerical simulations. It is worth noting that, as found by [13], if we consider a more general non-alfvénic outflow $v_{\rm out}$, Eq. (6) then modifies to $L/a = S^{1/3}(v_{\rm out}/c_A)^{2/3}$, granted that the current sheet is sufficiently thin to host the most unstable tearing eigenmode. Also in this case, [13] demonstrated that it is not possible to form a SP current sheet.

In the following sections we further investigate the nonlinear regime of the fast magnetic reconnection in the MHD case, and extend the study to include the effect of Hall currents, which may be relevant at reconnection sites where the characteristic scales of the tearing instability become comparable to the ion inertial length. Finally, we will present the results of a recent work [1] which generalizes the ideal tearing instability to include relativistic regimes.

2. Including the Hall term in the *ideal* tearing instability

We first discuss the tearing instability in the case of weakly collisional plasmas. At spatial ion scales, the electron and the ion velocities decouple. Since the electrons rather than the ions drive the magnetic field, the velocity term in the induction equation (1) is replaced by the electron velocity $\mathbf{v}_e = \mathbf{v} - \mathbf{J}/(en_e)$, where n_e is the numerical density of electrons, e is the unsigned fundamental electrical charge, and $\mathbf{J} = c/(4\pi)\nabla \times \mathbf{B}$ is the current density. Neglecting the effects of the electron inertia and of the electron pressure, there is only one new term in the induction equation, with respect to its MHD counterpart, that is the Hall term:

$$-\nabla \times \left[\frac{\mathbf{J}}{en} \times \mathbf{B}\right] = -\nabla \times \left[\frac{cm_i}{4\pi e\rho}(\nabla \times \mathbf{B}) \times \mathbf{B}\right], \tag{7}$$

where we used charge neutrality $n_e = n_i = n$ and approximated the plasma density with the ion density $\rho = m_i n$. To understand how the Hall term changes the behavior of the standard MHD tearing instability we need to identify the characteristic scale at which it becomes important. The fundamental parameter is the ion inertial scale $d_i = c/\omega_{pi} = c\sqrt{m_i/(4\pi n_i e^2)}$, which depends on the ion plasma frequency ω_{pi}. We renormalize the induction equation using the Alfvén speed $c_A = B_0/\sqrt{4\pi\rho_0}$ and time $\tau_A = L/c_A$, where B_0 and ρ_0 are the asymptotic values of the magnetic field and the density respectively, far from the current sheet, and L is the length of the current sheet. We then obtain

$$\partial_t \mathbf{B} = \nabla \times (\mathbf{v} \times \mathbf{B}) + \frac{1}{S}\nabla^2 \mathbf{B} - \eta_H \nabla \times \left(\frac{1}{\rho}(\nabla \times \mathbf{B}) \times \mathbf{B}\right) \tag{8}$$

where all quantities and operators are now renormalized with respect to L, τ_a, B_0, and ρ_0, and we introduced the Hall resistivity

$$\eta_H = \frac{d_{i0}}{L} = \frac{cm_i}{L\sqrt{4\pi e^2 \rho_0}} \tag{9}$$

with d_{i0} being the asymptotic value of the ion inertial length. The Hall term is not negligible when the ion inertial length becomes comparable to the width δ of the inner resistive layer of the tearing instability [14]. For the fastest growing mode, the inner width δ is described by the equation $\delta/a \simeq S_a^{-3/10}\Delta'^{1/5}$[15], where S_a is given in Eq. (3), and Δ' is an instability parameter which depend on the configuration considered for the equilibrium magnetic field. Δ' may depend on ka, which for the fastest growing mode scales as $ka \sim S_a^{1/4}$. For instance, in the equilibrium configuration considered by [5] Δ' is indipendent of S_a, while for the Harris sheet configuration considered in this work (see next Section) $\Delta' = 2\,(1/ka - ka)$. Thus, at high Lundquist numbers $S_a \gg 1$ and for the fastest growing mode, $ka \sim S_a^{1/4}$, we have $\Delta' \simeq 2S_a^{1/4}$, and $\delta/a \simeq S_a^{-1/4}$. Therefore, by rescaling to the macroscopic length L, the ratio

$$d_{i0}/\delta = \eta_H\,(a/L)^{-3/4}\,S^{1/4} \tag{10}$$

determines whether Hall effects are important in the dynamics of reconnection. For a current sheeet with an aspect ratio $L/a = S^\alpha$ that scales with a power of S, one obtains $d_{i0}/\delta = \eta_H S^{(3\alpha+1)/4}$. Finally, by considering a current sheet of ideal aspect ratio ($\alpha = 1/3$), we obtain

$$d_{i0}/\delta = \eta_H S^{1/2}. \tag{11}$$

We can identify three regimes: a MHD regime ($\eta_H \ll S^{-1/2}$) where the Hall term does not play a relevant role, a mild Hall regime ($\eta_H \lesssim S^{-1/2}$), where the ion inertial scale is comparable to the thickness of the inner layer, and a strong Hall regime ($\eta_H \gg S^{-1/2}$), where reconnection is dominated by the Hall effect and the classic theory of the tearing instability is no longer valid. The linear phase of the tearing instability in these three regimes has already been investigated by [14] in the case of a Harris current sheet in pressure equilibrium. They verified the existence of these regimes, and showed that the linear growth rate start increasing for, e.g., values $d_i/\delta \sim 3$ at $S = 10^6$. However this threshold is likely smaller, since Hall currents may affect the subsequent nonlinear evolution, where thinner current sheets formed between the ejected plasmoids may eventually host secondary reconnection events, as we will show in the next section.

3. Hall effects in the nonlinear evolution of magnetic reconnection

3.1. Numerical setup

Building on our previous works [11, 12, 16], we investigate the nonlinear evolution of a reconnecting current sheet, by means of 2D resistive Hall-MHD simulations. Togheter with the induction equation (8), we integrate numerically the compressible Hall-MHD equations in the form:

$$\partial_t\rho + \nabla\cdot(\rho\mathbf{v}) = 0\,, \tag{12}$$

$$\rho\,(\partial_t + \mathbf{v}\cdot\nabla)\,\mathbf{v} = -\nabla P + (\nabla\times\mathbf{B})\times\mathbf{B}\,, \tag{13}$$

$$(\partial_t + \mathbf{v}\cdot\nabla)\,T = (\Gamma-1)\left[-(\nabla\cdot\mathbf{v})T + \frac{1}{S}\frac{|\nabla\times\mathbf{B}|^2}{\rho}\right] \tag{14}$$

where $\Gamma = 5/3$ is the adiabatic index and all quantities retain their obvious meaning. Velocities, lengths, and time are expressed in units of the Alfvén speed c_A, the current sheet length L, and the Alfvén time τ_A respectively. The pressure is normalized to $B_0^2/4\pi$ and the energy equation is written in terms of the normalized temperature $T = P/\rho$ and expressed in units of $T_0 = (m_i/k_B)c_A^2$. With the chosen normalization the Lundquist number is the inverse of the magnetic diffusivity $S = \eta^{-1}$. For the Lundquist number we use $S = 10^6$ as our reference value. We focus on a initial force-free equilibrium and consider a Harris current sheet

$$\mathbf{B} = B_0\tanh(x/a)\hat{\mathbf{y}} + B_0\,\mathrm{sech}(x/a)\hat{\mathbf{z}}, \tag{15}$$

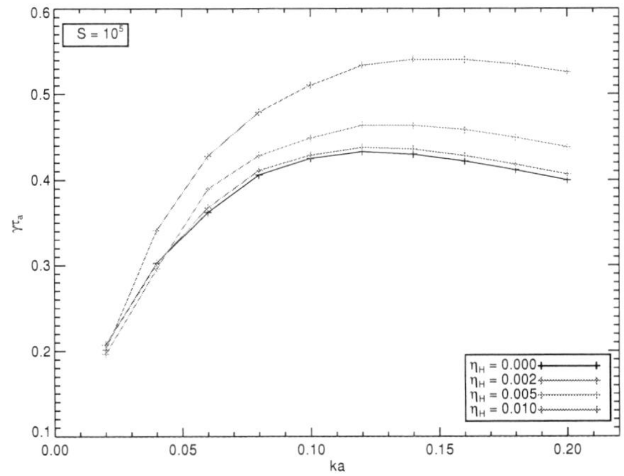

Figure 1. Growth rates $\gamma\tau_A$ vs. ka in the linear phase of Hall-MHD simulations with $S = 10^5$ and different values of η_H. The first ten wavenumbers with ka from 0.02 to 0.2 have been excited.

with homogeneous density $\rho_0 = 1$ and pressure $P_0 = \beta_0/2$, and with no velocities $\mathbf{v}_0 = 0$. The asymptotic field magnitude is $B_0 = 1$ and the plasma beta is set to $\beta_0 = 0.5$. The thickness of the current sheet is set to have an ideal aspect ratio $a/L = S^{-1/3} = \eta^{1/3}$. As it is defined, the current sheet is centered at $x = 0$ and is directed along the y-direction.

We consider a rectangular domain of size $[-L_x, L_x] \times [0, L_y]$. The value of L_y is chosen such that the lowest wavenumber resolved for the tearing instability is $ka = 2\pi a/L_y = 0.02$. This is more than sufficient to resolve the fastest growing mode of the tearing instability for $S = 10^6$. In the x-direction we set $L_x = 20a$ in order to have sufficiently far boundaries from the reconnection region while retaining the high resolution required inside the current sheet with a reasonable amount of grid points.

The Hall-MHD equations are numerically solved by means of the same MHD code we used in [11], modified to include the Hall term. We impose periodic boundary conditions in the y-direction and open boundary conditions in the x-direction using the method of projected characteristics. Spatial derivatives are calculated using Fourier decomposition along the periodic direction and a fourth-order compact scheme [17] across the current sheet. Time integration is performed with a third-order Runge-Kutta scheme. For further details we refer to [18].

The grid employed consists of $N_x \times N_y = 4096 \times 512$ points, which allows to resolve secondary reconnection events in both the x- and y-directions.

3.2. Results

We run several numerical simulations employing different values of the Lundquist number ranging from $S = 10^5$ to 10^7 and for different values of the Hall diffusivity. To trigger the tearing instability, the initial equilbrium was perturbed by sinusoidal magnetic perturbations of amplitude $\epsilon \sim 10^{-4}$, localized in the current sheet. For the perturbations we used the same analytical form of [11] but for the magnetic field instead of the velocity. We excited the first ten modes with ka ranging from 0.02 to 0.2.

In all the simulations, a linear tearing instability develops at the beginning and with the same qualitative behavior, but with quantitatives differences. As an example, Fig. 1 shows the dispersion relation calculated in the linear phase for four simulations with the same value of $S = 10^5$ and different values of $\eta_H = 0$ (MHD case), $0.002, 0.005$, and 0.01, corresponding to $d_i/\delta = 0, 0.6, 1.6,$ and 3.2 respectively. The dispersion relation has the same shape in all cases. However, the linear growth rate of each mode increases when d_i exceeds the thickness of the inner resistive layer δ, up to about 25% more than the MHD case for $ka = 0.14$ and $\eta_H = 0.01$. This is in qualitative and quantitative agreement with [14] even though the initial equilibrium

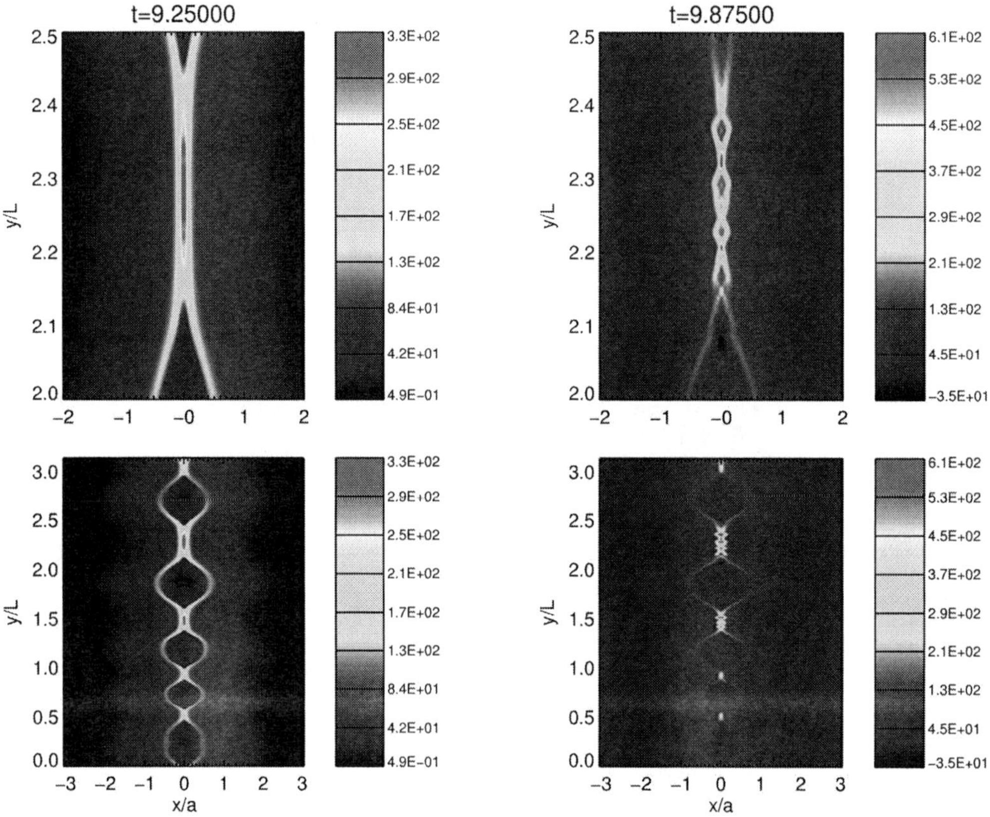

Figure 2. Filled Contour of J_z in the MHD case with $S = 10^6$ and at two different times, before a tearing instability is triggered in the secondary current sheets (left), and at the time when the secondary reconnection events reach the nonlinear phase (right). Bottom panels show J_z for the full length of the current sheet, top panels zoom into one of the secondary sheets. Note that x and y are normalized on different scales.

here is different and therefore the linear evolution may also be different, due to the additional terms present in the Hall-MHD linearized equations.

At the beginning of the nonlinear phase, that is at $t = 9.25\ \tau_A$ in our reference run (bottom left panel of Fig. 2), the size of the plasmoids is comparable to the current sheet thickness, and some of them already merged. Between plasmoids secondary current sheets have been formed, with a thickness of roughly one tenth of the original thickness. One of them is shown on the top left panel of Fig. 2. The subsequent evolution is characterized by the coalescence and nonlinear growth of plasmoids, but the most important feature is the onset of secondary reconnection events in the newly formed current sheets, which then drive the dynamics and eventually lead to the disruption of the whole system. These secondary tearing instabilities are indeed very fast, since at time $t = 9.875\ \tau_A$ (right panels of Fig. 2) they already fully developed, in less than 0.5 (macroscopic) Alfvén times.

To quantitatively support the above statements, we calculated the averaged reconnection rate $\gamma_{\rm rec}$ of the current sheet, shown in the left panel of Fig. 3 for the MHD reference run. The

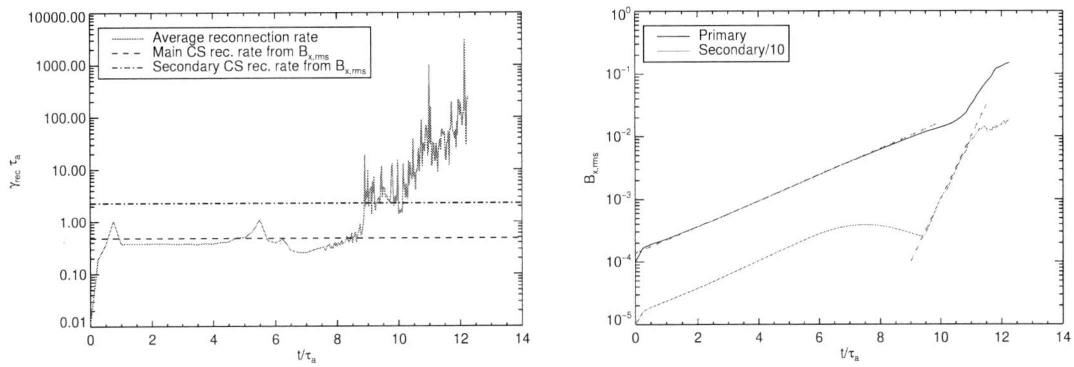

Figure 3. **Left:** Averaged reconnection rate $\gamma_{\rm rec}\tau_A$ vs. time for the MHD case $S = 10^6$. Horizontal lines are the linear growth rate as calculated by an exponential fit of $B_{x,\rm rms}$ (right panel). **Right**: $B_{x,\rm rms}$ vs. time of the primary current sheet (black solid line) and rms divided by ten of the secondary sheet of top panels of Fig. 2 (blue solid line). Red dashed lines denoted the exponential fit for the growth rates of the primary and secondary reconnection events.

averaged reconnection rate

$$\gamma_{\rm rec} = \frac{1}{N}\sum_{i=1}^{N}\frac{1}{\Phi_i}\frac{{\rm d}\Phi_i}{{\rm d}t}, \tag{16}$$

is calculated by taking, at each time, the logarithmic time derivative of the reconnected flux Φ_i between the i-th pair of X- and O-points and then averaging over all the pairs, N being the total number of pairs in the current sheet. In Fig. 3, after the initial perturbations reorganized to select the tearing modes, a clear plateau is present up to $t \simeq 9\tau_A$ and with a value $\gamma_{\rm rec}\tau_A \simeq 0.4$. In the same plot it is also possible to identify a second, more noisy plateau, corresponding to the average reconnection rate of the secondary current sheets. To better estimate the reconnection rates we performed a logarithmic fit of the rms of the x-component of the magnetic field $B_{x,\rm rms}$, shown in the right panel of Fig. 3 for $B_{x,\rm rms}$ calculated over the whole current sheet (black line) and by restricting to the region of the secondary current sheet in top panels of Fig. 2 (blue line). In the latter we clearly see the onset of the secondary tearing instability at around $t \simeq 9.5\tau_A$. The horizontal dashed and dot dashed lines in the left panel, corresponding to the values $\gamma_{\rm rec}\tau_A = 0.48$ and 2.30 given by the two fits, nicely match the two plateaux we identified. Indeed, the measured reconnection rate of the secondary current sheet is strongly superalfvenic and consistent with the preliminary estimate. In the final stage of the evolution (not shown here) the secondary reconnection events drive the dynamics, with new plasmoids being ejected by superalfvenic outflows and feeding the huge plasmoids generated by the first reconnection event. Eventually, the entire current sheet is disrupted.

Even though for values of $d_i/\delta \lesssim 1$ the Hall term does not affect the reconnection rates of the primary current sheet, as we already shown in Fig. 1, the changes in the secondary reconnection events are drammatic. Figure 4 shows a plot of $\gamma_{\rm rec}$ for the primary and the secondary reconnection events for three simulations with $S = 10^6$ and different values of the ratio d_i/δ. There, the reconnection rate of the secondary events (red lines) increases by 17% for a value $d_i/\delta = 0.01$, and almost doubles for $d_i/\delta = 1.4$. As a consequence, the evolution of the overall reconnection process is faster, and leads to the disruption of the current sheet in a shorter time.

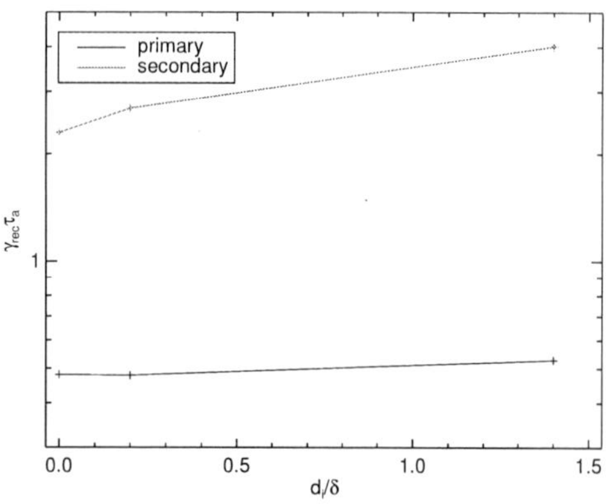

Figure 4. Reconnection rates, calculated with an exponential fit of $B_{x,\rm rms}$, of primary (black) and secondary (red) current sheets, for three simulations with $S = 10^6$ and $d_i/\delta = 0, 0.2$, and 1.4.

Going to larger ratios d_i/δ we enter in the strong Hall regime: the primary reconnection events becomes more and more violent, and the formation of secondary current sheets seems to be inhibited.

4. Ideal tearing instability in relativistic plasmas

What we presented so far applies to classical plasmas. In what follows we review the main results of our recent work [1], which extend the study of the ideal tearing instability to plasmas in extreme astrophysical environments. There, magnetic fields are so strong that the Alfvén speed approaches the speed of light and the plasma becomes relativistic. It is convenient to introduce the plasma magnetization $\sigma_0 = B_0^2/\rho_0$ and the enthalpy density $w_0 = \rho_0 + 4P_0$, where we let $c \to 1$ and $4\pi \to 1$. With this normalization the plasma beta is $\beta_0 = 2P_0/B_0$. At relativistic energies, the Alfvén speed takes the form $c_A = B_0/\sqrt{w_0 + B_0^2} = (1/\sigma_0 + 2\beta_0 + 1)^{-1/2}$, that verifies $c_A < c$. Here the enhanced inertia is due to all the energetic contributions, as it must be in a relativistic case.

To study the relativistic tearing instability we use the same equilibrium configuration (15) of classical and Hall MHD with no equilibrium velocities and no electric fields. We then introduce perturbations with small non-relativistic velocities ($v_1 \ll 1$). With the above assumptions, the linearized induction equation has the same form as in the MHD case, while the linearized momentum equation reads

$$\partial_t \left(w_0 \mathbf{v}_1 + \mathbf{E}_1 \times \mathbf{B}_0\right) = -\nabla(P_1 + \mathbf{B}_0 \cdot \mathbf{B}_1) + \mathbf{B}_0 \cdot \nabla \mathbf{B}_1 + \mathbf{B}_1 \cdot \nabla \mathbf{B}_0 \tag{17}$$

where the subscripts zero and one denote equilibrium and perturbed quantities respectively. By assuming $\mathbf{E}_1 \times \mathbf{B}_0 \simeq B_0^2 \mathbf{v}_1$ (which was numerically verified), we obtain

$$\partial_t \left(w_0 \mathbf{v}_1 + \mathbf{E}_1 \times \mathbf{B}_0\right) \simeq (w_0 + B_0^2)\partial_t \mathbf{v}_1 \tag{18}$$

and the linearized equations for the tearing instability take the same form of the MHD case, provided all quantities are normalized with respect to the enhanced inertia $w_0 + B_0^2$ instead of ρ_0. The linear analysis then proceeds as in the classic case, and we obtain the linear growth rate of Eq. (3) for the relativistic tearing instability, and the growth rate $\gamma\tau_A \simeq 0.67$ for a current sheet of ideal aspect ratio $L/a = S^{1/3}$. Differences between the relativistic and the classic tearing mode become evident if we renormalize the time scales with respect to the light time $\tau_c = L/c$ instead of the Alfvén time. Then one obtains that the maximum reconnection rate for a relativistic ideal tearing is

$$\gamma\tau_c \simeq 0.6(c_A/c). \tag{19}$$

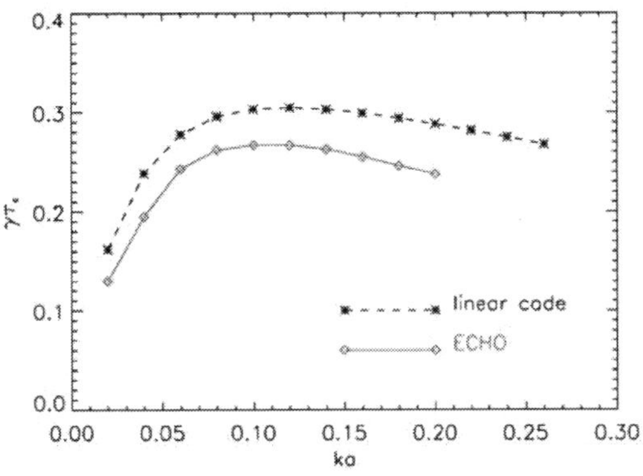

Figure 5. Growth rates $\gamma\tau_c$ vs. ka in the linear phase of the ideal relativistic tearing instability for a run with $\sigma_0 = 1$, $\beta_0 = 1$ (i.e., $c_A/c = 0.5$), and $S = 10^6$. Red solid lines and diamonds are obtained with the ECHO code, while black dashed lines and asterisks are calculated by means of a classical linear MHD code were we set $\rho_0 = 4$ to obtain $c_A = 0.5$ to mimic the linear relativistic behavior. The curve of the linear code exactly matches the theoretical dispersion relation for the ideal tearing instability.

This is a novel result, which suggests that the relativistic magnetic reconnection triggered by the tearing instability may occur on light crossing time scales in extremely thin current sheets when $c_A \to c$, compatible with the bursty events observed in many astrophysical objects.

5. Numerical simulations of relativitic fast magnetic reconnection

To corroborate our theoretical findings and to study the full evolution of relativistic magnetic reconnection, we performed numerical simulations by means of the Eulerian conservative High-Order (ECHO) code [19], which solves the resistive relativistic MHD equations combining shock-capturing properties and accuracy. In ECHO, time integration is performed through a 3rd-order, 4 stages Strong stability Preserving IMplicit-EXplicit Runge-Kutta (SSP3-IMEX-RK) scheme [20, 21], while spatial integration is performed with an MP5 limited reconstruction combined with the HLL two-wave Riemann solver. For further details on the ECHO code, see [19]. For the setup we used the same initialization of the MHD case for both the equilbrium and the initial perturbations. The size of the rectangular domain is the same. However, here we employed a different grid size of 1024×512 cells.

Figure 5 shows the dispersion relation $\gamma(k)$, calculated with ECHO (red solid curve) by performing a single-mode analysis as done in [11], for $S = 10^6$ and $c_A = 0.5$. For comparison the same dispersion relation (black dashed curve) has been calculated by means of the linear pseudospectral MHD code employed in [11], by setting for the latter $B_0 = 1$ and $\rho_0 = 4$ in order to obtain the same value for the Alfvén speed. The curve from the linearized code exactly matches the theoretical prediction, with the fastest growing mode having $\gamma\tau_c = 0.6c_A/c = 0.3$. Here the discrepancy between the linear code and ECHO are likely due to the diffusion of the magnetic field as the simulation evolves. Overall, the results confirm that the relativistic linear tearing modes grow independently of S and on time scales of the order of the light crossing time. Additional numerical simulations performed going to highly magnetized plasmas ($\sigma_0 \gg 1$) and small beta values ($\beta_0 \ll 1$), have confirmed that the tearing instability gets faster as the Alfvén speed approaches the speed of light, and eventually occurs on time scales comparable to the global light crossing time.

To investigate the nonlinear regime of the relativistic tearing instability, we performed a set of five simulations for the same value of the Lundquist number $S = 10^6$ but different values of plasma magnetization and plasma beta, in order to obtain for the Alfvén speed different values in the range from $0.5c$ up to $0.98c$. For these simulations we set $L_y = 2L$. In all simulations, the nonlinear phase is characterized by the growth and coalescence of the plasmoids, as in the classic MHD scenario. The final evolution brings to the formation of a single and well defined X-point, where a secondary reconnection event occurred, and to the coalescence of all plasmoids

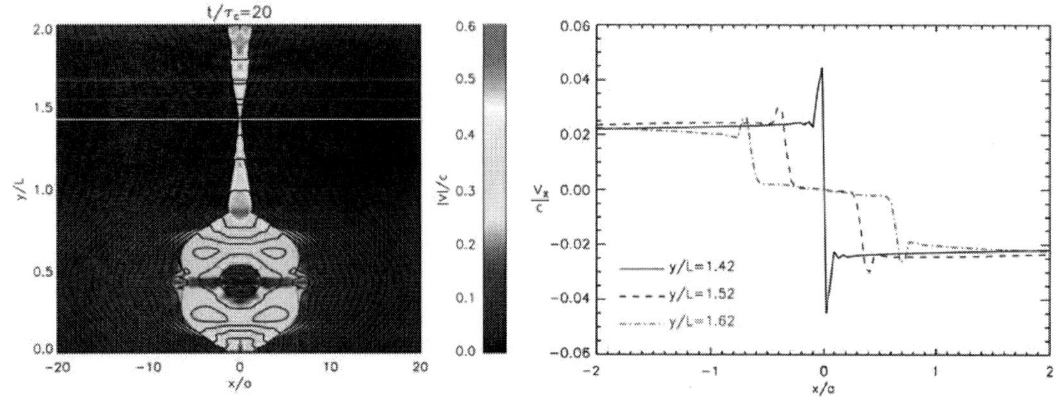

Figure 6. **Left**: color contour of the plasma velocity amplitude $|\mathbf{v}|$ at $t = 20\tau_c$. Black curves denote magnetic field lines. Horizontal white, blue, and red lines indicate the location of the cuts of v_x shown in the right panel. **Right**: cuts of v_x across the current sheet, at the X-point and at two different locations along one of the jets.

into a single, large one. Given the amplitude of the initial perturbations ($\sim 10^{-4}$), the whole reconnection process last within $\sim 20\tau_c$.

The final state (shown in the left panel of Fig. 6) is most interesting, as it shows a Petschek-like configuration [22]. Here superalfvenic jets originating from the X-point are channeled into an exhaust and feed the plasmoid through a fast magnetosonic shock. The jets are mildly relativistic, with a peak velocity of $\simeq 0.55c$ at the termination shock. These results are in agreement with previous works (see, e.g., [23, 24]). Additional simulations (not shown here) were performed by starting from a current sheet in pressure equilibrium. All confirm that the Petschek configuration is the natural final state of this tearing instability, independently on the initial equilibrium considered.

To compare our results with the Petschek model more quantitatively, we calculated the reconnection rate

$$\mathcal{R} \equiv M_A = |v_x|/c_A = \pi/(4 \ln S) \tag{20}$$

for a steady Petschek model. By using $S = 10^6$, we predict a reconnection rate of $\mathcal{R} \simeq 0.057$. Right panel of Fig. 6 shows a roughly constant inflow velocity $|v_x| \simeq 0.025$ at different distances from the X-point in the exhaust, which gives a value of $M_A \simeq 0.05$ for the Alfvénic Mach number, very close to the value predicted by Petschek and in agreement with Lyubarsky [25], who generalized the Petschek model to the relativistic case and predicted a value of $\mathcal{R} \sim 0.1$ for $c_A \to c$, i.e., twice the value we used. We also measured the angle θ between the magnetic field and the plane of the shock, that is, the inclination of the magnetic field lines with respect to the current sheet at the slow shock in the exhaust. We find a value of $\theta \sim 0.04$, compatible with the maximum of the inflow velocity at the X-point, $v_{in} = \tan\theta \simeq \theta$ and again in agreement with [25].

6. Conclusions

In this work we presented a unified study on the ideal tearing instability in classic, Hall, and relativistic plasmas, carried out by means of 2D fully nonlinear numerical simulations. Our results confirm that magnetic reconnection via the ideal tearing instability is indeed an efficient mechanism of energy conversion, which is as fast as ideal Alfvén timescales in MHD and Hall plasma, and up to light timescales in relativistic environments.

In the frame of MHD and Hall-MHD, after the ideal tearing instability saturated and the nonlinear phase has begun, we observed the onset of secondary reconnection events in newly formed current sheets, thinner of one order of magnitude than the initial current sheet. These secondary tearing instabilities are stronlgly superalfvénic, with reconnection rates $\gamma_{rec}\tau_A \simeq 2.3$, i.e., five time faster than the main instability for the MHD case. The net result is a much more violent reconnection process and a speed up in the disruption of the current sheet. Moreover, numerical simulations performed with increasing values of $\eta_H = d_i/L$ have showed that even though the Hall term is negligible in the linear phase, it considerably affects the secondary reconnection events in the nonlinear phase, by increasing the reconnection rates up to about 100% (for $\eta_H = 0.0014$) with respect to the pure MHD case and about ten times the reconnection rate of the linear phase. The last stage, before the current sheet disrupts, is very turbulent and characterized by enhanced reconnection rates, in agreement with [26].

Within the relativistic resistive MHD, the tearing instability behaves exactly as in the classical MHD case, provided the Alfvén speed is replaced by its relativistic counterpart and the rest mass density ρ_0 is replaced by the enhanced inertia term $\rho_0 + 4P_0 + B_0^2$. Moreover, when rescaled to the light crossing time scale $\tau_c = L/c$, the linear growth rate of the fastest reconnecting mode in a current sheet of ideal aspect ratio $L/a = S^{1/3}$ is found to be $\gamma\tau_c \simeq 0.6c_A/c$, indeed very fast and comparable to 1 when the Alfvén speed approaches the speed of light. In the subsequent nonlinear phase we see the merging of the plasmoids and the onset of secondary reconnection events. The final stage is dominated by superalfvénic Petschek-like jets originating from a single X-point and ending into a huge plasmoid through a fast magnetosonic shock. Finally, we retrieve the Petschek reconnection rate $\mathcal{R} \simeq (\ln S)^{-1}$, in agreement with [25]. The full evolution of the reconnection process, from the linear phase to the final Petschek state, only last for $t \simeq 10\tau_c$ when $c_A \to c$. All simulations performed show that the evolution of the ideal tearing reconnection is the same independently of the initial conditions and only depends on the Alfvén speed c_A, at least in the parameter range we explored for the plasma magnetization (up to $\sigma_0 = 50$) and the plasma beta (down to $\beta_0 = 0.01$).

Our results on the ideal tearing have potential applications in space and astrophysical plasmas, e.g., in modeling solar flares and coronal mass ejections, or in improoving our understanding of turbulence in the solar wind. For instance, recent numerical simulations retaining kinetic effects [27] assessed the role of reconnection in driving the turbulent cascade at sub-ion scales, through the destabilization of current sheets of thickness $a \simeq d_i$. Moreover, several theoretical works [28, 29, 30] strongly suggest that the characteristic spatial and temporal scales introduced by reconnecting current sheets embedded in turbulent environments are crucial in shaping the spectral and statistical properties of turbulence. Other astrophysical application in relativistic environments include giant flares observed in the magnetosfere of magnetars [31], and gamma-ray flares observed, e.g., in the Crab nebula [32].

Although not included in our study, we finally remark that in a full 3D system, different geometrical configurations and intrinsically 3D instabilities, both in MHD and kinetic regimes, may either enhance [33] or inhibit [34] magnetic reconnection.

Acknowledgments

The authors wish to acknowledge valuable exchanges of ideas with T. Tullio. EP thanks L. Franci for useful discussions. SL and LDZ acknowledge support from the PRIN-MIUR project prot. 2015L5EE2Y *Multi-scale simulations of high-energy astrophysical plasmas*. This research was conducted with high performance computing (HPC) resources provided by the CINECA ISCRA initiative (grant HP10B2DRR4).

References

[1] Del Zanna L, Papini E, Landi S, Bugli M and Bucciantini N 2016 *MNRAS* **460** 3753–3765

[2] Sweet P A 1958 *Electromagnetic Phenomena in Cosmical Physics* (*IAU Symp.* vol 6) ed Lehnert B p 123–134
[3] Parker E N 1957 *J. Geophys. Res.* **62** 509–520
[4] Furth H P, Killeen J and Rosenbluth M N 1963 *Physics of Fluids* **6** 459–484
[5] Loureiro N F, Schekochihin A A and Cowley S C 2007 *Physics of Plasmas* **14** 100703–100703
[6] Lapenta G 2008 *Phys. Rev. Lett.* **100** 235001
[7] Samtaney R, Loureiro N F, Uzdensky D A, Schekochihin A A and Cowley S C 2009 *Phys. Rev. Lett.* **103** 105004
[8] Bhattacharjee A, Huang Y M, Yang H and Rogers B 2009 *Physics of Plasmas* **16** 112102
[9] Cassak P A, Shay M A and Drake J F 2009 *Physics of Plasmas* **16** 120702
[10] Pucci F and Velli M 2014 *ApJLett* **780** L19
[11] Landi S, Del Zanna L, Papini E, Pucci F and Velli M 2015 *ApJ* **806** 131
[12] Del Zanna L, Landi S, Papini E, Pucci F and Velli M 2016 *J. of Physics Conference Series* vol 719 p 012016
[13] Uzdensky D A and Loureiro N F 2016 *Phys. Rev. Lett.* **116** 105003
[14] Pucci F, Velli M and Tenerani A 2017 *ApJ* **845** 25
[15] Biskamp D 1993 *Nonlinear magnetohydrodynamics*
[16] Landi S, Papini E, Del Zanna L, Tenerani A and Pucci F 2017 *Plasma Physics and Controlled Fusion* **59** 014052
[17] Lele S K 1992 *Journal of Computational Physics* **103** 16–42
[18] Landi S, Velli M and Einaudi G 2005 *ApJ* **624** 392–401
[19] Del Zanna L, Zanotti O, Bucciantini N and Londrillo P 2007 *A&A* **473** 11–30
[20] Del Zanna L, Bugli M and Bucciantini N 2014 *8th International Conference of Numerical Modeling of Space Plasma Flows (ASTRONUM 2013)* (*Astronomical Society of the Pacific Conference Series* vol 488) ed Pogorelov N V, Audit E and Zank G P p 217
[21] Bugli M, Del Zanna L and Bucciantini N 2014 *MNRAS* **440** L41–L45
[22] Petschek H E 1964 *The Physics of Solar Flares, Proceedings of the AAS-NASA Symposium* vol 50 ed Hess W N p 425
[23] Zenitani S, Hesse M and Klimas A 2010 *ApJLett* **716** L214–L218
[24] Zanotti O and Dumbser M 2011 *MNRAS* **418** 1004–1011
[25] Lyubarsky Y E 2005 *MNRAS* **358** 113–119
[26] Smith D, Ghosh S, Dmitruk P and Matthaeus W H 2004 *Geophys. Res. Lett.* **31** L02805
[27] Franci L, Cerri S S, Califano F, Landi S, Papini E, Verdini A, Matteini L, Jenko F and Hellinger P 2017 *ApjLett* **850** L16
[28] Mallet A, Schekochihin A A and Chandran B D G 2017 *Journal of Plasma Physics* **83**
[29] Boldyrev S and Loureiro N F 2017 *ApJ* **844** 125
[30] Loureiro N F and Boldyrev S 2017 *ApJ* **850** 182
[31] Elenbaas C, Watts A L, Turolla R and Heyl J S 2016 *MNRAS* **456** 3282–3295
[32] Cerutti B, Werner G R, Uzdensky D A and Begelman M C 2014 *Physics of Plasmas* **21** 056501
[33] Che H 2017 *Physics of Plasmas* **24** 082115
[34] Gingell P W, Burgess D and Matteini L 2015 *ApJ* **802** 4

A simple and efficient solver for self-gravity in the DISPATCH astrophysical simulation framework

J P Ramsey and T Haugbølle and Å Nordlund

Centre for Star and Planet Formation, the Niels Bohr Institute and the Natural History Museum of Denmark, University of Copenhagen, Øster Voldgade 5-7, DK-1350 Copenhagen, Denmark

E-mail: jramsey@nbi.ku.dk; haugboel@nbi.ku.dk; aake@nbi.ku.dk

Abstract. We describe a simple and effective algorithm for solving Poisson's equation in the context of self-gravity within the DISPATCH astrophysical fluid framework. The algorithm leverages the fact that DISPATCH stores multiple time slices and uses asynchronous time-stepping to produce a scheme that does not require any explicit global communication or sub-cycling, only the normal, local communication between patches and the iterative solution to Poisson's equation. We demonstrate that the implementation is suitable for both collections of patches of a single resolution and for hierarchies of adaptively resolved patches. Benchmarks are presented that demonstrate the accuracy, effectiveness and efficiency of the scheme.

1. Introduction

Gravity is at the root of the formation of cosmological structure, galaxies, stars and planets. It also dominates the dynamics of many astrophysical systems, not least planets in orbit around a star or the spiral arms of a galaxy. Determining the strength of gravity at any given location requires solving for the gravitational potential due to a particular mass distribution via Poisson's equation, a second-order partial differential, elliptical equation with Dirichlet boundary conditions. In general, the solution to Poisson's equation for an arbitrary mass distribution is not analytically tractable. As such, given the importance of gravity in the universe, the accurate (and efficient) numerical solution of Poisson's equation has received considerable attention in computational astrophysics.

There exists many different approaches to numerically solve Poisson's equation. The most popular, basic methods include fast Fourier transforms (FFT), direct N-body calculations [1], multi-pole expansions [2], relaxation methods (e.g. Gauss-Seidel and successive over relaxation; SOR) and multi-grid techniques [3]. Many problems in astrophysics, meanwhile, have an intrinsically large dynamic spatial range (e.g. galaxy and star formation), which makes these problems expensive to simulate. In order to keep the cost of simulations down, considerable effort has been spent developing adaptive techniques that focus numerical resolution only when and where it is needed (e.g. adaptive mesh refinement; AMR; [4]). Naturally, the combination of these adaptive techniques with the solution of Poisson's equation is desirable and indeed required in certain contexts. Examples of adaptive resolution techniques for Poisson's equation include variants or combinations of the above listed methods (e.g. FFT + multi-grid; [5]) that have been modified for a specific adaptive framework or data structure (e.g. AMR multi-grid; [6, 7, 8]; tree gravity [9]; sink particles [10]).

What all of these methods share in common is the need to globally synchronise information whenever a solution to Poisson's equation is undertaken. Indeed, in a large-scale, domain-decomposed simulation this usually requires substantial communication of information across domains (most commonly, MPI ranks[1]) as the Poisson solver iterates, and a synchronisation point whereby all domains agree on the global solution. Not surprisingly, the details of how the Poisson equation is solved and how updated information is propagated between domains can have a substantial effect on the performance and scalability of a code or simulation framework.

In this contribution, we present a simple and efficient method for solving self-gravity in the DISPATCH framework. DISPATCH is a new framework for grid-based astrophysical fluid simulations [11, 12, 13] which incorporates novel algorithms to produce a framework with potentially unlimited parallel scaling and excellent performance even at the node level. The salient features of DISPATCH can briefly be described as follows: 1) We decompose the computational domain into small, semi-independent patches, which readily fit into cache and can be efficiently vectorised; 2) we store multiple time slices of the conservative variables, which then permits us to use individual, asynchronous time-stepping in each patch; 3) we ensure there are many more patches than hardware threads (e.g. OpenMP) available, thus ensuring threads keep busy; 4) work scheduling is task-based and patches most in need of updating are picked first; 5) each MPI rank only communicates with its nearest neighbours, whether they be geometric neighbours or causally-coupled ones; 6) load and communication imbalances are handled by trading entire patches to MPI neighbours on-demand and continuously. Taken together, these features prevent a very small fraction of the computational domain from dominating the cost of a simulation, they permit continuous and low-cost load balancing and they result in linear OpenMP and MPI scaling to $> 10^5$ cores [11].

The modular design of DISPATCH allows for different solvers to be used depending on the set of partial differential equations being solved and the needs of the user. Currently available solvers include the RAMSES hydrodynamical (HD) and magnetohydrodynamical (MHD) solvers [14, 15], STAGGER solvers of 2nd, 4th and 6th order [16], the ZEUS-3D solver [17, 18] and even the PPcode solver for particle-in-cell applications [19]. In this work, we choose to apply the hydrodynamical variants of the ZEUS-3D and 2nd-order STAGGER solvers. The self-gravity algorithm presented herein is, however, independent of the choice of hydrodynamical solver.

What makes this particular algorithm novel is that it exploits the multiple time slots and asynchronous time-stepping features of DISPATCH, plus the existing infrastructure for prolongation and restriction, leading to a simple but effective means for solving Poisson's equation. Indeed, the algorithm requires no global communication or blocking synchronisation across domains.

2. Solving Poisson's Equation

Poisson's equation for Newtonian gravity is a prototypical example of a second-order partial differential, elliptic equation constrained by boundary conditions:

$$\nabla^2 \Phi = 4\pi G \rho, \tag{1}$$

where ρ is the gas density, G is the universal gravitational constant, Φ is the gravitational potential and $\nabla^2 = \nabla \cdot \nabla$ is the Laplacian operator. The term on the right hand side is commonly referred to as the "source" and, in the case of gravity, this is the distribution of mass.

In periodic cases, one must offset the source term by its global average. Doing this renormalises the problem, effectively cancelling the effect of the mass in the periodic neighbours, which could otherwise lead to a divergent solution.

[1] MPI: "Message Passing Interface"; currently, the most common approach to distributed scientific computing.

To wit, the form of the Poisson equation that must be solved in the periodic case is:

$$\nabla^2\Phi = 4\pi G(\rho - \langle\rho\rangle) = 4\pi G\rho', \tag{2}$$

where $\langle\rho\rangle$ is the average value of the density over the entire domain. It is sufficient to determine this value from the initial conditions and therefore does not require a global summation. In cases with fixed boundary conditions, there is no need for such a renormalisation, and would indeed result in an incorrect solution.

As already discussed, many different methods are available to solve eq. 2. Direct methods (e.g. FFT, N-body) would be preferred because they provide an "exact" solution, but these approaches are often impractical or even intractable in many cases. This is particularly true when the mass is discretised onto a hierarchy of grids, e.g., when using AMR. Instead, iterative relaxation methods have proven to be the most cost-effective in this situation. We choose to solve eq. 2 iteratively via the form:

$$4\pi GR = \nabla^2\Phi - 4\pi G\rho', \tag{3}$$

where R is the residual and is a direct measure of the adjustment in mass that would be required to make Φ the exact solution for a given set of boundary conditions. While the goal is to reduce the residual to zero iteratively, in practice, one instead aims to decrease its magnitude below a given tolerance level. In our case, the quantity used to measure convergence is:

$$\varepsilon = \frac{R}{\rho + \rho_0}, \tag{4}$$

where ρ_0 is a 'floor' value characterising the desired tolerance in regions with very low density. When $\varepsilon < \varepsilon_0$, typically set to 10^{-4}, convergence is signalled and the Poisson solver exits.

For solving Poisson's equation on a single DISPATCH patch, we have implemented both successive over-relaxation (SOR) with Chebyshev acceleration [3] and a conjugate gradient (CG) solver [20] with preconditioner [21]. One could, of course, also apply Gauss-Seidel relaxation, but since SOR is always faster there is no reason to. An important consideration for SOR is that a red-black scheme should be employed, otherwise initial symmetries that exist in a problem will not be preserved. In fact, preserving exact symmetry, in any case, may require the use of an odd number of points in non-periodic problems, and an even number of points in periodic problems.

The cost per point of the SOR and CG (with preconditioner) solvers are relatively similar for a typical number of cells per patch. For example, the average cost per cell update on an Intel Core i5 Haswell-based laptop is $\sim$0.015 μs/point/iteration. Of course, as the number of cells increases, the CG solver becomes more efficient than SOR (c.f. the tests of Section 4.1).

Once we have obtained the gravitational potential, the gravitational acceleration, $\vec{g} = \nabla\Phi$, can be calculated and applied as a source term within the hydrodynamical solver. The Poisson solver is implemented independently of the hydrodynamical solver; the feedback of gravity on the gas is accomplished via accelerations. For example, when using 2nd-order staggered differences (e.g. ZEUS-3D and STAGGER solvers), the x-component of the gravitational acceleration at the cell face indexed by (i, j, k) may be written as:

$$g_{x,i} = -\partial_x\Phi = -\frac{\Phi_i - \Phi_{i-1}}{\Delta x} + \mathcal{O}(\Delta x^2), \tag{5}$$

where Δx is the grid spacing in the x-direction. Since the gravitational potential is zone-centred, this simple form gives accelerations that are face-centred and co-spatial with staggered velocity/momentum components. In the case of a centred solver, such as RAMSES, we instead apply the following centred difference formula [22]:

$$g_{x,i} = -\partial_x\Phi = -\frac{4}{3}\frac{\Phi_{i+1} - \Phi_{i-1}}{2\Delta x} + \frac{1}{3}\frac{\Phi_{i+2} - \Phi_{i-2}}{4\Delta x} + \mathcal{O}(\Delta x^4). \tag{6}$$

3. Multiple patch and multiple resolution algorithm

When applying adaptive grid techniques to the solution of the (M)HD equations (e.g. [4]), small errors are readily introduced at interfaces between grids of different resolutions. In the case of quantities which are relatively smooth, these errors can be compensated for by applying so-called "flux corrections". In the more extreme case that a discontinuous wave passes through a change in resolution, however, errors are unavoidably generated (see, e.g., Sect. 7.1 of [18]). However, these errors are rarely ever consequential in complex, large-scale numerical simulations and, as such, they are frequently tolerated.

Although self-gravity is clearly a global problem, the gravitational potential is typically smooth and slowly-varying slowly in time. With this in mind, in DISPATCH, we choose to take a different approach to self-gravity: We solve Poisson's equation locally, on a per patch basis, and then export the updated information to neighbouring patches. This potentially admits errors after a single call to the Poisson solver but, independent of the strategy to reduce such errors, their magnitude can always be monitored by computing the residual (eq. 3). If the errors become large, it is typically because the density distribution changes too much per time step, and a simple remedy is then to reduce the local time step. This effectively leads to more Poisson solutions per unit code time, and hence more frequent exchange of updated information between neighbours. Importantly, frequent exchange between neighbours propagates information about changes in the mass distribution across the domain at a rate proportional to the inverse of the (M)HD time step. Furthermore, Poisson's equation is iterated until convergence on all patches within the domain.

The first step to solving for gravity in DISPATCH is restriction and prolongation, which we term here as 'downloading'. Before the Poisson solver is invoked, up-to-date information about the mass distribution and gravitational potential in neighbouring patches is downloaded. Depending on the details of the patch arrangement, this will either take the form of temporal and spatial interpolation of guard zone data from neighbours (i.e. prolongation) or the conservative averaging of finer resolution data from overlying child patches to the current patch (i.e. restriction). In the first regard, we employ linear interpolation in both time and space; higher order interpolation in space is possible, but more costly, and linear interpolation has proven more than sufficient for the current application. In the second regard, we use linear averaging. Via experimentation, we have found that using the finest resolution data available when prolonging values of Φ provides the best precision. The downloading procedure for the gravitational potential is the same as for any of the hydrodynamical variables. Indeed, that is one of advantages of our scheme: no special considerations are needed for the gravitational potential relative to the other variables.

The next step is to actually solve Poisson's equation on a patch (see above). This is accomplished after the download, but before the (M)HD equations are solved. The resulting gravitational potential is then applied as a source term in the (M)HD solver (eqs. 5, 6).

A critical part of the algorithm is to now call the Poisson solver again for the same patch but using the recently time-updated values of the density. This solution typically only takes 1 or 2 iterations to converge and the resulting gravitational potential is placed into the *next* time slot. When using local time-stepping, a neighbour patch will generally not be at exactly the same time as the current patch and will therefore require some interpolation or extrapolation in time to fill its ghost cells before it can take a step forward in time; by providing a forward *prediction* for Φ, the neighbour can fill its ghost cells and proceed with its update.

Since we have access to multiple time slices of Φ for each patch in DISPATCH (see [11]), it is possible to make a (e.g. linear, quadratic, or higher order) extrapolation in time for Φ from these time slices that can then serve as an initial guess for the next invocation of the Poisson solver on a given patch. Experimentation shows that a linear extrapolation in time for a single, isolated patch can reduce the number of iterations in a subsequent Poisson solve by up to a

IOP Conf. Series: Journal of Physics: Conf. Series **1031** (2018) 012021 doi:10.1088/1742-6596/1031/1/012021

factor of three. Quadratic temporal extrapolation, meanwhile, typically reduces the iteration count by a further ~10%. This feature is most effective when the potential changes significantly between time steps (e.g. a quickly moving massive clump). In certain pathological cases, this feature results in non-convergence of the Poisson solver, however, and thus it is implemented as an optional feature.

Finally, in our algorithm, if an initial guess for Φ is not available, a short "burn-in" phase may be desirable to obtain a globally consistent Φ before the dynamics are activated. This is accomplished by running DISPATCH with the dynamics deactivated and only solving for the gravitational potential (e.g. Section 4.1). The procedure followed is the same as when dynamics are activated, and so no special modifications are necessary for this phase. Furthermore, it remains a local process; no explicit global communication of information is required.

4. Verification

4.1. Static gravitational potential

As a first test of our self-gravity algorithm, we investigate a slightly modified version of the static, analytical gravitational potential of [22]. This test employs a 2D, analytic, static and purely radial gravitational potential given by:

$$\Phi_{\rm e}(r) = \ln\left[\left(\frac{r}{r_0}\right)^2 + 1\right] + 1, \tag{7}$$

where $r^2 = x^2 + y^2$ is the radial coordinate and r_0 is the parameter that determines the radial "concentration" of the potential. We have slightly modified this expression relative to [22] so that, at the centre of the domain ($r = 0$), it has a non-zero value. This has no effect on the resulting density distribution or the quality of the solution, but simplifies the comparison of numerical and analytical solutions near $r = 0$.

The resulting density distribution, obtained by applying Equation (1) to $\Phi_{\rm e}$, is:

$$\rho_{\rm e}(r) = \frac{4r_0^2}{\left(r^2 + r_0^2\right)^2}. \tag{8}$$

The test is set up by initialising the density to eq. (8) everywhere and setting the gravitational potential to eq. (7) in the boundaries of the computational domain, but zero elsewhere. We set the domain size to $(x, y) \in [-0.5, 0.5]$ and the concentration parameter to $r_0 = 0.1$. We then proceed to solve for Φ iteratively on every patch. A convergence tolerance of 10^{-4} was employed for the Poisson solver. As described above, in between calls to the Poisson solver, updated boundary information is retrieved from neighbouring patches. We alternate between Poisson solves and downloading until the global value of the $L_\infty(\Phi)$-norm converges. As this is a time-independent test, time extrapolation in the gravitational potential was not used.

Furthermore, because this test does not require hydrodynamics or even a time step, we have exploited the modular and hierarchical structure of DISPATCH to implement a shell 'solver' for Poisson's equation that leverages the framework without needing to include any code for a hydrodynamical solver.

We consider three configurations: First, a single patch with varying resolution, from 16^2 to 2048^2 cells. Second, the domain is split into either 3^2 or 9^2 patches of equal resolution and tests run for 4^2 to 256^2 cells per patch. Third, a set of nested, refined grids with a maximum refinement level[2] of 3 or 4 but a constant number of cells per patch. Each nested level is comprised of 3^2 patches centred on the origin and refined relative to the previous level by a

[2] The "root" grid, which encompasses the entire domain, has level = 1.

IOP Conf. Series: Journal of Physics: Conf. Series **1031** (2018) 012021 doi:10.1088/1742-6596/1031/1/012021

factor of three. Tests were run with 5^2 up to 129^2 cells per patch. In the nested patches set up, a maximum refinement level of 3 (4) is equivalent to covering the domain with 9^2 (27^2) patches.

Figure 1 shows the resulting L_∞-norm values in both the gravitational potential and acceleration for the three scenarios at different effective resolutions: a single patch of uniform resolution (circles), a domain sub-divided into multiple patches (squares), and nested sets of patches of increasing resolution (triangles). With respect to the gravitational potential, all three scenarios show that the L_∞-norm decreases proportional to the number of cells squared, demonstrating that our algorithm converges at second-order, and that the L_∞-norm does not vary significantly between configurations. As for the gravitational acceleration, $\vec{g}$ our algorithm retains second-order convergence when all patches are of the same resolution. However, once patches of different resolution are introduced, we observe that the convergence in the gravitational acceleration degrades to first-order. As noted in other works (e.g. [22]), this is a result of using linear interpolation for filling guard zones at changes in resolution, and adopting a higher-order interpolation in Φ would improve the convergence rate.

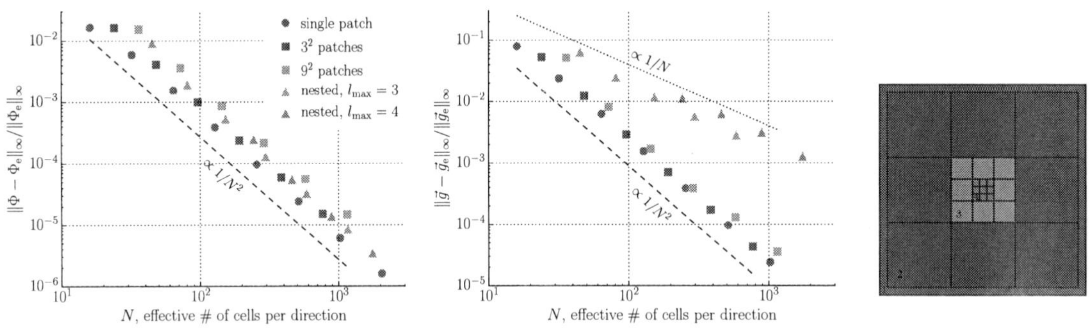

Figure 1. *Left:* The L_∞-norm in the gravitational potential for the static potential problem of Sect. 4.1 as a function of (effective) resolution. The straight line denotes an L_∞-norm which is proportional to $1/N^2$, where N is the effective number of cells. *Middle:* The L_∞-norm in the gravitational acceleration given by eq. 5, as a function of resolution. *Right:* A schematic representation of the nested patch hierarchy used for $l_{\max} = 4$. Numbers denote level; black lines denote patch boundaries.

4.2. The Truelove experiments

The collapse of a gaseous clump under its own gravity is critical to understanding star, disk and planet formation and its dynamics has therefore been well-studied under a wide range of conditions (e.g. [23, 24, 25, 26, 27, 10]).

The criterion for runaway gravitational collapse is originally due to Jeans [28], who found that density perturbations on scales larger than:

$$\lambda_{\rm J} = \sqrt{\frac{\pi c_{\rm s}^2}{G\rho}} \tag{9}$$

are unstable. With respect to numerical simulations of gravitational collapse, [29, 6] further determined that, in order to avoid numerical fragmentation, one must ensure that the so-called Jeans number,

$$J = \frac{\Delta x}{\lambda_{\rm J}}, \tag{10}$$

IOP Conf. Series: Journal of Physics: Conf. Series **1031** (2018) 012021 doi:10.1088/1742-6596/1031/1/012021

where Δx is the cell size, is kept below a value of $J_{\rm max} = 1/4$.

To test the combination of self-gravity and hydrodynamics, we follow [6] and place a massive clump of radius R_0 that is, by definition, gravitationally unstable, at the centre of a 3D Cartesian domain. The density of the clump is given by $\rho_0(1 + A\cos(m\vartheta))$ for $r \leq R_0$ and by ρ_0/χ_ρ, otherwise, where ρ_0 is the uniform density of the clump, A is the amplitude of a possible perturbation, m is the mode of the perturbation, ϑ is the polar angle in the $x - y$ plane and $\chi_\rho = 100$ is the density contrast between clump and ambient medium. In addition to the density, these clumps are characterised by their mass, $M = 1M_\odot$, and their thermal and rotational energies with respect to gravity:

$$\alpha = \frac{E_{\rm therm}}{E_{\rm grav}} = \frac{5}{2}\left(\frac{3}{4\pi\rho_0 M^2}\right)^{1/3}\frac{c_{\rm iso}^2}{G}; \tag{11}$$

$$\beta = \frac{E_{\rm rot}}{E_{\rm grav}} = \frac{1}{4\pi}\frac{\Omega^2}{G\rho_0}. \tag{12}$$

For a given initial clump density, the gravitational free-fall time,

$$t_{\rm ff} = \sqrt{\frac{3\pi}{32G\rho_0}}, \tag{13}$$

can be used to estimate the time scale of collapse.

The criteria for dynamic refinement applied here is that the Jeans number (eq. 10) must remain below 1/4 in every cell. If this criteria is violated, a patch is split into 3^3 child patches, each with the same number of cells as the parent, and the parent data is prolonged onto the new patches.

We attempted three different configurations drawn directly from [6]: i) a non-rotating clump of initially uniform density; ii) a uniformly rotating clump of uniform density; iii) a slightly perturbed (10% amplitude in the $m = 2$ mode) but uniformly rotating clump. The parameters employed can be found in Table 1.

We employed both the ZEUS-3D solver and the 2nd-order STAGGER solver. In the current configuration, both employ the internal energy equation and use a Courant factor of 0.2. The solver-specific artificial viscosity parameters employed for ZEUS-3D are `qlin` = 0.1 and `qcon` = 1 [17], while the relevant hyper-viscosity parameters for STAGGER are $\nu_1 = 0.2$ (viscous stress tensor coefficient), $\nu_2 = 1.0$ (artificial pressure coefficient) [16]. We furthermore used 17^3 cells per patch, a refinement ratio of three and a convergence tolerance of 10^{-4} for the Poisson solver.

As can be seen in Figure 2, our benchmark results compare favourably with those of [6]. There are differences in the rate of collapse, but the density and velocity structures are very similar. In the uniform, non-rotating case, by $t = 1.036t_{\rm ff}$, the density has increased from its initial value by 4.3 orders of magnitude to $\rho_{\rm max} = 10^{-10.7}$ g cm^{-3} and we reach a maximum effective resolution of 459 cells ($l_{\rm max} = 4$). Similarly, for the uniform, rotating case, the density increases by 5.2 orders of magnitude to $\rho_{\rm max} = 10^{-12.6}$ g cm^{-3} after $1.239t_{\rm ff}$ and we reach an effective resolution of 1377 cells ($l_{\rm max} = 5$). Finally, in the 10% perturbation + rotating test, by $1.321t_{\rm ff}$, the maximum density has reached $\rho_{\rm max} = 10^{-11.4}$ g cm^{-3}, an increase of 6.04 orders of magnitude, and we attain an effective resolution of 4131 cells ($l_{\rm max} = 6$). The bottom right panel of Figure 2 shows a slice of the x-component of the gravitational acceleration through the center ($y = z = 0$) of the perturbed Truelove experiment. The profile appears smooth across patch boundaries, indicating that our algorithm is approximating the gravitational force reasonably. However, we do observe changes in the slope of the acceleration (at the <10% level) at jumps in resolution. Therefore, we are currently investigating the use of higher-order interpolations for Φ in the boundaries.

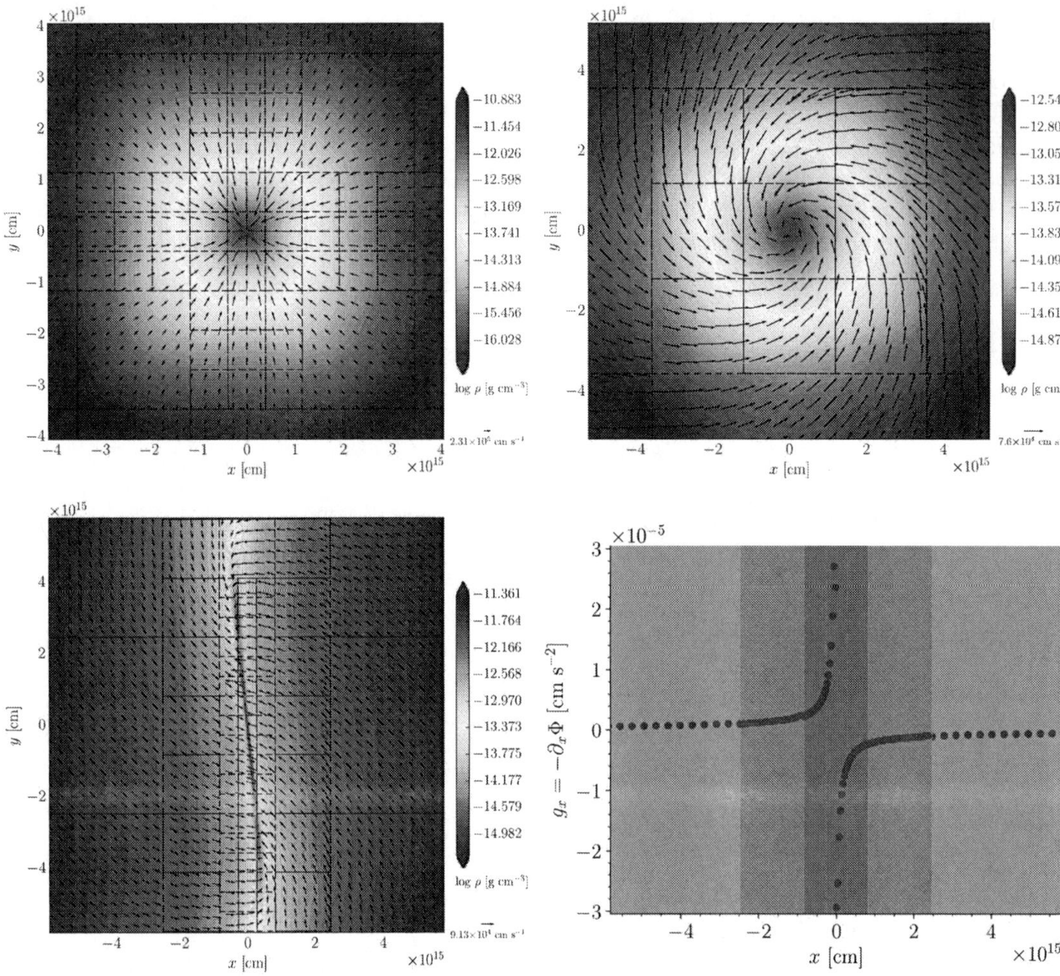

Figure 2. Truelove collapse tests. Plotted is an $x - y$ slice of the logarithm of density at $z = 0.0$. Black arrows denote velocity vectors. Dashed lines indicate patch boundaries. *Top left:* Uniform, non-rotating collapse at $t = 1.036t_{\rm ff}$. *Top right:* Uniform, rotating collapse at $t = 1.239t_{\rm ff}$. *Bottom left:* Perturbed ($m = 2$, 10% amplitude), rotating collapse at $t = 1.321t_{\rm ff}$. *Bottom right:* The x-component of the gravitational acceleration (eq. 5) through $y = z = 0.0$ in the perturbed Truelove experiment. Successively darker shading indicates levels 4–6 while vertical dashed lines denote patch boundaries.

Differences in our results with respect to [6] can be attributed to two factors: i) We are employing fixed rather than periodic boundary conditions and ii) we are using a box size of $8R_0$ versus $4R_0$. The absence of 'mirror' images due to periodic boundaries, combined with a larger box size, implies that our set up should collapse somewhat faster than the simulations presented in [6]. Indeed, with the exception of the uniform, rotating collapse, our simulations attain higher densities than at the times stated in [6].

5. Summary

Herein, we have presented a method for solving Poisson's equation on adaptive resolution grids for self-gravity that does not require any explicit global communication. The approach is

	Non-rotating	Rotating	Rotating + perturbed
ρ_0 [g cm^{-3}]	10^{-15}	$10^{-17.9}$	$10^{-17.4}$
α	0.0475	0.54	0.26
β	0	0.08	0.16
$\chi_p = \frac{P_{\rm clump}}{P_{\rm ambient}}$	1	1	10
R_0 [cm]	7.80×10^{15}	7.23×10^{16}	5.0×10^{16}
$t_{\rm ff}$ [s]	6.65×10^{10}	1.87×10^{12}	1.05×10^{12}

Table 1. Parameters and derived values used in the Truelove tests. In all cases, we employ a box size of $8R_0$, $\gamma = 1.0001$, $\chi_\rho = \rho_{\rm clump}/\rho_{\rm ambient} = 100$ and a cloud mass of $1M_\odot$. See the text for additional details.

built around the concept of multiple individual patches that only exchange information via prolongation and restriction operations; each patch solves gravity locally and then exchanges updated information with its neighbours. The approach exploits the local time-stepping and time slice storage features of DISPATCH to provide time interpolated and extrapolated values of the gravitational potential. There is also the option to increase efficiency by using forward time prediction and by using OpenMP to solve Poisson's equation simultaneously on more than one patch.

Using the static potential benchmarks of [22] and the gravitational collapse benchmarks of [6], we have demonstrated that Poisson's equation can be efficiently solved on a collection of semi-independent patches, and that a global communication or synchronisation step is not required to obtain an accurate solution.

In the Truelove collapse experiments, the DISPATCH self-gravity algorithm accounts for roughly 20% of the total cost during a simulation. Slightly less than half of the remaining cost is due to the hydrodynamical solver, while ~25-30% is due to downloading. A portion of the downloading cost originates with the self-gravity algorithm because we make an intermediate download after the (M)HD step, but before the predictive solution of Poisson's equation; in experiments without self-gravity, the relative cost of downloading is reduced to ~15%.

The first scientific application of this self-gravity algorithm is towards models of planet formation in protoplanetary disks. More specifically, we are currently investigating the self-gravity of combined gas+dust fluids and their potential for gravitational instability. We are also currently implementing additional experiments to test this algorithm under a wider range of conditions including, for example, an equal-mass binary with a moving centre-of-mass. The results of these benchmarks will be featured in future work. Finally, DISPATCH, and the self-gravity algorithm presented here, will be made open-source and publicly available in the near future.

Acknowledgments

The work of TH was supported by the Sapere Aude program of the Danish Council for Independent Research (DFF). The work of ÅN was supported by grant 1323-00199B from the DFF. The Centre for Star and Planet Formation is funded by the Danish National Research Foundation (DNRF97). Storage and computing resources at the University of Copenhagen HPC centre, funded in part by Villum Fonden (VKR023406), were used to carry out some of the simulations presented here.

References

[1] Evrard A E 1988 MNRAS **235** 911–934

[2] Couch S M, Graziani C and Flocke N 2013 *The Astrophysical Journal* **778** 181
[3] Press W H, Teukolsky S A, Vetterling W T and Flannery B P 1992 *Numerical recipes in FORTRAN. The art of scientific computing*
[4] Berger M J and Colella P 1989 *Journal of Computational Physics* **82** 64–84
[5] Collins D C, Xu H, Norman M L, Li H and Li S 2010 ApJS **186** 308–333 (*Preprint* 0902.2594)
[6] Truelove J K, Klein R I, McKee C F, Holliman II J H, Howell L H, Greenough J A and Woods D T 1998 ApJ **495** 821–852
[7] Ricker P M 2008 ApJS **176** 293-300 (*Preprint* 0710.4397)
[8] Almgren A S, Beckner V E, Bell J B, Day M S, Howell L H, Joggerst C C, Lijewski M J, Nonaka A, Singer M and Zingale M 2010 ApJ **715** 1221–1238 (*Preprint* 1005.0114)
[9] Wünsch R, Walch S, Whitworth A P and Dinnbier F 2017 *ArXiv e-prints* (*Preprint* 1708.06142)
[10] Federrath C, Banerjee R, Clark P C and Klessen R S 2010 ApJ **713** 269–290 (*Preprint* 1001.4456)
[11] Nordlund Å, Ramsey J P, Popovas A and Küffmeier M 2018 MNRAS (*Preprint* 1705.10774)
[12] Popovas A, Nordlund Å, Ramsey J P and Ormel C W 2018 *ArXiv e-prints* (*Preprint* 1801.07707)
[13] Ramsey J P, Nordlund Å and Popovas A *in preparation*
[14] Teyssier R 2002 A&A **385** 337–364 (*Preprint* astro-ph/0111367)
[15] Fromang S, Hennebelle P and Teyssier R 2006 A&A **457** 371–384 (*Preprint* astro-ph/0607230)
[16] Kritsuk A G, Nordlund Å, Collins D, Padoan P, Norman M L, Abel T, Banerjee R, Federrath C, Flock M, Lee D, Li P S, Müller W C, Teyssier R, Ustyugov S D, Vogel C and Xu H 2011 ApJ **737** 13 (*Preprint* 1103.5525)
[17] Clarke D A 2010 ApJS **187** 119–134
[18] Ramsey J P, Clarke D A and Men'shchikov A B 2012 ApJS **199** 13 (*Preprint* 1112.4390)
[19] Haugbølle T, Frederiksen J T and Nordlund A 2013 *Physics of Plasmas* **20** 062904 (*Preprint* 1211.4575)
[20] Ghysels P and Vanroose W 2014 *Parallel Computing* **40** 224 – 238 ISSN 0167-8191 7th Workshop on Parallel Matrix Algorithms and Applications URL http://www.sciencedirect.com/science/article/pii/S0167819113000719
[21] Ament M, Knittel G, Weiskopf D and Strasser W 2010 *Parallel, Distributed and Network-Based Processing (PDP), 2010 18th Euromicro International Conference on* (IEEE) pp 583–592
[22] Guillet T and Teyssier R 2011 *Journal of Computational Physics* **230** 4756–4771 (*Preprint* 1104.1703)
[23] Hoyle F 1953 ApJ **118** 513
[24] Hunter C 1962 ApJ **136** 594
[25] Nordlund Å, Haugbølle T, Küffmeier M, Padoan P and Vasileiades A 2014 *Exploring the Formation and Evolution of Planetary Systems* (*IAU Symposium* vol 299) ed Booth M, Matthews B C and Graham J R pp 131–135
[26] Burkert A and Bodenheimer P 1993 MNRAS **264** 798
[27] Bate M R and Burkert A 1997 MNRAS **288** 1060–1072
[28] Jeans J H 1902 *Philosophical Transactions of the Royal Society of London Series A* **199** 1–53
[29] Truelove J K, Klein R I, McKee C F, Holliman II J H, Howell L H and Greenough J A 1997 ApJ **489** L179–L183

IOP Conf. Series: Journal of Physics: Conf. Series **1031** (2018) 012022 doi:10.1088/1742-6596/1031/1/012022

MHD Simulation of Laboratory Jets

Olga Toropina, Gennadiy Bisnovatyi-Kogan, Sergei Moiseenko

Space Research Institute, 84/32 Profsoyuznaya Str, Moscow, Russia, 117997

E-mail: toropina@iki.rssi.ru

Abstract. We performed numerical MHD simulation of the laboratory experiment for creating plasma jets on a NEODIM laser installation. In this experiment, the plasma ejection is formed as a result of the action of a powerful laser on the target. To describe these processes and simulate plasma flow, we chose a numerical method, boundary and initial conditions. We investigated the picture of the flow and compared it with the experiment. We found the distribution of the density of matter at various distances from the target and at different time, and investigated the possible structures of matter on the surface of the detector.

1. Introduction

Astrophysical jets are observed at various astronomical scales and are found in active galactic nuclei, quasars, compact objects in binary systems and young stellar objects. The emergence of jets is possible with the explosion of collapsing supernovae. The processes of formation and evolution of astrophysical jets are one of the most interesting problems of astrophysics [1].

The study of astrophysical jets is conducted in two directions: using observations in the optical, X-ray and radio bands and using multidimensional numerical MHD modeling. Such a simulation must take into account the entire complex nature of relativistic jets, including the gravitational field and the magnetic field of the central object. Recently, the methods of investigating relativistic jets have been supplemented with methods of laboratory astrophysics. They allow the creation of plasma jets with the help of lasers and make it possible to investigate in the laboratory conditions the structure, evolution, and features of the structure of jets.

The equations of magnetic hydrodynamics describing the formation and evolution of jets are nonlinear even in the one-dimensional case. For the theoretical description of the processes occurring in jets, it is necessary to apply methods of multidimensional numerical simulation. Multidimensional numerical calculations make it possible to obtain a detailed picture of the flow of matter in the jet and the shock-wave structure of the flow. Numerical simulation allows calculations to be performed both for initial data corresponding to astrophysical jets, and for a set of parameters from a laboratory experiment.

Laboratory research is conducted by several groups of scientists. Fua, Liang, Tzeferacos and Lamb work on the creation of a magnetized supersonic laboratory jet. In parallel they perform numerical simulation of the experiment using the two-dimensional FLASH code [2]. Experiments on laboratory modeling of astrophysical jets were carried out at the Kurchatov Institute on the installation of the plasma focus PF-3 by Krauz at al. [3]. Pikuz at al. [4] presented the results of a laboratory laser experiment and the numerical simulation of a magnetized jet expiring from a young stellar object. The collimated jet in these calculations is obtained due to the presence of a poloidal magnetic field.

IOP Conf. Series: Journal of Physics: Conf. Series **1031** (2018) 012022 doi:10.1088/1742-6596/1031/1/012022

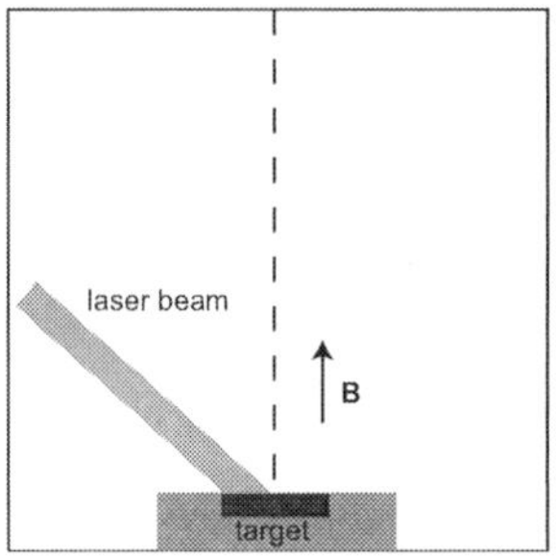

Figure 1. Experimental scheme

2. Experiment

The purpose of the study is numerical modeling and analysis of the results of a laboratory experiment simulating the formation of cosmic relativistic plasma jets on the 10 TW picosecond NEODIM laser installation (TsNIIMASH, Korolev, Russia) [5].

The laser unit has the following parameters of the laser pulse: energy up to 10 J, wavelength 1.055 mkm, duration 1.5 ps, laser radiation contrast about 10^7. The focusing system based on an off-axis parabolic mirror with a focal length of 20 cm provides a concentration of at least 40% of the energy of the laser beam into a 15 μm spot and, correspondingly, a peak intensity of $2 \times 10^{18} W/cm^2$.

The laboratory experiment consists of the following: a cylindrical chamber is placed as a target a foil of Cu or Ta with a thickness of 30 μm and 50 μm. On the opposite side are track detectors CR-39. The energy $10\,J = 6.24 \times 10^{19}$ eV is uniformly and instantaneously applied to the central region of the target with a diameter of 10 m. As a result of heating the target, two directional plasma ejections appear in the form of symmetrical jets. In our numerical simulation, only one of them is considered. The scheme of the experiment is shown in Figure 1.

In the experiment, clearly distinct ring structures formed on track detectors by protons with energies in the range of 0.8 ÷ 1.7 MeV are observed. An example of images of proton beams is shown in Figure 2.

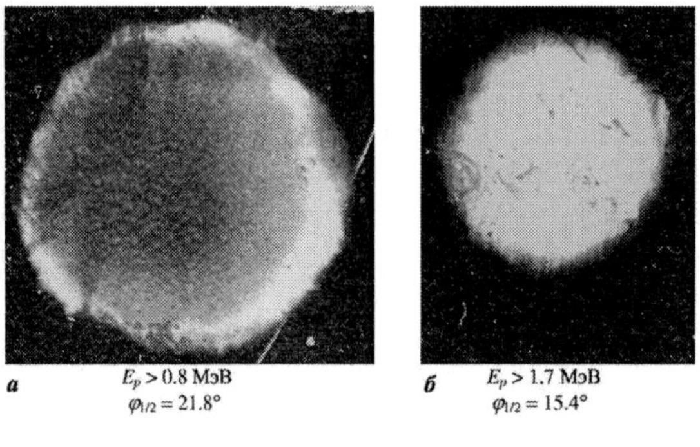

Figure 2. Images of proton beams with different energies on track detectors CR-39 for a target of Cu thickness 50 m.

IOP Conf. Series: Journal of Physics: Conf. Series **1031** (2018) 012022 doi:10.1088/1742-6596/1031/1/012022

3. Similarity criteria

The scale of physical processes occurring in astrophysical jets is very different from those in jets that occur when a laser pulse is applied to a target. However, the application of similarity criteria makes it possible to obtain a correspondence between a number of characteristics of an astrophysical and laboratory jet. The use of such similarity criteria is convenient both for direct comparison of laboratory and observational data, and for multidimensional numerical simulation. Investigation of the similarity criteria for comparing the results of laboratory experiments and astrophysical processes has been devoted to many works, see, for example, [6]. Despite the significant difference in the physical parameters of laboratory and astrophysical jets, there are scaling criteria that allow to connect these quantities.

Laboratory jet on NEODIM installation [5]	Laboratory jet after scaling	Jets from young stars
$x = 0.1 - 1cm$,	$(3 \div 30) \times 10^{17} cm$,	$\sim 3 \times 10^{17} cm$,
$t = 10^{-8} \div 10^{-7} s$,	$3 \times (10^{10} \div 10^{9}) s$,	$\sim 3 \times 10^{10} s$,
$v = 10^{7} cm/s$,	$10^{7} cm/s$,	$(2 \div 5) \times 10^{7} cm/s$,
$\rho = 10^{-4} g/cm^{3}$,	$10^{-23} g/cm^{3}$,	$10^{-23} g/cm^{3}$,
$n = 2 \cdot 10^{19} cm^{-3}$,	$2 cm^{-3}$,	$(1 \div 100) cm^{-3}$,
$T = 10^{4} K$,	$10^{4} K$,	$(1 : 6) \times 10^{4} K$
$H = 10^{8} \div 10^{9} Gs$,	$1 \div 10 Gs$	$1 \div 100 Gs$

As a result of scaling, we get a satisfactory similarity of a number of jets parameters expiring from the young stars and the jet obtained on the NEODIM laser installation.

4. Numerical model

During the simulation, the processes of laser heating, melting, and evaporation of the target material were not considered in detail. A considerable amount of work has been devoted to the study of laser ablation processes. For example, Mazhukin et al. investigated nanosecond laser ablation in subcritical and transcritical modes [7]. They also conducted a simulation of laser explosive effervescence of metal [8]. We believe that the ablation process has already happened, the substance has evaporated from the surface of the target and is taken into account in determining the initial and boundary conditions.

For numerical simulation of the experiment, the system of equations of magnetohydrodynamics [9] was used. In the future, to more accurately describe the processes occurring with the target material as a result of high-power laser irradiation, it seems necessary to combine MHD and the kinetic approach. But at the initial stage of modeling, we can confine ourselves to the system of MHD equations.

In the simulation, two options were considered. In the first of these, the external magnetic field was not included. In the second variant, an external constant magnetic field was taken into account in different configurations: a poloidal magnetic field directed along the normal to the plate, and a toroidal field. Simulations were carried out for different values of the magnetic field. In both cases, the emission from the plasma target was simulated under the influence of laser radiation. A single-component plasma flow was studied-the region and shape of the flow at various distances from the target and at different times, the spatial distribution of plasma density and energy, and also the magnetic field in the second version.

4.1. Simulation region

The analysis of the experiment showed that the formation of the jet is associated with the heating of the plate, and the geometry of the incident laser beam plays no part. Because of this,

IOP Conf. Series: Journal of Physics: Conf. Series **1031** (2018) 012022 doi:10.1088/1742-6596/1031/1/012022

the mathematical model of jet formation in the experiment in Figure 1 can be constructed in axisymmetric approximation, assuming that the spot heated by the laser has the form of a circle. To simulate the problem, an inertial cylindrical coordinate system (r, ϕ, z) is used. The center of the coordinate system is placed in the middle of the target. The orientation of the coordinate system was chosen so that the z axis is parallel to the vector of the poloidal magnetic field H and perpendicular to the target. The problem assumes axial symmetry in the distribution of all the macroscopic quantities ρ, T, v, H, $\partial/\partial\phi = 0$, but all three components of the velocity v and magnetic fields H.

Simulation is carried out in half of the coordinate plane r - z. The problem uses a uniform difference grid, the coordinates of the grid nodes lie in the range $0 \le r \le R_{max}$, $0 \le z \le Z_{max}$. Simulation is performed on a uniform grid (r, z) with the number of cells $N_R \times N_Z = 257 \times 513$, or $N_R \times N_Z = 513 \times 1026$.

The target (the central region of the plate to which the energy is applied) is a homogeneous cylinder of radius R_d, of thickness Z_d. In this case, $R_d << R_{max}$, Z_{max} and $Z_d << R_{max}$, Z_{max}. The scheme of the modeling domain is shown in Figure 3.

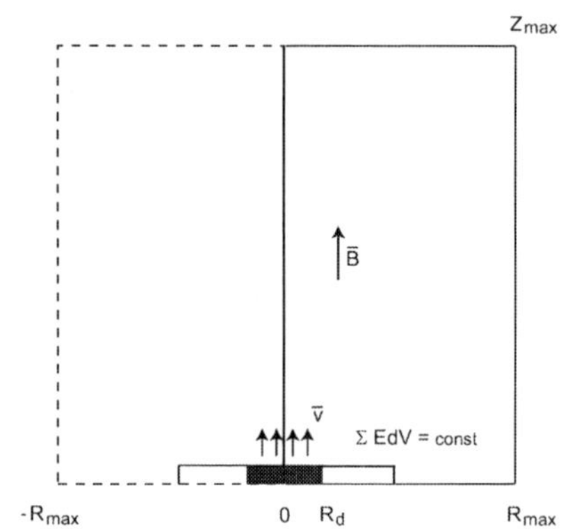

Figure 3. Simulation region

4.2. The system of equations

We use an axisymmetric, resistive MHD code. The code incorporates the methods of local iterations and flux-corrected-transport developed by Zhukov, Zabrodin, Feodoritova [10]. The flow is described by the resistive MHD equations in the absence of gravity:

$$\frac{\partial \rho}{\partial t} + \nabla \cdot (\rho\ \mathbf{v}) = 0\ , \tag{1}$$

$$\rho \frac{\partial \mathbf{v}}{\partial t} + \rho(\mathbf{v} \cdot \nabla)\mathbf{v} = -\nabla p + \frac{1}{c}\mathbf{J} \times \mathbf{H} \tag{2}$$

$$\frac{\partial \mathbf{H}}{\partial t} = \nabla \times (\mathbf{v} \times \mathbf{H}) + \frac{c^2}{4\pi\sigma}\nabla^2 \mathbf{H}\ , \tag{3}$$

$$\frac{\partial(\rho\varepsilon)}{\partial t} + \nabla \cdot (\rho\varepsilon\mathbf{v}) = -p\nabla \cdot \mathbf{v} + \frac{\mathbf{J}^2}{\sigma}\ , \tag{4}$$

$$\frac{4\pi}{c}\mathbf{J} = \nabla \times \mathbf{H}\ . \tag{5}$$

The variables in the system of equations have the following values: c is the speed of light, ρ is the density of matter, v is the velocity, H is the magnetic field strength, ε is the internal energy of the mass unit of matter, $\mathbf{J}$ is the electric current density, σ is the effective electrical conductivity of the medium. To describe the thermodynamic properties of a substance, the adiabatic equation of state of an ideal gas is used: $p = (\gamma - 1)\rho\varepsilon$ where γ is the adiabatic exponent. For current calculations, $\gamma = 5/3$ is chosen, which corresponds to the standard adiabatic exponent for monatomic gases. When writing the system of equations (1) - (5), Ohm's law was taken into account in the form: $\mathbf{J} = \sigma(\mathbf{E} + \mathbf{v} \times \mathbf{H}/c)$, where σ is the electric conductivity. The associated magnetic diffusivity, $\eta_{\rm m} \equiv c^2/(4\pi\sigma)$, is assumed to be a constant. The gravitational force, $\mathbf{F}^g = 0$.

The magnetic field has one of three configurations: the poloidal field (H_z, H_r), the toroidal field H_ϕ or their superposition. In order to fulfill exactly the condition $\nabla \cdot \mathbf{H} = 0$, in the calculations instead of the poloidal component of the magnetic field strength H_z and H_r, the toroidal component A_ϕ vector potential $\mathbf{A}$ is used, $\mathbf{H} = \nabla \times \mathbf{A}$. Early we used this MHD code in the simulations of accretion onto stars with dipole magnetic fields [11, 12].

4.3. Boundary and initial conditions

The system of equations (1) - (4) is solved numerically in the cylindrical region ($0 \le z \le Z_{max}$, $0 \le r \le R_{max}$). It is assumed that there is a magnetic field at the outer boundaries, but there are no currents. For this, the so called free boundary conditions are used, i.e. for any value f, $\partial f/\partial n = 0$. Here $\partial f/\partial n$ denotes the derivative along the normal to the surface, that is, on the outer boundary $f(r, z) = const$. In accordance with the axial symmetry assumption, the boundary conditions on the axis z ($r = 0, 0 \le z \le Z_{max}$) are written as: $v_r = v_\phi = 0, \quad A_\phi = H_\phi = 0$.

On the boundary ($z = Z_{min}$, $R_d \le r \le R_{max}$) fixed boundary conditions are given, $\rho = \rho_0$ $v_0 = 0$. The flow of matter through this boundary into the modeling area is prohibited.

On the opposite boundary ($z = Z_{max}$, $0 \le r \le R_{max}$), the free boundary conditions are given, $\partial/\partial n = 0$. This is allow matter flow out of the field. The flow of matter through this boundary into the simulation region is also prohibited.

On the cylindrical boundary ($Z_{min} \le z \le Z_{max}$, $r = R_{max}$) the free boundary conditions $\partial/\partial n = 0$ are given. To verify the influence of boundary conditions, a series of calculations was performed for various dimensions of the calculation region.

The vector potential of the magnetic field $A = A_\phi$ was determined and defined throughout the calculation region. For a homogeneous magnetic field H co-directed with the axis z, the vector potential is written as follows: $A_\phi = rH$, $A_r = 0$, $A_z = 0$, the toroidal component of the field $H_\phi = 0$. For a toroidal magnetic field $A_\phi = 0$ in the whole region, the toroidal component of the field is written as follows: for $0 \le r \le R_d$ $H_\phi = H_0 r/R_d$, for $R_d \le r \le R_{max}$ $H_\phi = H_0 R_d/r$, where H_0 is constant.

At the initial moment $t = 0$, the magnetic field was a pure poloidal or toroidal field. The initial density of the substance was assumed to be homogeneous throughout the calculation region, with the exception of the target, and had a small value: $\rho = \rho_0$. The initial velocity of the substance in the calculation region, with the exception of the target, was zero $v_z = v_r = 0$. The substance has no angular momentum, $v_\phi = 0$.

In the target region $0 \le r \le R_d$, $0 \le z \le Z_d$ the value of the density $\rho >> \rho_0$ is given. The velocity of the particles emitted from the target, which was determined from the condition: $\rho v^2/2 = E$, $v = \sqrt{2E/\rho}$, where E is the energy at the initial time.

After reducing the system of equations (1) - (4) to a dimensionless form, a set of characteristic dimensionless parameters is appeared. First parameter

$$\beta = \frac{8\pi P_{0jet}}{H_0^2} = \frac{2}{\gamma}\frac{c_{s0}^2}{V_{A0}^2}\,, \tag{6}$$

determines the relation between the characteristic values of the pressure of the matter $P_{0jet} = \rho_{0jet} c_{s0}^2/\gamma$ and the pressure of the magnetic field. Here ρ_{0jet} is the density of the inflowing substance at the boundary of the calculated region, P_{0jet}, c_{s0} is the pressure and sound speed in the incoming matter, H_0 is the magnetic field strength in the region, $V_{A0} = \frac{H_0}{\sqrt{4\pi\rho_{0jet}}}$ is the Alfven velocity of the inflowing substance at the boundary of the calculated region.

The second parameter of the numerical model is the characteristic dimensionless magnetic viscosity

$$\tilde{\eta}_M = \frac{\eta_M}{L_0 V_{A0}} = \frac{c^2}{4\pi\sigma L_0 V_{A0}} = \frac{1}{Re_M}, \tag{7}$$

where σ is the constant effective conductivity of the plasma; L_0 is the characteristic size and V_{A0} is the characteristic Alfven velocity used in determining the Reynolds magnetic number Re_M. As the characteristic size L_0 the scale of the problem R_{max} was taken. As the characteristic velocity V_0 the Alfven velocity of the inflowing matter is taken at the boundary of the calculated region V_{A0}.

5. Results of numerical simulation

5.1. *Hydrodynamic approximation*

As a test of the numerical MHD model, the hydrodynamic approximation with $H = 0$ is usually used. It allows us to evaluate the correctness of the choice of the dimensions of the modeling domain, parameters and boundary conditions. The initial density of matter in the whole calculation region, with the exception of the target, is homogeneous and has a small value: $\rho = \rho_0$. The initial velocity of the substance in the calculation region, with the exception of the target, is zero $v_z = v_r = 0$. The substance has no angular momentum, $v_\phi = 0$.

In the target region $0 \leq r \leq R_d$, $0 \leq z \leq Z_d$, the density $\rho \approx 300\rho_0$ and the supersonic velocity of the particles emitted from the target $v \geq c_s$ are set, here $c_s = \sqrt{\gamma \frac{P_0}{\rho_0}}$ is the sound speed in the unperturbed region. For simplicity of test calculations, the velocity of the emitted matter at the initial instant of time has only the z-component v_z; $v_r = 0$.

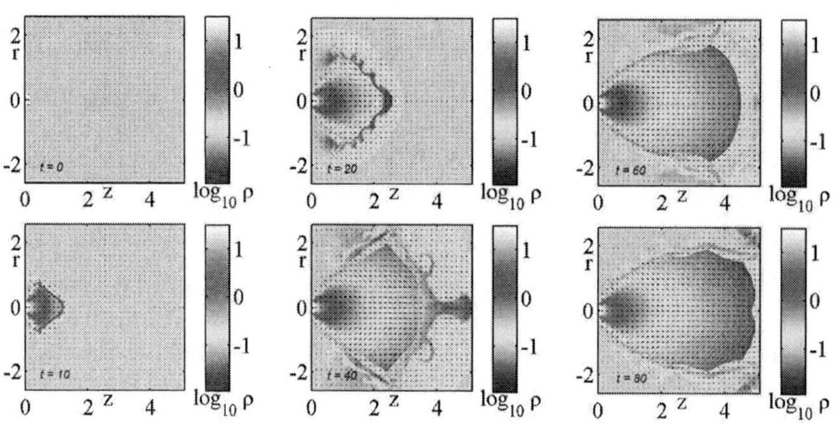

Figure 4. Picture of the matter flow for the hydrodynamic case at different moments.

The initial distribution of density and velocity and the results of modeling the ejection of matter in the hydrodynamic approximation are shown in Figure 4. The color background corresponds to the logarithm of the density, the arrows indicate the velocity vectors, the length of the arrow is proportional to the velocity value. It can be seen that the target substance under the influence of the thermal energy of the laser beam begins to fly along the axis z to the opposite wall of the chamber, expanding in the direction of the r axis along the way. A shock wave moves from the target to the boundary of the chamber, leaving a steady outflow of a cone-like shape. There is a small collimation of the substance due to the presence of an external (background) gas.

5.2. The poloidal magnetic field with $\beta = 10^{-3}$

At the next step of the simulation we included an external constant poloidal magnetic field directed along the normal to the target H. We performed a series of calculations for different values of the parameter β. The characteristic parameters for this case are: $\beta = 10^{-3}$ and $\tilde{\eta}_M = 10^{-5}$

The initial setups for the background density and velocity are the same as those of the simulation described in Section 5.1. The magnetic field is defined as follows: $A_\phi = rH$, $A_r = 0$, $A_z = 0$, $H_\phi = 0$.

The results of modeling the release of matter for the case $\beta = 10^{-3}$ are shown in Figure 5. The color background corresponds to the logarithm of the density, the arrows indicate the velocity vectors, the length of the arrow is proportional to the velocity value. Solid lines represent magnetic field lines. It can be seen that the target substance under the action of the thermal energy of the laser beam begins to fly along the axis Z to the opposite wall of the chamber. A shock wave of elongated shape is formed, which moves from the target to the boundary of the chamber, leaving behind a steady outflow. Due to the strong magnetic field, the flux is well-collimated, the substance does not expand in the direction R. The collimation by the external (background) gas in the presence of a strong field does not play a role. Thus, in the presence of a sufficiently strong poloidal magnetic field, a sufficiently sharp continuous spot can be observed on the detector.

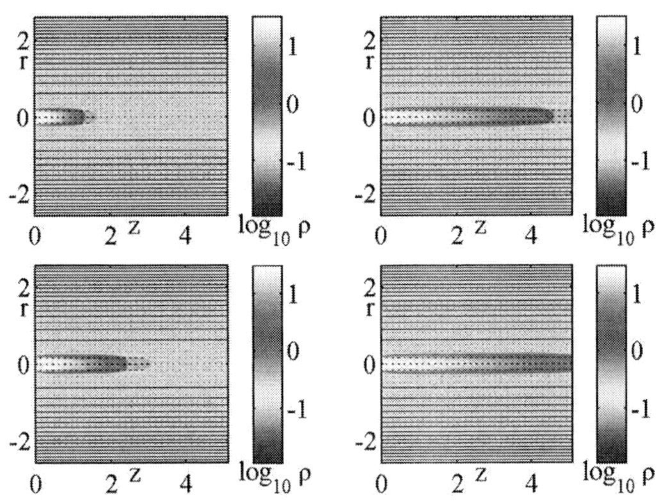

Figure 5. Picture of the matter flow for MHD case $\beta = 10^{-3}$ at different moments

5.3. The poloidal magnetic field with $\beta = 10^{-1}$

In the following calculation, we reduced the magnetic field. The characteristic parameters for this case are: $\beta = 10^{-1}$ and $\tilde{\eta}_M = 0^{-5}$ The initial setups for the background density, velocity, and magnetic fields are the same as those of the simulation described in Section 5.2 except β. The results of modeling the release of matter for the case $\beta = 10^{-1}$ are shown in Figure 6.

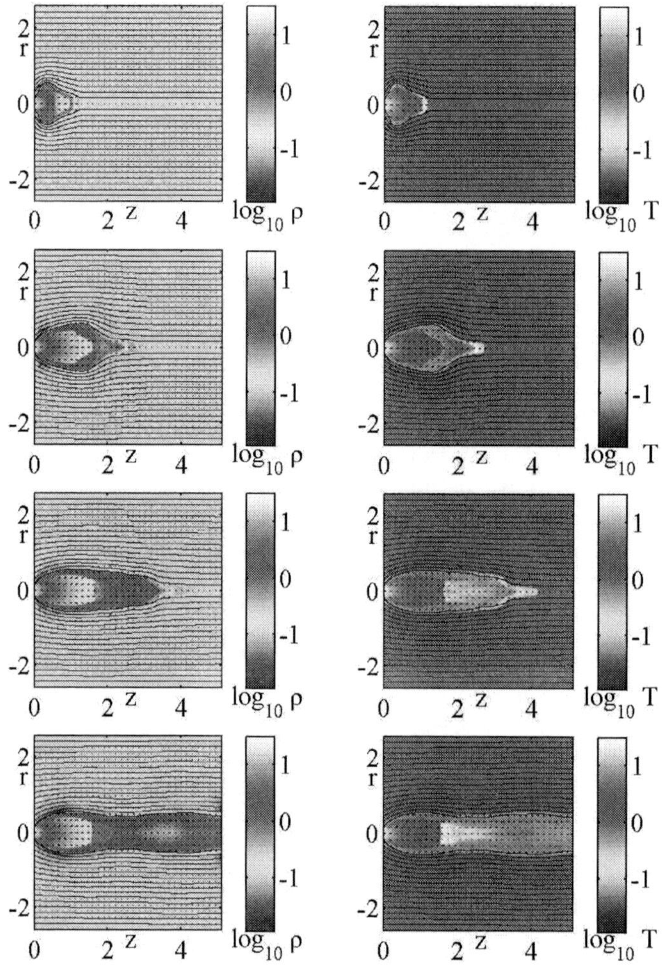

Figure 6. Picture of the matter flow for MHD case $\beta = 10^{-1}$ at different moments.

The color background of the left column corresponds to the logarithm of the density. The color background of the right column corresponds to the logarithm of temperature. The arrows indicate the velocity vectors, the length of the arrow is proportional to the velocity value. Solid lines represent magnetic field lines. It can be seen that after the release of matter it begins to expand in all directions, but then the flow is collimated by a magnetic field. The influence of the external gas on the collimation of the flow in this case is insignificant. Inside the flow, a shock wave is formed at a distance of $z \approx 1.6$.

Figure 7 shows the Mach number $M = v/c_s$ as a function of the distance from the target. Figure 8 shows the Mach number $M = v/c_s$ as a function of the distance from the r-axis. Both figures correspond to the moment of the forth epoch in Figure 6. We can see that the velocity of matter rises to a distance of $z \approx 1.6$ and reaches a value of $M \approx 10$. After the passage of the shock wave, the velocity of matter falls sharply to $M \approx 0.5$, and then again begins to increase to $M \approx 2$ at a distance of $z \approx Z_{\max}$.

IOP Conf. Series: Journal of Physics: Conf. Series **1031** (2018) 012022 doi:10.1088/1742-6596/1031/1/012022

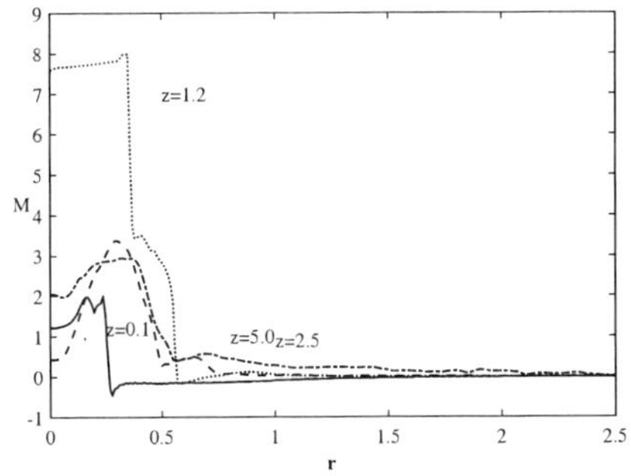

Figure 7. The dependence of the Mach number $M = v/c_s$ on the distance from the target, $\beta = 10^{-1}$.

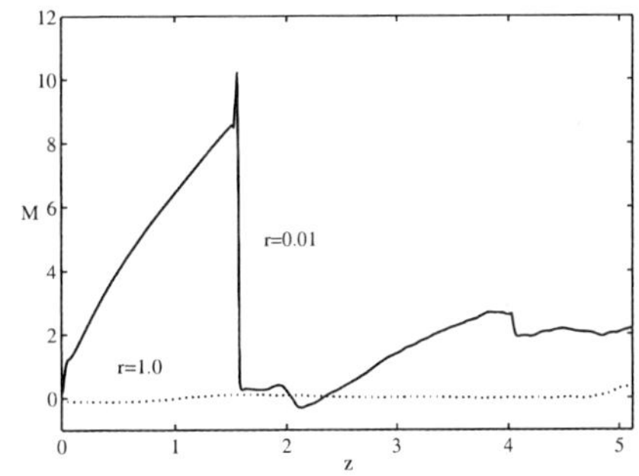

Figure 8. The dependence of the Mach number $M = v/c_s$ on the distance from the r-axis $\beta = 10^{-1}$.

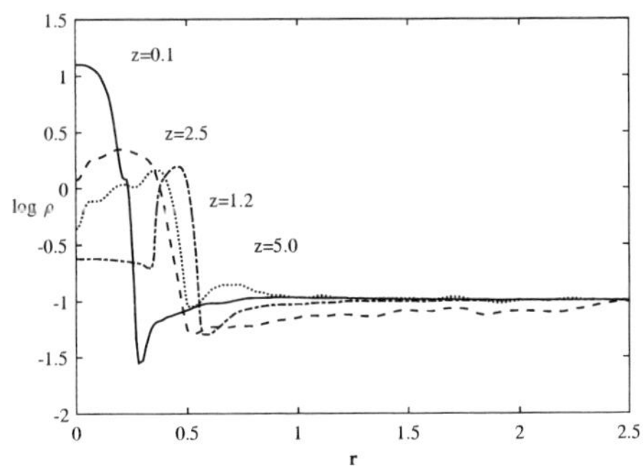

Figure 9. The flux density along r as a function of the distance from the target at different distances z from the target surface for the case $\beta = 10^{-1}$.

Figure 9 shows the flux density along R as a function of the distance from the target at different distances z from the target surface for the case $\beta = 10^{-1}$. A thin solid line corresponds to the flux density at the boundary of the target. The dot-dashed line corresponds to the distance $z = 1.2$, the dashed line corresponds to the density in the middle of the chamber, $z = 2.5$, the dotted line corresponds to the density near the camera boundary, $z = 5.0$. It can be seen that the density distribution in the plasma flow depends on the distance from the target. If at the beginning of the flow the maximum plasma flux density falls on a narrow range comparable to the target dimensions, then in the shock wave region, $z = 1.2$, the density distribution forms a clear ring structure. After the passage of the shock wave, the density distribution becomes more uniform and the ring structure is washed out. From Figures 7 and 9, we can determine the angle of the cone of the flow at the distance of the shock wave. It is approximately equal to $\theta \approx 40$.

It is seen from Figure 6 that near the target a cavity is formed, bounded by a shock wave, the size of which depends on the magnitude of the poloidal magnetic field. Similar results were obtained in the work [4]. This leads to the formation of ring structures, which is observed in the experiment. The angle of the solution of these structures depends on the magnitude of the magnetic field. For a weak field (and also in the absence of magnetic field, see Fig. 4), this angle is large. With increasing field strength, the dimension of the cavity and the opening angle of the ring structure decreases, and then a solid spot is observed instead of the ring (Figure 5).

6. Conclusions

The paper presents the results of numerical simulation of magnetized and non-magnetized supersonic jets. It is shown that in the absence of a magnetic field, a substantial expansion of the jet occurs. In the presence of a strong poloidal magnetic field, the jet expands only slightly, which confirms the possibility of a magnetic collimation mechanism for astrophysical jets. Calculations have shown that under certain values of the poloidal magnetic field ring structure may form. The opening angle of these structures approximates the results of the laboratory experiment ($\approx 40^\circ$).

Thus, a comparison of the results of the laboratory experiment and the numerical calculations of magnetized jets in this paper, as well as the results of the work [4], confirm the possibility of the formation of ring structures whose characteristics depend on the magnitude of the magnetic field. Further theoretical and experimental research is needed in this field.

Formation and development of jet in a laboratory laser experiment is a complex physical phenomenon involving a large number of different physical processes. Numerical simulation of such an experiment, allowing to account for all possible physical processes, is a complex problem and is currently difficult to implement. We used the approximation of magnetohydrodynamics, as the first step in the study of this problem.

7. Acknowledgements

We are grateful to V.V. Saveliev (IPM RAS) for provided the original version of the MHD simulation program. This work was partially supported by RFBR grant 17-02-00021, 17-02-00760, Leading scientific school grant NS-6579.2016.2, and by Russian Academy of Science program I.7P.

References

[1] Bisnovatyi-Kogan G.S. Proc. 6th Int. Workshop of the Astronomical Observatory of Capodimonte (OAC 6), Capri, Italy, September 18-21, 1991, Dordrecht: Kluwer. Eds. L. Errico, and Alberto A. Vittone. Astrophys. Space Sci. Library, **186**, 369, 1993
[2] Fua W., Liang E.P., Tzeferacos P., Lamb D.Q., High Energy Density Physics, **17**, 42, 2015
[3] Mitrofanov, K. N. et.al. Astronomy Reports, 61, 138, 2017
[4] Albertazzi B. et al. Science, **346**, 325, 2014
[5] Belyaev V. S., Vinogradov V. I. , Matafonov A. P. Nuclear physics, **71**, 466, 2008
[6] Ryutov D.D., Drake R.P., Remington B.A. ApJS, **127**, 465, 2000.
[7] Mazhukin V.I., Shapranov A.V., Demin M.M., Samokhin A.A., Zubko A.E. Mathematica Montisnigri, **38**, 3, 2017
[8] Mazhukin V.I., Demin M.M., Shapranov A.V. Appl. Surf. Sci., **302**, 6, 2014
[9] Landau, L. D., Lifshitz, E. M. 1960, Electrodynamics of Continuous Media (New York: Pergamon), chap. 8
[10] Zhukov, V. T., Zabrodin, A. V., Feodoritova, O. B. 1993, Comp. Maths. Math. Phys., 33, No. 8, 1099
[11] Toropina, O.D., Romanova, M.M., Toropin, Yu.M.,& Lovelace, R. V. E. ApJ, **561**, 964, 2001
[12] Toropina, O.D., Romanova, M.M., & Lovelace, R. V. E. MNRAS, **420**, 810, 2012

IOP Conf. Series: Journal of Physics: Conf. Series **1031** (2018) 012023 doi:10.1088/1742-6596/1031/1/012023

Generating Automatic Network Reduction Module for Chemical Hydrodynamics Simulations

Jeongkwan Yoon
Ulsan National Institute of Science and Technology (UNIST), Ulsan, South Korea
E-mail: logicyoon@unist.ac.kr

Kyujin Kwak
Ulsan National Institute of Science and Technology (UNIST), Ulsan, South Korea
E-mail: kkwak@unist.ac.kr

Abstract. Since molecules exist in the interstellar cloud and they affect the hydrodynamic evolution through their formation and destruction, physical states of the actual interstellar cloud are different from calculations of conventional hydrodynamics simulations. Furthermore, in the case of running star-forming simulations, conventional hydrodynamics models are not enough to explain molecular lines emitted from clouds such as those detected from the ALMA observatory. In order to simulate the chemical evolution of a hydrodynamic cloud, building an efficient chemical network that contains relevant chemical reactions is crucial for cutting down the computation cost. A key factor for generating an efficient chemical network is to avoid using an abnormally small network that contains only a few reactions because using too small a network does not simulate the effects of molecules accurately. There already exist a few chemical hydrodynamics simulation codes, which provide pre-built reduced networks. Although those pre-built networks make the simulations simple and light, *they cannot be used universally for various clouds under diverse* initial chemical compositions and environmental conditions. Therefore, it is necessary to build a more flexible network according to the conditions of the model. In this study, we *propose to make* an automatic *network reduction* module which builds an optimized closed network corresponding to the specific simulation conditions. *As a preliminary result,* we test our module with *simple primordial test clouds* by comparing our results with those obtained with the full network. *In the future study, we will validate our reduction module for primordial cloud models and expand its usage to various physicochemical cases such as AGB stars.*

1. Introduction

Hydrodynamics (HD) simulations play a crucial role in studying numerous celestial phenomena because many astronomical objects can be treated as fluid. However, there are some situations where HD simulations are not sufficient alone. For example, magnetohydrodynamics (MHD) is often required in order to explain astronomical phenomena that occur in the plasma states of the material [1]. Another recent example is chemical-hydrodynamics (CHD)[1] simulations which trace the change of chemical species as a function of time while running HD simulations. The need for CHD simulations emerged in recent years due to the ALMA observations of tremendous

[1] *CHD is generally known as physicochemical hydrodynamics (PCHD) as well.*

molecular lines, but the CHD simulations have not been actively performed compared to HD and/or MHD simulations.

A crucial reason for lacking the CHD simulations is the limitation of computational resources. In order to trace the change of all the relevant chemical species, the abundance of each molecule (or atom) must be treated as a variable in the code. Because tracing a decent number of species is usually required in the CHD simulations in order to explain the observed molecular lines, the realistic CHD simulations need to take at least a few tens of variables. For the conventional three dimensional (3D) HD simulations, it is sufficient to use six or seven variables (density, three components of velocity, internal energy, pressure and/or temperature depending of equation of state). Therefore, the 3D CHD simulations require more computational resource than the 3D HD simulations by at least a factor of three. Furthermore, the more species are included, the more reactions need to be solved. This also increases the computing time. As a result, CHD simulations are generally more challenging than HD and MHD ones.

It is possible, however, to perform these challenging simulations efficiently by using a selected network (a network is composed of species and reactions) which is optimized for the problem-specific situations. The optimization can be done by reducing the number of chemical species and reactions from the large original network which includes all the known species and reactions. This approach has been tried already and there are some CHD simulators which use a pre-built reduced network [2]. *Most CHD simulators which use pre-built networks select the species and reactions manually and pre-built networks made in this way may not guarantee accurate simulations under diverse physical and/or chemical conditions. Meanwhile, Wiebe et al. [3] suggested a sturdy method to make an automatic reduction. They calculated priorities of each species and/or reaction to determine which species and reactions are reducible. Although their method is well-organized, it shows significant errors during CHD simulations, especially for the case of the diffuse gas cloud.* Therefore, it is *still* necessary to build a proper automatic module that makes an optimized closed network for a given physicochemical condition by reducing the large original network. *In this study, we propose to develop such an automatic module and present our preliminary algorithm and results obtained with simple tests.*

In the following section, we explain how the reduction module works and the algorithm of the module. *We also test our reduction module.* We choose two simple test models (Section 3). We evolve them with both the reduced network built by our reduction module and the full (i.e., original) network and compare two results (Section 4). We summarize our study and discuss the future work in the final section.

2. Reduction Module

2.1. Role of Reduction Module

A grid-based hydrodynamics simulator is designed to update the physical states of gas (or any fluid) over time at the segmented cells which are combined together to cover the entire computational domain. As mentioned earlier, on top of the hydrodynamics part, the CHD simulator additionally updates the abundances of chemical species by calculating the rate equations that contain the chemical reactions among the chemical species. Therefore, the CHD simulator requires the input information on the chemical composition of the model cloud and the list of chemical reactions related to the chemical composition. Because the abundances of the chemical species are also affected by environmental conditions such as cosmic rays, background ultraviolet (UV) radiation, and visual extinction, it is also necessary to implement these environmental parameters into the CHD simulator.

Figure 1 shows schematically how the CHD simulator works. Both the initial composition and the physical parameters determining the environmental conditions, which are specified by the problem, are the input not only to the simulator itself but also to the network reducer (i.e., our automatic reduction module). The network reducer reduces both species and reactions from the full (original) network and builds a reduced network which is used now as input to the CHD simulator. Without the reduction process, the full network is used as the input to the simulator, but in this case the simulator does unnecessary calculations for the species and reactions that are not relevant to the problem. Because these dummy calculations are just a waste of the computing resources, the reduction module can increase the computing efficiency.

Since the currently available full network includes more than 400 chemical species and 6000 reactions, it is not a trivial task to build an efficient automatic reduction module which works for all the various cases. In the following subsection, we introduce the algorithm that we use for the reduction. Note that the current version of our algorithm is still under development. In the final section, we will briefly mention what will be required to improve the current algorithm.

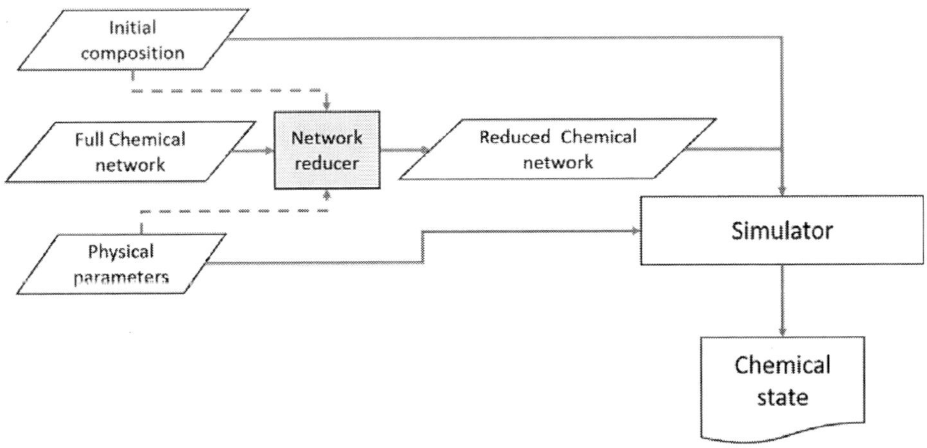

Figure 1. The schematic figure showing how the CHD simulator works. Our network reduction module is indicated by a yellow box.

2.2. Algorithm of Reduction Module

The flowchart of our reduction algorithm is shown in Figure 2. Before explaining the details, it is worth mentioning an important rule for the reduction process. A network both full and reduced must be closed. That is, all the species in the network must interact with the others through the reactions in the network. Without violating this, our algorithm reduces the full network according to the following steps.

First, by using the information on the composition of the problem, our module identifies zero-abundance reactants *which do not appear at all* during the period of simulations and removes the species of zero-abundance reactants and the reactions which involve the zero-abundance reactants.

The second step eliminates reactions from the full network according to the physical conditions. In this step, the reduction module determines whether the model cloud has the proper conditions such as temperature and environments for certain reactions to occur. For example, when the cloud has neither background radiation nor cosmic ray, most of photo-dissociation reactions are removed.

In the final step, we tried to reduce the network further by checking out individual reaction rates. When the rates of certain reactions are very small *(i.e., smaller than the threshold value of 10^{-20})*, our reduction module ignores them. But care needs to be taken in order not to violate the rule that the reduced network must be closed. When used in the chemical evolution, the reduction module with this step may produce different amounts of some chemical species in comparison with those calculated with the full network. However, as long as the difference is not significant, the reduced network can be used in the place of the full network reducing the computing resources and time in the chemical evolution calculations.

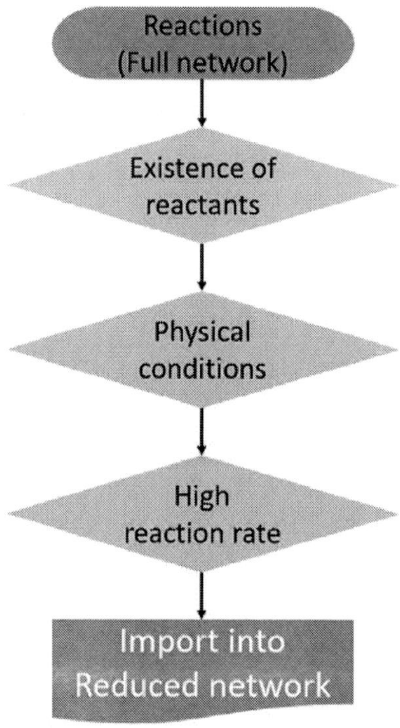

Figure 2. The schematic algorithm of the reduction module

3. Test Models

We tested our reduction module for a simple static situation. In this case, the dynamical motion of the fluid can be ignored and the abundance change of chemical species can be traced by solving the rate equations only. In other words, it is sufficient to use a chemical evolution code as the simulator in Figure 1 without calculating any hydrodynamics update. We used *Astrochem* (version 0.7) for the calculation of the chemical evolution [4]. *Astrochem* numerically solves the rate equations which are in fact ordinary differential equations. When we consider all of formation and destruction reactions for an arbitrary molecule AB, the ordinary differential equation for the abundance evolution of this molecule is

$$\frac{\mathrm{d}}{\mathrm{d}t}N(\mathrm{AB}) = \sum_i k_i \prod_a N(X_a) - N(\mathrm{AB}) \sum_j k_j \prod_b N(Y_b) \, , \tag{1}$$

where $\mathrm{X_a}$ is an a–th reactant of the i–th formation reaction and $\mathrm{Y_b}$ is a b–th reactant of the j–th destruction reaction of molecule AB. A set of coupled equations for participating species are solved together by *Astrochem*.

Astrochem can also handle the effects of various physicochemical environments such as cosmic ray and photon by considering reactants or products in the above equation as cosmic ray or photon. *Astrochem* uses the full network based on the Ohio State University (OSU) astrochemistry database (version `osu_01_2009`) which contains 6046 reactions and 468 species[2]. Note that we also used the OSU database as the full network from which our reduction module builds the reduced network.

We considered two simple model clouds for the test of the static situation. The first cloud contains only hydrogen molecule and the second cloud is the simplest two-element cloud which contains helium and molecular hydrogen. In both cases, all hydrogen atoms are bounded as diatomic molecular state at the initial time. So, we set the abundance of H_2 as 0.5 in both test models (Table 1) because the abundances of H, H_2, H_3^+ have a relation, $N(H) + 2 \times N(H_2) + 3 \times N(H_3^+) = n(H)$, where $N(X)$ and $n(H)$ indicate the *number density* of a given species X and hydrogen nuclei, respectively, in unit of cm^{-3} [4]. We set the abundance of helium for the second model cloud as 0.14 which is selected from [6].

Species	Abudance(/n(H))
H_2	0.5
He	0.14

Table 1. Initial abundances in our test models

For our test, we also include the physical and environmental effects. Table 2 shows the physical parameters and their values used in our test. These values are the input parameters to *Astrochem*.

Density	CRIR	UV intensity	Temperature	Visual Extinction
n(H)=500 cm^{-3}	1.3×10^{-17}	1.0 Draine[2]	100 K	5.0 magnitude

[2] See [7] for a description of Draine units.

Table 2. Physical and environmental parameters in our test models

4. Results

By using the initial composition and the physical conditions of our two test models as the input information, our reduction module builds the reduced networks that correspond to each model, respectively. We calculate the chemical evolution of two test models with the reduced network and compare the results with those evolved with the full network. *Table 3 shows the size of the reduced network and the computational gain for two simulations. The list of species in each reduced network is posted at Tables 4 and 5. When we used the reduced network for the two tests, the computing time got decreased as small as 1/80th of that run with the full network.*

Model	Size of reduced network[2]	Gain
H-only	7 (24)	~ 80
He+H	9 (30)	~ 80

[2] number of species (number of reactions)

Table 3. Size of reduced network and computational gain for our test models

[2] See http://faculty.virginia.edu/ericherb/research.html and [5].

4.1. Test Model 1: Hydrogen-only Cloud

Table 4 lists the species in the reduced chemical network for the first test model, the hydrogen-only cloud. The reduced network contains 24 reactions only among these seven species that make it a closed network.

The upper-left and upper-right panel in Figure 3 show the evolution of the seven species obtained with the reduced network and the full network, respectively. At a glance, there seems no significant difference between them, but the fractional difference defined as $[N(X)_{red} - N(X)_{full}]/N(X)_{full}$, where $N(X)_{red}$ and $N(X)_{full}$ are the abundance of species X calculated with the reduced and full network, respectively, looks apparent for two species, tri-hydrogen cation (H_3^+) and electron (e^-) although the maximum errors for these two species are 4.3% (H_3^+) and 0.2% (e^-) throughout the entire evolution of 10^9 years (see lower panel in Figure 3). Note that the amounts of produced species from the initial hydrogen molecule are still very small compared to that of hydrogen molecule. Hydrogen atom (H) was produced most abundantly, but its amount is still less than 1% of H_2.

We find that these errors are not based on the third step in the algorithm of the reduction module which removes the reactions with small rates in the full network because the relevant reactions in the reduced network are identical to those in the full network, i.e., the third step did not operate in this case. We speculate that the fractional errors in two species might be caused by the numerical precision of the machine because the maximum values of absolute difference are approximately 10^{-18} for H_3^+ and 10^{-9} for e^-. We are now investigating the exact cause for these errors, but the overall results show that our reduction module builds the properly reduced network.

H	H^+	H^-	H_2
H_2^+	H_3^+	e^-	

Table 4. Chemical species obtained with our reduction module for the first test model, hydrogen-only cloud

4.2. Test Model 2: Helium-containing Hydrogen Cloud

Table 5 shows the species in the reduced network for the second test model, the helium-containing hydrogen cloud. This model cloud containing molecular hydrogen and helium is the simplest two-element model because both elements do not directly react with each other. Although there *is* a helium hydride cation molecule (hydro-helium, HeH^+), its unique production reaction requires the existence of formyl-radical (HCO). *Our reduction module recognizes this information and does not import* HeH^+ *into our reduced network* which contains nine species only *because there is no HCO in the initial condition.* Again, we check out that our reduced network having nine species and 30 reactions is a closed one.

Figure 4 shows the results for the second test model. As test model 1 does (see Figure 3), *the results of this test model show* that the chemical evolution of nine species simulated with the full network can be reproduced with the reduced network. However, there is an error in the fractional difference defined as $[N(X)_{red} - N(X)_{full}]/N(X)_{full}$ for tri-hydrogen cation (H_3^+) (see the bottom panel of Figure 4). As in the test model 1, we find that the third step did not operate either in this case. Thus we also speculate that the fractional errors shown in this test case are caused by the numerical precision of the machine. Note that the evolution pattern of the error looks similar in both test models for tri-hydrogen cation (H_3^+) (compare the bottom panel of Figure 3 and 4) and the maximum errors for this cation also have a similar value of 4.3%. The exact reason for this error shown in both test models is under investigation now.

IOP Conf. Series: Journal of Physics: Conf. Series **1031** (2018) 012023 doi:10.1088/1742-6596/1031/1/012023

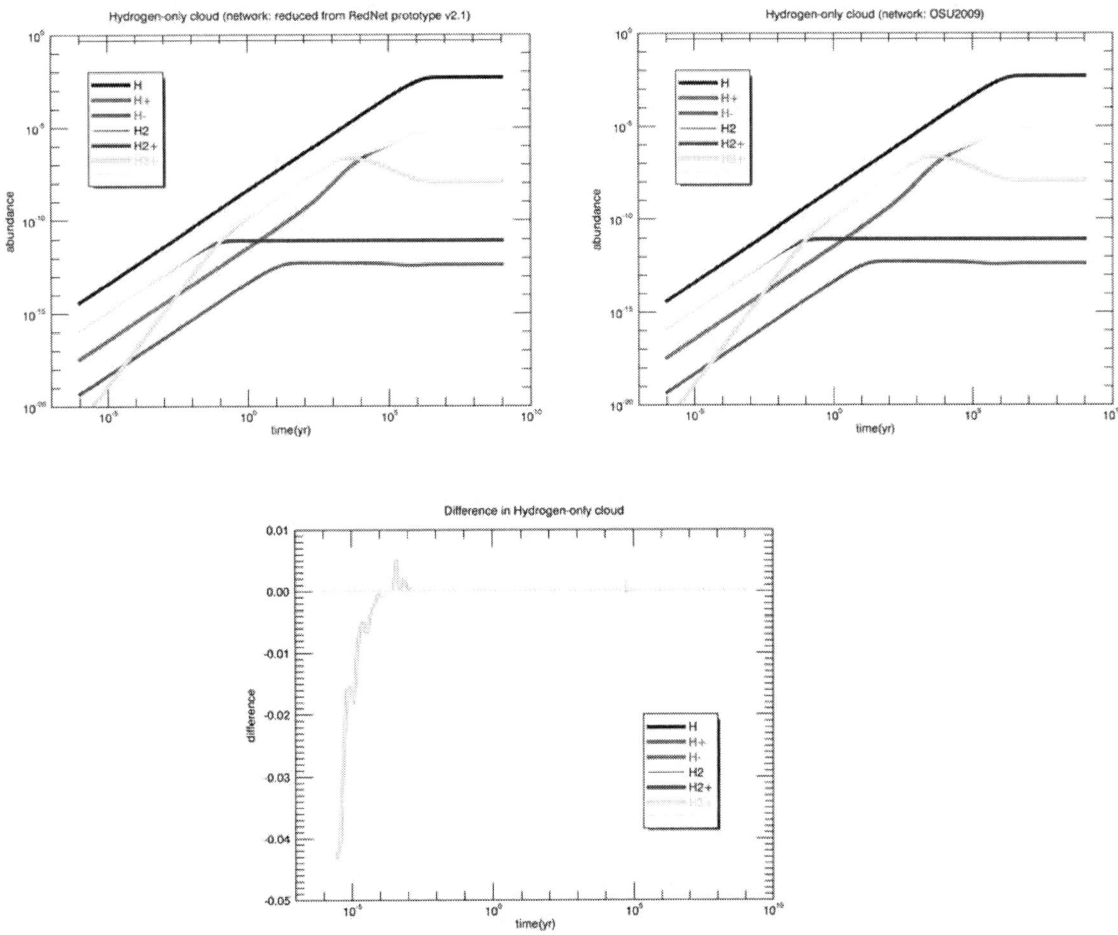

Figure 3. The evolution of seven species for test model 1, the hydrogen-only cloud simulations obtained with the reduced network (upper-left) and the full network (upper-right). Note that the abundance of H_2 (thin black lines close to the top boundary) does not change much throughout the evolution and that some curves are overlapped. The lower panel shows the fractional difference defined as $[N(X)_{red} - N(X)_{full}]/N(X)_{full}$ as a function of time.

H	H^+	H^-	H_2	H_2^+
H_3^+	He	He^+	e^-	

Table 5. Chemical species obtained with our reduction module for the second test model, helium-containing hydrogen cloud

5. Summary & Future work

In order to reduce the computing resources and time in the CHD simulations, we develop a reduction module which builds a reduced chemical network from the full network. The reduction is done by using the initial composition and the physical conditions of the problem-specific model which are also the initial input to the simulations. We *test* our reduction module with two simple static clouds by using *Astrochem*, a chemical evolution code, as a simulator.

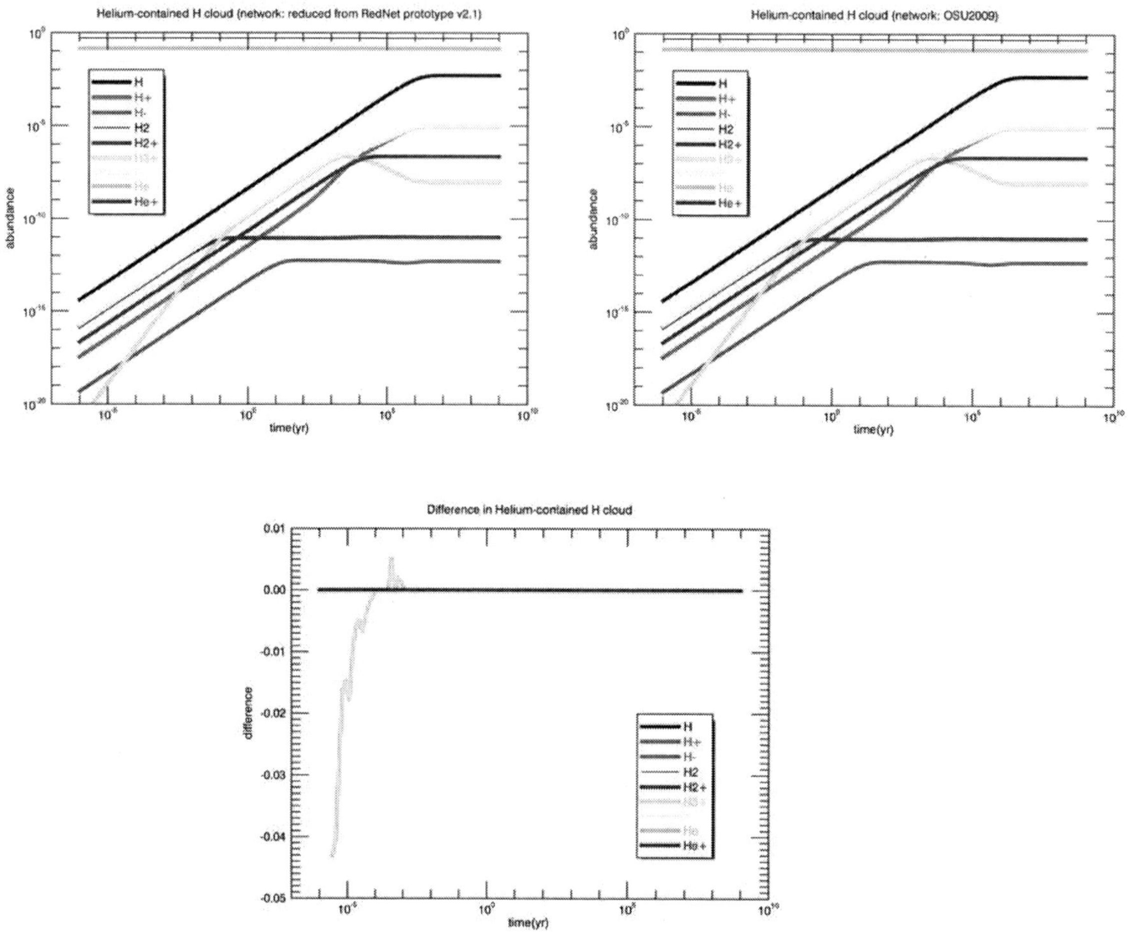

Figure 4. The evolution of nine chemical species for test model 2, the helium-containing hydrogen cloud simulated with the reduced network (top-left) and the full network (top-right). As in Figure 3, the abundance of H_2 (thin black lines close to the top boundary) does not change much throughout the evolution and some curves are overlapped. The bottom panel shows the fractional difference defined as $[N(X)_{red} - N(X)_{full}]/N(X)_{full}$ as a function of time.

Our two test model clouds contain only hydrogen or both hydrogen and helium, which are relevant for the primordial chemistry simulations. In the future work, we will include more species like deuteron for the more accurate primordial chemistry. We will also test our reduction module which works with more complex chemistry such as a two-element system composed of carbon and oxygen and a three-element system composed of hydrogen, carbon, and oxygen. These models can address the chemical evolution that can occur in the outer part of asymptotic giant branch(AGB) stars as well as in the star-forming region.

Acknowledgments
This article is based on the poster presentation given at ASTRONUM 2017 – the 12th International Conference on Numerical Modeling of Space Plasma Flows at Saint Malo, France, June 26-30, 2017. This work is supported by the National Research Foundation of Korea (NRF)

grant funded by the Korean government, No. 2016R1A5A1013277, 2016R1D1A1B03936169, and 2014R1A1A1006050.

References

[1] Alfvén H 1946 *Stockholms Observatoriums Annaler* **14** 9–1
[2] Grassi T, Bovino S, Schleicher D, Prieto J, Seifried D, Simoncini E and Gianturco F 2014 *Monthly Notices of the Royal Astronomical Society* **439** 2386–2419
[3] Wiebe D, Semenov D and Henning T 2003 *Astronomy & Astrophysics* **399** 197–210
[4] Maret S and Bergin E A 2015 Astrochem: Abundances of chemical species in the interstellar medium Astrophysics Source Code Library (*Preprint* 1507.010)
[5] Wakelam V, Herbst E, Loison J C, Smith I, Chandrasekaran V, Pavone B, Adams N, Bacchus-Montabonel M C, Bergeat A, Béroff K *et al.* 2012 *The Astrophysical Journal Supplement Series* **199** 21
[6] Mookerjea B 2016 *Monthly Notices of the Royal Astronomical Society* **459** 2822–2826
[7] Draine B 1978 *The Astrophysical Journal Supplement Series* **36** 595–619
[8] Harada N, Herbst E and Wakelam V 2010 *The Astrophysical Journal* **721** 1570

IOP Conf. Series: Journal of Physics: Conf. Series **1031** (2018) 012024 doi:10.1088/1742-6596/1031/1/012024

Meeting the Challenges of Modeling Astrophysical Thermonuclear Explosions: Castro, Maestro, and the AMReX Astrophysics Suite

M. Zingale[1], A. S. Almgren[2], M. G. Barrios Sazo[1], V. E. Beckner[2], J. B. Bell[2], B. Friesen[3,2], A. M. Jacobs[4], M. P. Katz[5], C. M. Malone[6], A. J. Nonaka[2], D. E. Willcox[1], and W. Zhang[2]

[1]Department of Physics and Astronomy, Stony Brook University, Stony Brook, NY 11794-3800 USA
[2]Center for Computational Sciences and Engineering, Lawrence Berkeley National Lab, Berkeley, CA 94720 USA
[3]National Energy Research Scientific Computing Center, Lawrence Berkeley National Lab, Berkeley, CA 94720 USA
[4]Department of Physics and Astronomy, Michigan State University, East Lansing, Michigan 48824 USA
[5]NVIDIA Corporation, 2788 San Tomas Expressway, Santa Clara, CA, 95050 USA
[6]Los Alamos National Laboratory, Los Alamos, NM, 87545 USA

Abstract. We describe the AMReX suite of astrophysics codes and their application to modeling problems in stellar astrophysics. Maestro is tuned to efficiently model subsonic convective flows while Castro models the highly compressible flows associated with stellar explosions. Both are built on the block-structured adaptive mesh refinement library AMReX. Together, these codes enable a thorough investigation of stellar phenomena, including Type Ia supernovae and X-ray bursts. We describe these science applications and the approach we are taking to make these codes performant on current and future many-core and GPU-based architectures.

1. Introduction

Astrophysical explosions come in many flavors: gravitational and thermonuclear supernovae, unstable burning on the surface of compact objects, and explosive ignition of burning stages in stellar evolution. Accurate modeling of these events requires the coupling of hydrodynamics, gravity, thermonuclear reactions, and in some cases, radiation and magnetic fields. Further, these environments are characterized by a wide range of length scales, from the size of the star or binary system down to the burning zone width and dissipation scales. Temporal scales are equally impressive—stellar evolution occurs over 10s of millions to billions of years, the simmering phase leading up to explosions lasts hours or days to millenia, and the explosion can be over in seconds to hours. The radiation leakage, which leads to the observables we see lasts from hours to months.

No single algorithm meets all of the demands imposed by these events. Instead, we advance our understanding of these events by piecing together simulations of different phases of the evolution from different codes. Here we discuss our simulation codes, Maestro and Castro, designed to perform three-dimensional models of the early subsonic evolution leading to runaway and the subsequent explosion, respectively. Together this suite of codes allows us to address many problems in stellar and nuclear astrophysics. We describe some of the design details, the current architecture of the code, and some applications below.

2. Science drivers and challenges

Our interests are thermonuclear explosions, including Type Ia supernovae (SNe Ia), X-ray bursts (XRBs), and novae. The basic ingredients for these events are thermonuclear energy release and a degenerate equation of state that allows a runaway to build without a pressure response. Most of the current models for these events are characterized by a long timescale "simmering" phase where reactions heat the star or layer and drive convection. Eventually, reactions become vigorous enough that a runaway takes place, perhaps with an accompanying burning front that spreads through the star.

2.1. Type Ia supernovae

Among the most significant open questions for SNe Ia is the identity of the progenitor. About 20 years ago, the community had mostly converged upon the single-degenerate scenario—a Chandrasekhar-mass C/O white dwarf that accretes from its companion, eventually leading to runaway at the center that burns through the star (see [1] for the state of the field at that time). Since then, a wealth of observations has indicated that there is a lot of diversity in SNe Ia, and searches for progenitor systems have strongly suggested that Chandra-mass white dwarfs cannot explain most SNe Ia. Today, a pair of merging white dwarfs (the double degenerate scenario) has perhaps become the most popular model. Other progenitors, like He burning on the surface of a sub-Chandra white dwarf, have also seen interest in explaining some of the observed diversity. See [2] for a review.

There are open questions in all of these scenarios that can be addressed through simulation. For the Chandra and sub-Chandra models, what is the distribution (spatial and temporal) of the hotspots set up by turbulent convection that give rise to burning fronts? A longstanding question with the Chandra model is whether a deflagration can transition to a detonation during the burning front propagation through the star. For the sub-Chandra double detonation model, it is not clear whether it is possible to create a detonation in the thin surface He layer. For double degenerates, it is still unresolved whether the burning takes place promptly or after some delay. However it proceeds, we need to avoid an accretion-induced collapse to a neutron star. For many of these investigations, we need to address numerical issues such as whether it is possible to accurately model the ignition of a detonation with the spatial resolution we can attain. These are some of the questions we seek to answer.

2.2. X-ray Bursts

X-ray bursts—the burning of accreted H/He on the surface of a neutron star—can be important probes of neutron star structure. Interpreting observations requires that we understand what we are seeing, which can be influenced by the products of the burning and how the burning spreads across the star. Many efforts have focused on different aspects of these events. One-dimensional models capture the energetics well and inform us about the nucleosynthesis [3]. Global models show the importance of rotation in confining the burning [4], while models inspired from atmospheric science can explore the vertical structure [5]. However, the resolution differences from the scale of the burning to a reasonable fraction of the neutron star surface has prevented detailed explorations of the burning and how it feeds back on the flame structure and propagation in resolved calculations. Algorithms and computing architectures are starting to reach the point where we can span the gaps in spatial scales between these calculations to

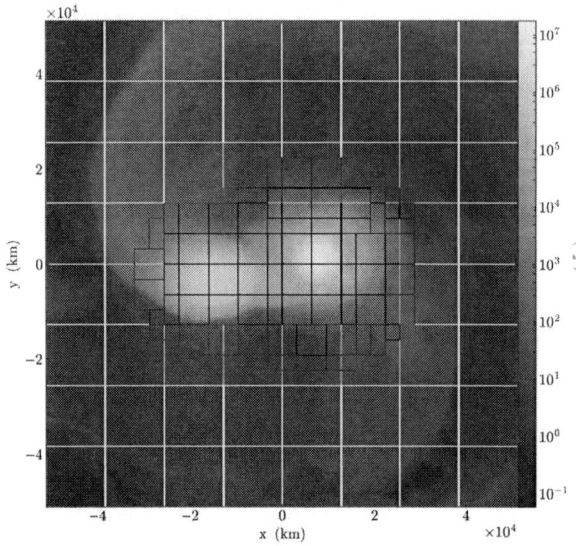

Figure 1. A slice through a Castro three-dimensional simulation of merging white dwarfs showing the 2-level grid structure. The boxes making up the base grid are drawn in white and the boxes making up the finer grid are drawn in black. Note that the finer boxes can overlap multiple coarse boxes, and not all boxes are the same size.

provide a better understanding of the ignition and propagation of the burning front, and the nucleosynthesis produced.

2.3. Requirements

These problems share common algorithmic requirements: strong coupling between hydrodynamics and burning, support for a general equation of state, self-gravity, including isolated boundary conditions, and long timescale evolution for the convective phases. All of these problems are inherently three-dimensional, as turbulence, fluid instabilities, and rotation affect the dynamics. Conservation is important as well, suggesting approaches that implement gravitational and rotation sources conservatively, and methods for improving angular momentum conservation.

3. AMReX Astrophysics Suite

Our suite of application codes is built on the AMReX block-structured adaptive mesh refinement (AMR) library. The basic programming model has AMReX managing the grid data-structures and parallel communications, and it calls the computational kernels on a patch-by-patch basis. The core library is written in C++ with computational kernels written in Fortran[1] —this allows us to take advantage of the strengths of both languages. AMReX supports subcycling in time, which we use in Castro.

AMReX uses a hybrid MPI + OpenMP approach to parallelism. Distribution of grid patches to nodes using MPI provides a natural coarse-grained approach to distributing the computational work, while subdividing patches into tiles and using threads to parallelize over tiles using OpenMP provides effective fine-scale parallelization and amortizes thread overhead over large units of work. Additionally, tiling does not incur the large increase in metadata associated with using smaller patches in a flat MPI mode [6] since we can use fewer, larger patches. This strategy is especially important for many-core architectures like the Intel Xeon Phi. Ongoing development is being done to support GPU offloading, using managed memory provided by the latest generations of GPUs. Figure 1 shows an example of a 2-level grid.

[1] Currently Maestro is written in pure-Fortran, but will be ported to the updated C++ AMReX framework this coming year.

There are many application codes built on AMReX, including those in combustion, multiphase flow, accelerator design, and microfluidics. In astrophysics, these include Maestro and Castro for stellar and nuclear astrophysics applications and Nyx [7] for cosmological applications. Here we focus on the former two.

3.1. *Maestro*

Maestro [8] is a low Mach number stellar hydrodynamics code designed for efficiently modeling convection in stars. Maestro decomposes the state variables into a one-dimensional hydrostatic base state and a three-dimensional Cartesian state that models the deviation from hydrostatic equilibrium. A constraint equation is derived by requiring that the pressure everywhere is close to the background hydrostatic pressure. The constraint acts to enforce instantaneous acoustic equilibration, effectively filtering soundwaves from the system, while retaining the compressibility effects due to the background stratification of the star and local heat release, as well as the hydrostatic adjustment of the star. In this fashion, it is more general than traditional anelastic methods. The timestep constraint for these equations depends only on the fluid velocity, not the sound speed, enabling much larger timesteps than compressible codes for highly subsonic flows.

The state is advanced using a second-order accurate projection method. Fluid quantities are advected using an unsplit Godunov method, with reactions incorporated via operator splitting. The provisional velocities are then projected onto the space that satisfies the divergence constraint. The projections require solving a variable coefficient elliptic equation, which is done numerically using a multigrid algorithm. A number of recent advances in low Mach number modeling [9, 10] have been incorporated into Maestro.

Maestro has been applied to convection in the Chandrasekhar-mass model for SNe Ia [11–13], the sub-Chandra model for SNe Ia [14, 15], XRBs [16–18], and convection in massive stars [19].

3.2. *Castro*

Castro [20–22] is a fully-compressible radiation hydrodynamics code that supports arbitrary equations of state, nuclear reaction networks, and Poisson gravity using geometric multigrid. The main hydrodynamics scheme in Castro is an unsplit piecewise parabolic method. The radiation solver in Castro uses the flux-limited diffusion approximation for gray or multigroup radiation. The integration algorithm on the grid hierarchy is a recursive procedure in which coarse grids are advanced in time, fine grids are advanced multiple steps to reach the same time as the coarse grids and the data at different levels are then synchronized. The synchronization for self-gravity is similar to the algorithm introduced by [23]. Recent developments in Castro include a spectral-deferred correction method of coupling hydrodynamics and reactions, a conservative gravity and rotation source formulation [24], and a retry mechanism to redo a step based on criteria evaluated during the integration.

Castro has been applied to core-collapse supernovae [25], radiative shock breakout in supernovae [26], population III pair-instability supernovae [27], the Chandra model for SNe Ia [28], the sub-Chandra SNe Ia model [29], and white dwarf mergers as a model for SNe Ia [24, 30]. For Maestro simulations that evolve from the subsonic regime to the sonic regime, we have demonstrated the ability to restart the calculations in Castro to continue the evolution into the sonic regime [31, 32] (in this case, for Chandra model SNe Ia).

3.3. *StarKiller Microphysics*

Maestro and Castro share the same microphysics, available as the StarKiller Microphysics GitHub

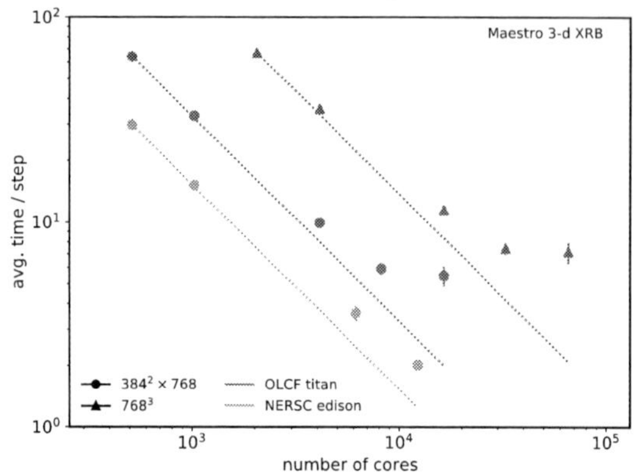

Figure 2. Maestro strong scaling on NERSC Edison and OLCF Titan for a 3-d XRB problem. Two different problem sizes are shown. We see excellent strong scaling to high core counts for this problem.

project[2]. This includes equations of state and nuclear reaction networks[3]. The reaction networks are written such that the rates and integration strategy are decoupled, allowing us to change the integration strategy for a given set of rates. They are also written to be threadsafe and with GPUs in mind (more on that below). The goal of StarKiller is to create a set of community microphysics solvers that can be used in a variety of nuclear astrophysics codes, not just those discussed here.

3.4. Open source and reproducibility

All of our simulation codes are open source and follow a fully open development model—the development git repos are hosted on GitHub[4], available for anyone to see and contribute to using issues and pull-requests. Additionally, we have mailing lists for discussions and asking for help. Several branches are used in our workflow. New changes are put into the `development` branch in each repo. Nightly regression testing is used to ensure that no new bugs were introduced. Once a month, `development` is merged into `master`. Finally, *all source files, model files, input parameters, etc. for any published science results are also available in the code repos.* When feasible, the git hashes for the published results are included in paper acknowledgments.

4. Parallel Performance and GPUs

A key design goal of our application codes is performance portability. We want the same kernels to run on clusters, manycore machines (e.g. Intel Xeon Phi), and GPU-based machines. Our development has balanced this need with architecture-specific optimizations to maximize code reuse.

Figure 2 shows strong scaling for Maestro on the 3-d XRB problem. This is a typical Maestro application [18], where burning can be a significant part of the overall evolution. We ran on both OLCF Titan and NERSC Edison. We see that the code scales well to $\mathcal{O}(10^4)$ processors. The upturn at the end of the scaling curves on Titan reflect the change from 1 MPI task / 8 OpenMP threads per NUMA node to 1 MPI task / 16 OpenMP threads per compute node (2 NUMA nodes). The main limitation to the scaling at the moment is the multigrid solves used to enforce the projection (in particular the nodal solver). Also, Maestro currently uses a simpler fine-grained approach to parallelism where planes in the z-direction are divided among OpenMP

[2] `https://github.com/StarKiller-astro/microphysics/`

[3] Several of the reaction network righthand sides and the EOS were provided from Frank Timmes' *cococubed* software instruments page `http://cococubed.asu.edu/code_pages/codes.shtml`. We thank him for making them available.

[4] `https://github.com/AMReX-Astro/`

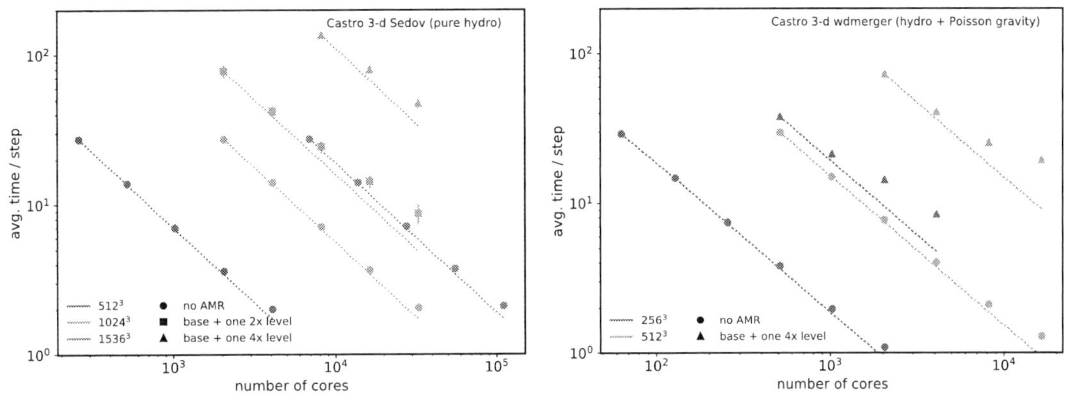

Figure 3. (left) Castro strong scaling on OLCF Titan for a pure hydro (Sedov w/ real EOS) problem. The different colors represent different base resolutions. We see excellent strong scaling for the single level runs. For the 1024^3 run, we also ran with one level of refinement by a factor of 2 (triangles) or a factor of 4 (squares), and see good strong scaling. The variability (shown by the error bars) at high core counts shows we are becoming work-starved. These runs used the PGI 17.7 compilers. (right) Castro strong scaling on OLCF Titan for a hydro + Poisson gravity (wdmerger) problem. This problem uses a multipole solver to determine Dirichlet boundary conditions representing an isolated mass distribution, and then geometric multigrid to solve for the potential in the interior. Two coarse grid sizes are shown, and demonstrate great strong scaling. We also look at a single level of refinement (by a factor of 4) on top of this coarse grid. These runs used the Cray 8.5.7 compilers.

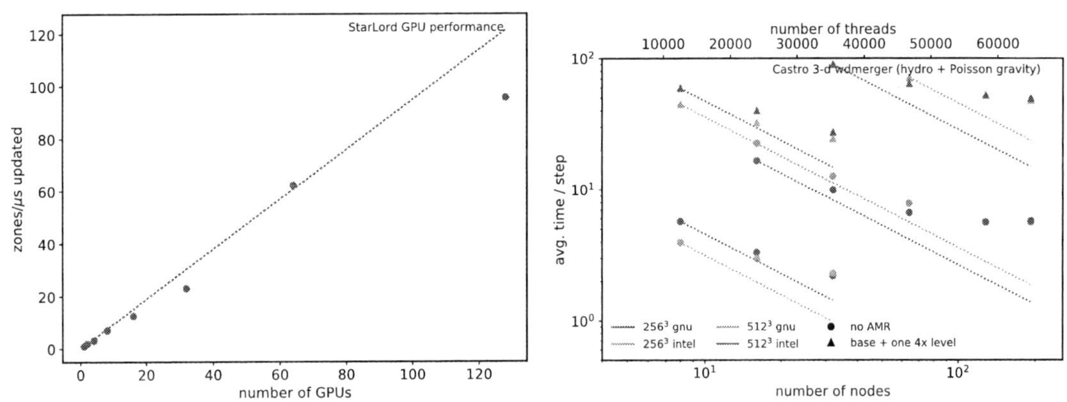

Figure 4. Castro scaling on new architectures. The left figure shows the Castro GPU proxy app (StarLord) on OLCF Summitdev. We see good scaling behavior to 64 GPUs in this test of a pure hydro problem (Sedov w/ real EOS). On the right, we explore performance on NERSC Cori (an Intel Xeon Phi platform).

threads. As we port Maestro to the C++ AMReX code base we will take advantage of ongoing development for increased tile size control.

Figure 3 shows Castro scaling behavior for a pure hydro problem with a real EOS and a hydro + self-gravity problem with a real EOS. All tests were run on OLCF Titan. The AMD processors in Titan can be used in an 8 or 16 cores / node configuration. For all these runs, we ran in the 16 cores / node mode. We see excellent scaling behavior for the pure hydro problem to $\mathcal{O}(10^5)$ cores. For the self-gravity problem, we scale very well with a uniform grid, but with a factor of 4 refinement for the AMR level on top of this, we observe degraded performance at higher core counts. This is a challenging problem, as only 3% of the domain volume is refined. For a given core count, there are a number of different combinations of MPI ranks and OpenMP threads we can use. In general, with multilevel problems, we found the best performance with fewer MPI ranks and more OpenMP threads.

Our latest focus has been on GPU ports of our application codes. A small proxy app, StarLord, was created from Castro with just the hydrodynamics and stellar equation of state. It uses a simple method-of-lines formulation of the hydrodynamics and advects 13 nuclear species in addition to the hydrodynamics. To offload work on GPUs, GPU support was added directly into AMReX[5]. In AMReX, an iterator loops over all of the boxes at the same level of refinement and passes a data pointer into a Fortran kernel function where the work is done. For the simulation state data that resides in each box, we have modified the memory allocator so that it can use a CUDA allocator (mainly relying on managed memory). The domain iterator is configured to handle data motion to and from the device, so that the compute kernels can operate on data that is presumed to already be there, and the computation is decoupled from the memory management. Computation on the data can then be performed with OpenACC, CUDA Fortran, or (more recently) OpenMP 4.5. We have also built CUDA compute support into AMReX so that a compute kernel can be transparently operated on using CUDA Fortran without substantially modifying the kernels (and we anticipate using a similar strategy to use OpenACC and/or OpenMP in the future). This helps ensure that we can continue to maintain performance portability in our simulation codes, by decoupling the physics algorithms from the backend support used to implement them on various compute architectures.

Figure 4 shows the performance of StarLord on the Summitdev platform at OLCF[6]. The highest GPU count (128) corresponds to 32 nodes on Summitdev. We see a nearly linear speedup with the number of GPUs, indicating good weak scaling on this machine. A single P100 GPU achieves a performance approximately 2.5 times that of the 20 Power8 cores. Efforts are underway to complete the port of the GPU developments into Castro. This figure also shows our performance on the Intel Xeon Phi manycore processors (using NERSC Cori). This is for the same problem as the self-gravity test on Titan. The Intel Xeon Phi chips in Cori have 68 cores that can be run with 1, 2, or 4 hardware threads each. For all these runs, we ran with 4 threads / core (272 threads / chip). We have not yet focused on optimizing Castro for the Intel Xeon Phi architecture.

A parallel effort is porting our microphysics—in particular reaction networks—to GPUs. Our strategy is to do the entire ODE integration on the GPU, i.e., the data for a patch of zones is passed to the GPU, all righthand side and Jacobian evaluations and the timestepping itself is done on the GPU, and once the burning in all zones is completed, we access the data as needed on the CPU. To enable this, we ported our workhorse ODE integrator (VODE [33]) to CUDA Fortran. This required extensive rewrites of the internals of VODE, which was originally written using Fortran 77 syntax unsupported by CUDA Fortran. Accelerating VODE with CUDA Fortran has proven successful, and we now see significant performance gains with the CUDA version of our reaction networks, even for moderate-sized networks (a 13-isotope standard network). Figure 5 shows the speed-ups on a GPU vs. single CPU core. The main issue with scaling is running out of local stack memory per thread with larger networks. Reducing the memory footprint is a near-term goal for this work. This test problem also shows a lot of thread divergence due to the widely differing thermodynamic conditions in the zones that are burning.

[5] This support is currently on a feature branch in the git repo, awaiting merge into `development`.

[6] Summitdev consists of 2 IBM Power8 processors and 4 NVIDIA Pascal GPUs per node.

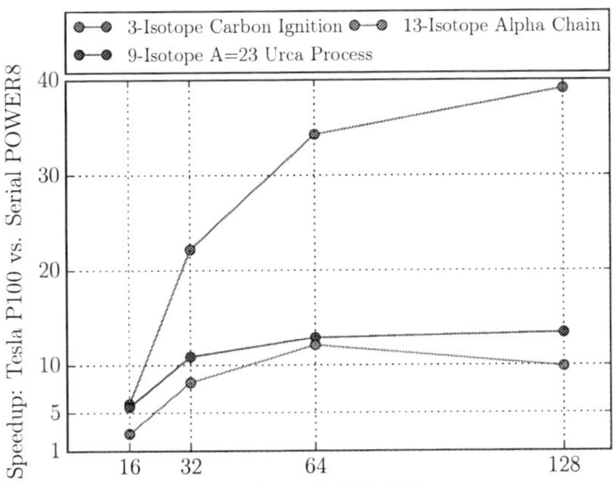

Figure 5. GPU speed up over single CPU core for three different reaction networks. reaction networks. The 13-isotope alpha-chain is a potential network for the sub-Chandra simulations; the 9-isotope Urca network is used for our Urca simulations; and the 3 isotope carbon network was used for our original Maestro white dwarf convection calculations. A variety of grid sizes were used, 16^3, 32^3, 64^3, and 128^3. In all cases, we see a good speed up on the GPU vs. a single core. These speed-up numbers are from the Summitdev machine, comparing a single Tesla P100 GPU to a single Power8 CPU core.

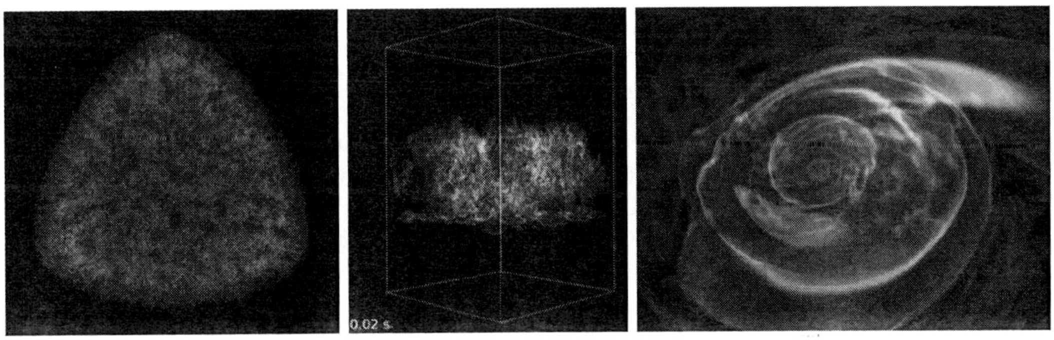

Figure 6. (left) Convective plumes in a Maestro sub-Ch calculation. (center) Vertical velocity showing the convective structure in a Maestro XRB calculation. (right) Snapshot of a Castro simulation of the merger of two white dwarfs, with 0.90 and 0.81 solar masses. The contours represent density levels.

5. Some science results

Figure 6 shows some of our recent science simulations. The left panel is an image of convection in the helium layer on a sub-Chandra white dwarf. This is part of a study of the early stages of the double detonation SNe Ia model. Using Maestro, we are able to model the convection in the He layer for many turnover times and saw a range of outcomes depending on the mass of the white dwarf and He layer, including a both nova-like behavior where the entire layer runs way together and localized burning ignited in a small region [15]. We are performing further studies to characterize the ignition.

The middle figure shows convection in a H/He layer on a neutron star, as a model of the early burning in an XRB. This Maestro model was the first 3D model of convection for this problem [18]. This study showed that the convective field became fully turbulent, achieving a Kolmogorov spectrum, and the overall dynamics was very different than our earlier 2-d simulations. This calculation acts as a bridge to our next set of studies that will look at larger scales.

The rightmost image is the coalesced remains of the merger of a 0.9 $M_\odot$ and 0.6 $M_\odot$ WD performed with Castro. This used the developments from [24]. Our primary focus with this suite of simulations is understanding the numerical sensitivity of mergers and collisions on the burning that takes place. This work is ongoing.

6. Summary and future development

We have described our suite of astrophysics codes built on the AMReX block-structured adaptive mesh refinement framework. These codes were developed to model problems in stellar astrophysics spanning from low speed convection to explosive burning. A major theme of the codes is the open development model, with the code development done on GitHub and all problem files needed to recreate any science results freely available. Future development efforts for Maestro include higher-order hydrodynamics and time-integration and rotation. For Castro, we are investigating stronger coupling between hydrodynamics and reactions, new solvers (including MHD), and finishing the GPU port.

Acknowledgments

The work at Stony Brook was supported by DOE/Office of Nuclear Physics grant DE-FG02-87ER40317 and NSF award AST-1211563. The work at LBNL was supported by the DOE Office of Advanced Scientific Computing Research under Contract No, DE-AC02-05CH11231. An award of computer time was provided by the Innovative and Novel Computational Impact on Theory and Experiment (INCITE) program. This research used resources of the Oak Ridge Leadership Computing Facility at the Oak Ridge National Laboratory, which is supported by the Office of Science of the U.S. Department of Energy under Contract No. DE-AC05-00OR22725. We appreciate the efforts of the OLCF (and in particular Fernanda Foertter) in organizing the GPU hackathons. This research used resources of the National Energy Research Scientific Computing Center, which is supported by the Office of Science of the U.S. Department of Energy under Contract No. DE-AC02-05CH11231. The authors would like to thank Stony Brook Research Computing and Cyberinfrastructure, and the Institute for Advanced Computational Science at Stony Brook University for access to the high-performance LIred and SeaWulf computing systems, the latter of which was made possible by a $1.4M National Science Foundation grant (#1531492). The GPU work benefited from a grant of an Titan X Pascal board from NVIDIA Corporation through the GPU Grant Program. Visualizations were done using yt [34]. This research has made use of NASA's Astrophysics Data System Bibliographic Services.

References

[1] Hillebrandt W and Niemeyer J C 2000 *Annu. Rev. Astron. Astrophys* **38** 191–230
[2] Maoz D, Mannucci F and Nelemans G 2014 *Annual Review of Astronomy and Astrophysics* **52** 107–170 (*Preprint* `https://doi.org/10.1146/annurev-astro-082812-141031`) URL `https://doi.org/10.1146/annurev-astro-082812-141031`
[3] Woosley S E, Heger A, Cumming A, Hoffman R D, Pruet J, Rauscher T, Fisker J L, Schatz H, Brown B A and Wiescher M 2004 *Astrophysical Journal Supplement* **151** 75–102
[4] Spitkovsky A, Levin Y and Ushomirsky G 2002 *Astrophysical Journal* **566** 1018–1038
[5] Cavecchi Y, Watts A L, Braithwaite J and Levin Y 2013 *Monthly Notices of the Royal Astronomical Society* **434** 3526–3541 (*Preprint* `1212.2872`)
[6] Zhang W, Almgren A, Day M, Nguyen T, Shalf J and Unat D 2016 *SIAM J. Scientific Computing* **38** S156–S172
[7] Almgren A S, Bell J B, Lijewski M J, Lukić Z and Van Andel E 2013 *Astrophysical Journal* **765** 39 (*Preprint* `1301.4498`)
[8] Nonaka A, Almgren A S, Bell J B, Lijewski M J, Malone C M and Zingale M 2010 *Astrophysical Journal Supplement* **188** 358–383 paper V
[9] Klein R and Pauluis O 2012 *Journal of Atmospheric Sciences* **69** 961–968
[10] Vasil G M, Lecoanet D, Brown B P, Wood T S and Zweibel E G 2013 *Astrophysical Journal* **773** 169 (*Preprint* `1303.0005`)
[11] Zingale M, Almgren A S, Bell J B, Nonaka A and Woosley S E 2009 *Astrophysical Journal* **704** 196–210
[12] Zingale M, Nonaka A, Almgren A S, Bell J B, Malone C M and Woosley S E 2011 *Astrophysical Journal* **740** 8

[13] Nonaka A, Aspden A J, Zingale M, Almgren A S, Bell J B and Woosley S E 2012 *Astrophysical Journal* **745** 73 (*Preprint* 1111.3086)
[14] Zingale M, Nonaka A, Almgren A S, Bell J B, Malone C M and Orvedahl R J 2013 *Astrophysical Journal* **764** 97 (*Preprint* 1212.4380)
[15] Jacobs A M, Zingale M, Nonaka A, Almgren A S and Bell J B 2016 *Astrophysical Journal* **827** 84 (*Preprint* 1507.06696)
[16] Malone C M, Nonaka A, Almgren A S, Bell J B and Zingale M 2011 *Astrophysical Journal* **728** 118 (*Preprint* 1012.0609)
[17] Malone C M, Zingale M, Nonaka A, Almgren A S and Bell J B 2014 *Astrophysical Journal* **788** 115
[18] Zingale M, Malone C M, Nonaka A, Almgren A S and Bell J B 2015 *Astrophysical Journal* **807** 60 (*Preprint* 1410.5796)
[19] Gilet C, Almgren A S, Bell J B, Nonaka A, Woosley S E and Zingale M 2013 *Astrophysical Journal* **773** 137
[20] Almgren A S, Beckner V E, Bell J B, Day M S, Howell L H, Joggerst C C, Lijewski M J, Nonaka A, Singer M and Zingale M 2010 *Astrophysical Journal* **715** 1221–1238 (*Preprint* 1005.0114)
[21] Zhang W, Howell L, Almgren A, Burrows A and Bell J 2011 *Astrophysical Journal Supplement* **196** 20 (*Preprint* 1105.2466)
[22] Zhang W, Howell L, Almgren A, Burrows A, Dolence J and Bell J 2013 *Astrophysical Journal Supplement* **204** 7 (*Preprint* 1207.3845)
[23] Miniati F and Colella P 2007 *Journal of Computational Physics* **227** 400–430
[24] Katz M P, Zingale M, Calder A C, Swesty F D, Almgren A S and Zhang W 2016 *Astrophysical Journal* **819** 94 (*Preprint* 1512.06099)
[25] Dolence J C, Burrows A and Zhang W 2015 *Astrophysical Journal* **800** 10 (*Preprint* 1403.6115)
[26] Lovegrove E, Woosley S E and Zhang W 2017 *Astrophysical Journal* **845** 103 (*Preprint* 1706.02440)
[27] Chen K J, Heger A, Woosley S, Almgren A and Whalen D J 2014 *Astrophysical Journal* **792** 44 (*Preprint* 1402.5960)
[28] Ma H, Woosley S E, Malone C M, Almgren A and Bell J 2013 *Astrophysical Journal* **771** 58 (*Preprint* 1305.2433)
[29] Moll R and Woosley S E 2013 *Astrophysical Journal* **774** 137 (*Preprint* 1303.0324)
[30] Moll R, Raskin C, Kasen D and Woosley S E 2014 *Astrophysical Journal* **785** 105 (*Preprint* 1311.5008)
[31] Almgren A, Bell J, Kasen D, Lijewski M, Nonaka A, Nugent P, Rendleman C, Thomas R and Zingale M 2010 *Proceedings of SciDAC 2010* (*Preprint* 1008.2801)
[32] Malone C M, Nonaka A, Woosley S E, Almgren A S, Bell J B, Dong S and Zingale M 2014 *Astrophysical Journal* **782** 11 (*Preprint* 1309.4042)
[33] Brown P N, Byrne G D and Hindmarsh A C 1989 *SIAM J. Sci. Stat. Comput.* **10** 1038–1051
[34] Turk M J, Smith B D, Oishi J S, Skory S, Skillman S W, Abel T and Norman M L 2011 *Astrophysical Journal Supplement* **192** 9 (*Preprint* 1011.3514)